CRYSTAL STRUCTURE DETERMINATION

The Role of the Cosine Seminvariants

CRYSTAL STRUCTURE DETERMINATION

The Role of the Cosine Seminvariants

Herbert A. Hauptman

Mathematical Biophysics Department
Medical Foundation of Buffalo
Buffalo, New York

Springer Science+Business Media, LLC

Library of Congress Catalog Card Number 72-80574

ISBN 978-1-4684-9956-8   ISBN 978-1-4684-9954-4 (eBook)

DOI 10.1007/978-1-4684-9954-4

© 1972 Springer Science+Business Media New York

Originally published by Plenum Press, New York 1972

Softcover reprint of the hardcover 1st edition 1972

A Division of Plenum Publishing Corporation

227 West 17th Street, New York, N. Y. 10011

United Kingdom edition published by Plenum Press, London

A Division of Plenum Publishing Company, Ltd.

Davis House (4th Floor), 8 Scrubs Lane, Harlesden, London, NW10 6SE, England

PREFACE

   The central theme of this monograph is that the cosine
seminvariants are the key to crystal structures. The cosine
seminvariants are the cosines of those linear combinations of the
phases (the so-called structure seminvariants) whose values, for
a given functional form for the geometric structure factor, are
uniquely determined by the crystal structure alone and are therefore independent of the choice of permissible origin. It follows
that the cosine seminvariants themselves are uniquely determined,
in general, by the observed magnitudes of the normalized structure
factors. The values of the cosine seminvariants in turn lead
unambiguously to the values of the individual phases and thus to
the crystal structure by means of the E-map (Fourier synthesis).
It is this property of the cosine seminvariants, that they serve
to link the observed magnitudes with the desired phases of the
normalized structure factors, which accounts for their importance
and explains the emphasis which is here placed on their role.

   In this monograph, both the theoretical and applied aspects
of this theme are developed, and it has been the author's
intention to do this in such a way that the reader will be able to
apply the principles described here with that understanding which
is necessary in order to tackle, with reasonable hope of success,
crystal structures of moderate to great complexity. Thus the
attempt has been made to motivate and make plausible the
theoretical arguments and to develop the typical mathematical derivations which are included here deliberately and carefully, without sacrificing mathematical rigor. Although details are not
always included, the serious reader can usually supply these by

working out the exercises or referring to the literature cited.
In all cases, whether or not the mathematical analyses are given,
great care has been taken in the precise statement of hypotheses
and conclusions so that there should be no question of the logical
basis or significance of the results obtained or claimed. It is
felt that only in this way will the desired clarity and understanding be achieved and a solid basis for intelligently making
the applications realized. The critical study of the logical
underpinning of the methods has yielded an additional bonus, for
it has thrown light on the limitations of some of the techniques
and suggested ways to strengthen them.

While the present work therefore stresses the importance of
a more than superficial understanding of the methods, the author
has felt, on the other hand, that in order to be most valuable to
the practising crystallographer, a major thrust of the monograph
should be on the applications. For this reason a great deal of
emphasis has been placed on the "case history" approach, on the
author's own experiences and those of his colleagues in implementing the various techniques, and in describing in sufficient detail
the actual applications which have been made. Postmortem studies
throw additional light on the several procedures employed, their
relative strengths and weaknesses, their limitations, the accuracy
of the phase determination, and the pitfalls to be avoided.
Finally, a significant portion of the book is devoted to bridging
the gap between the theoretical basis and the applications. It is
hoped that in these ways the monograph will provide the necessary
tools, and serve as a useful manual and guide, for the crystallographer interested in solving crystal structures and will enable
him rationally to attack the more difficult problems which may
confront him.

Although the foundations of the direct methods of crystal
structure determination were laid in the 1950's, it has only been
in the last several years that systematic and extensive applica-

tions of these techniques have been made. The success of these
methods has been such that there has been a renewal of interest
and activity in developing further their theoretical basis and a
number of significant advances have been made in the recent past
which have resulted in novel and improved techniques. An
additional goal of the author in writing the present monograph,
then, is to stimulate further this renewed activity and to suggest,
on the basis of both the recent theoretical work and the applications (much of which is included here, some for the first time)
that a more intensive study of the role of the structure
seminvariants in the solution of the phase problem will prove to
be fruitful.

It should be stressed that the methods described in this
monograph, in common with all methods of crystal structure determination, do not presently have unlimited applicability. Experience shows that one can have good expectation of success for
structures of moderate complexity, say 40-50 independent nonhydrogen atoms, and the straightforward solution of the valinomycin structure (Chapter XII) shows that even more complex
molecules are vulnerable. However, for the successful solution
of structures of much greater complexity, all methods presently
known would require excessive computing time. It is hoped that
the program described in this book will point the way for future
research and the methods described here improved and perfected.

Although these methods are applicable in the case that the
crystal is centrosymmetric or has one or several heavy atoms, it
is felt that, in general, given the present state of knowledge,
only the noncentrosymmetric structures having no heavy atom
present a real challenge. For this reason it is the latter case,
with one or two exceptions, which is treated here.

It is assumed that the reader has a good basic knowledge of
X-ray crystallography as is contained, for example, in the
recent book of M. M. Woolfson, "An Introduction to X-ray

Crystallography" (Cambridge, 1970).

It is now the author's pleasant task to make grateful
acknowledgement of his indebtedness to a number of people without
whose help and cooperation this book could not have been written.
Foremost among these are Drs. William Duax and Charles Weeks who
have not only shown great resourcefulness in making the initial
applications but have suggested novel approaches, particularly in
the applications, which have often, in turn, led to further
theoretical advances. They have generously permitted the author
to include in this book a number of crystal structure determinations not previously published (e.g. 9α-fluorocortisol, valinomycin
and $9-t$-butyl-9,10-dihydroanthracene). The author is still
further indebted to Dr. Weeks because he has written Chapters X,
XI, and XIII and, with the assistance of Mr. Steve Pokrywiecki
and Dr. James Edmonds, has also written most of the computer programs.

The author is particularly indebted to the Administration
of the Medical Foundation of Buffalo, especially Dr. Dorita Norton,
Research Director, and Dr. George Koepf, President, for their
continuing assistance and encouragement, and for this support the
author expresses his deep gratitude.

Finally, due acknowledgement is made to Miss Deanna Hefner,
Mrs. Vilma Kamysz, Mr. Donald Maracle, Mrs. Theodora Potter, Mrs.
Janice Stancliffe and Miss Melda Tugac who are responsible for the
excellence of the typescript from which this book was photo-
reproduced.

Buffalo, March 1972   Herbert Hauptman
                                       Medical Foundation of Buffalo

CONTENTS

Part A. THE THEORETICAL BASIS

Part C. THE APPLICATIONS

THE THEORETICAL BASIS

Part A is devoted to a description of the logical basis of
the direct methods of phase determination and the development of
the required mathematical apparatus. In Chapter I the idea of the
structure invariants and seminvariants is introduced in a natural
way and their fundamental importance, both with respect to origin
and enantiomorph specification and their potential use as a practical tool for phase determination, are stressed. Derived concepts such as linear dependence and independence, primitivity and
accessibility are introduced as needed.

In Chapter II the algebraic theory of the phase problem is
elaborated in great detail. Derivations of the explicit formulas
for the three kinds of cosine seminvariants, $\cos \phi$, $\cos(\phi_1 + \phi_2)$
and $\cos(\phi_1 + \phi_2 + \phi_3)$ are given in such a way as to exhibit their
similarities and to suggest the general conjecture that every
cosine seminvariant has a similar explicit representation in terms
of the magnitudes of normalized structure factors alone. The
algebraic analyses given here have the advantage that they are
shorter, more transparent, and require less mathematical machinery
than the probabilistic development given in the succeeding
chapters, but the latter is more powerful, is capable of greater
generality, and yields results having no algebraic counterpart
(e.g. probability measures and levels of significance).

Chapters III and IV contain the probabilistic theory of the
phase problem. Owing to widespread misconceptions concerned with
the proper use of probability theory as applied to this problem,
a great deal of attention is paid here to the rigorous development

of the theory. In particular, the primitive random variable is carefully defined and the various conditional distributions and expectation values derived in detail. It is necessary to stress this development since only in this way is it possible to justify rigorously the use of certain averaging procedures (notably the tangent formula, Chapter VII) over subsets of vectors in reciprocal space associated with large structure factor magnitudes. Although a more extensive mathematical machinery is required for these chapters than for the preceding one, the results obtained justify the additional effort.

CHAPTER I

THE PHASE PROBLEM

1. Introduction

Denote by $\phi_{\vec{h}}$ the phase of the crystal structure factor $F_{\vec{h}}$:

$$F_{\vec{h}} = |F_{\vec{h}}| \exp(i\phi_{\vec{h}}). \qquad (1.1)$$

The relationship between the (complex) structure factor $F_{\vec{h}}$ and the electron density function $\rho(\vec{r})$ in a crystal is given by the pair of equations

$$F_{\vec{h}} = \int_V \rho(\vec{r}) \exp(2\pi i\vec{h}\cdot\vec{r}) \, dv, \qquad (1.2)$$

3

$$\rho(\vec{r}) = \frac{1}{V} \sum_{\vec{h}} F_{\vec{h}} \exp(-2\pi i \vec{h} \cdot \vec{r}) =$$

$$\frac{1}{V} \sum_{\vec{h}} |F_{\vec{h}}| \exp(i\phi_{\vec{h}} - 2\pi i \vec{h} \cdot \vec{r}) \qquad (1.3)$$

in which V represents the unit cell or its volume. Only the
magnitudes $|F_{\vec{h}}|$ of a finite number of structure factors are
obtainable from experiment; the values of the phases $\phi_{\vec{h}}$ cannot
be determined experimentally. If one substitutes arbitrary values
for the phases $\phi_{\vec{h}}$ in (1.3) together with the observed values
for $|F_{\vec{h}}|$, an electron density function $\rho(\vec{r})$ is obtained which
satisfies (1.2) also. Of course the density functions $\rho(\vec{r})$
obtained in this way are, in general, not non-negative for all
values of $\vec{r}$ so that most sets of values for the $\phi_{\vec{h}}$ may be ruled
out by imposing the non-negativity requirement on $\rho(\vec{r})$. However, even if the values of the $\phi_{\vec{h}}$ are restricted in such a way
that the resulting electron density $\rho(\vec{r})$, as found from (1.3),
is non-negative for all $\vec{r}$, then there are, in general, still
infinitely many ways of choosing the $\phi_{\vec{h}}$. In other words there
are infinitely many non-negative electron density functions
$\rho(\vec{r})$ such that the values of the $|F_{\vec{h}}|$, as calculated from (1.2),
agree with their observed values. Clearly the phase problem,
to determine the values of the phases $\phi_{\vec{h}}$ of the structure
factors $F_{\vec{h}}$ when only the magnitudes $|F_{\vec{h}}|$ are given, is, in
principle, unsolvable when formulated in these terms.

Next, suppose that the real crystal, with continuous
electron density $\rho(\vec{r})$, is replaced by an idealized one, the unit

cell of which consists of N discrete, non-vibrating, point atoms. Then the structure factor $F_{\vec{h}}$ is replaced by the normalized structure factor $E_{\vec{h}}$ and (1.1) – (1.3) are replaced by

$$E_{\vec{h}} = |E_{\vec{h}}| \exp{(i\phi_{\vec{h}})} \qquad (1.4)$$

$$E_{\vec{h}} = \frac{1}{\sigma_2^{1/2}} \sum_{j=1}^{N} Z_j \exp{(2\pi i \vec{h}\cdot\vec{r}_j)} \qquad (1.5)$$

$$\left\langle E_{\vec{h}} \exp{(-2\pi i \vec{h}\cdot\vec{r})} \right\rangle_{\vec{h}} = \frac{1}{\sigma_2^{1/2}} \left\langle \sum_{j=1}^{N} Z_j \exp{\left[2\pi i \vec{h}\cdot(\vec{r}_j-\vec{r})\right]} \right\rangle_{\vec{h}}$$

$$\left. \begin{array}{l} = \dfrac{Z_j}{\sigma_2^{1/2}} \text{ if } \vec{r} = \vec{r}_j \\[2em] = 0 \quad \text{ if } \vec{r} \neq \vec{r}_j \end{array} \right\} , \qquad (1.6)$$

where Z_j is the atomic number, $\vec{r}_j$ is the position vector of the atom labeled j, and

$$\sigma_2 = \sum_{j=1}^{N} Z_j^{2} . \qquad (1.7)$$

In practice the magnitudes $|E_{\vec{h}}|$ of the normalized structure factors $E_{\vec{h}}$ are obtainable (at least approximately) from the observed magnitudes $\left|F_{\vec{h}}\right|$ while the phases $\phi_{\vec{h}}$, as defined by (1.4) and (1.5), cannot be determined experimentally. Since one now requires only the 3N components of the N position vectors $\vec{r}_j$, rather than the much more complicated electron density function $\rho(\vec{r})$, it turns out that, in general, the known magnitudes are more than sufficient. This is most readily seen from (1.5) which, if real and imaginary parts are equated, is in reality a system of 2n equations where n is the number of known magnitudes $\left|E_{\vec{h}}\right|$. The unknowns consist of the n phases $\phi_{\vec{h}}$ and the 3N components of the vectors $\vec{r}_j$, or n + 3N unknowns in all. Since, in general, the number of equations (1.5), 2n, exceeds by far the number of unknowns, n + 3N, the problem to determine the phases $\phi_{\vec{h}}$ when only the magnitudes $\left|E_{\vec{h}}\right|$ are given (the phase problem), is now greatly overdetermined. Thus the phase problem is, in principle, solvable when reformulated in terms of fixed, point atoms.

There is another point of view which throws additional light on this matter. Although only a finite number of magnitudes $\left|E_{\vec{h}}\right|$ are obtainable from experiment and the phases $\phi_{\vec{h}}$ cannot be measured, Eq. (1.6) can still be used to determine crystal structures since the prior knowledge that the unit cell of the crystal consists of N point atoms of known atomic numbers severely restricts the permitted values of the phases $\phi_{\vec{h}}$. In fact, it is plausible to suppose that, in general, the magnitudes $\left|E_{\vec{h}}\right|$ uniquely determine (for fixed origin and enantiomorph, as clarified later) those phases $\phi_{\vec{h}}$ having the property that the left side of (1.6), as a function of $\vec{r}$, is then zero everywhere except at N points where it has the value $Z_j/\sigma_2^{1/2}$. Naturally, owing to the finite number of data obtainable from experiment, Eq. (1.6) is non-vanishing in N discrete regions, not points, and is only approximately zero elsewhere. In practice therefore it is the maxima of Eq. (1.6) which yield the atomic positions. In any event then, one is led to the earlier conclusion that, if rigid, point atoms are assumed, the phase problem is not only solvable, at least in principle, but is actually overdeterminate.

2.   The Structure Invariants

The concept of the structure invariants and seminvariants was first introduced, for the primitive centrosymmetric space groups, in 1953. By 1956 the theory had been worked out for the primitive noncentrosymmetric space groups and, a few years later, for all the remaining space groups. The initial motivation for developing this theory was that the structure invariants and seminvariants are the natural tool for clarifying the nature of the relationship between the values of the individual phases and the choice of origin and enantiomorph and a proper understanding of this relationship is essential for the development of procedures for phase determination. Thus, by means of this concept, the identity of those phases whose values are uniquely determined by the crystal structure and are independent of the choice of origin, is readily found. Again, once the values of certain phases have been specified, the theory provides a simple answer to the question of which other phases are then uniquely determined by the structure. Finally the theory leads to procedures for choosing the origin and enantiomorph by fixing the values of certain phases.

In 1969, with the solution of the estriol structure as a consequence of improved techniques for calculating the values of the cosine invariants, the goal of utilizing the structure invariants as a practical tool for phase determination was at last realized. About sixteen non-centrosymmetric structures have since been solved by one variant or another of this basic theme and it appears likely that, as the method is refined, it will become an increasingly important technique for structure determination.

Eq. (1.6) implies that the normalized structure factors $E_{\vec{h}}$ determine the crystal structure. However (1.5) does not imply that, conversely, the crystal structure determines the values of the normalized structure factors $E_{\vec{h}}$ since the position vectors $\vec{r}_j$ depend not only on the structure but on the choice of origin as well. It turns out nevertheless that the magnitudes $|E_{\vec{h}}|$ of the normalized structure factors are in fact uniquely determined by the crystal structure and are independent of the choice of origin but that the values of the phases $\phi_{\vec{h}}$ depend also on the choice of origin. Although the values of the individual phases depend on the structure and the choice of origin, there exist certain linear combinations of the phases, the so-called structure invariants, whose values are determined by the structure alone and are independent of the choice of origin. Our next task is to determine those linear combinations of the phases which are structure invariants for all the space groups.

If the origin of coordinates is shifted to a different point whose position vector with respect to the initial origin is $\vec{r}_0$, then $\vec{r}_j$, the position vector of the j th atom with respect to the first origin, is replaced by

$$\vec{r}_j' = \vec{r}_j - \vec{r}_0 \tag{2.1}$$

and E_h of (1.5) is replaced by

$$E_{\vec{h}}' = \frac{1}{\sigma_2^{1/2}} \sum_{j=1}^{N} Z_j \exp (2\pi i \vec{h} \cdot \vec{r}_j') \tag{2.2}$$

$$E_{\vec{h}}' = \frac{1}{\sigma_2^{1/2}} \sum_{j=1}^{N} Z_j \exp \left(2\pi i \vec{h} \cdot (\vec{r}_j - \vec{r}_0)\right) \tag{2.3}$$

$$E_{\vec{h}}' = \frac{\exp\ (-2\pi i \vec{h} \cdot \vec{r}_0)}{\sigma_2^{1/2}} \sum_{j=1}^{N} Z_j \exp\ (2\pi i \vec{h} \cdot \vec{r}_j) \qquad (2.4)$$

or, in view of (1.5),

$$E_{\vec{h}}' = E_{\vec{h}} \exp\ (-2\pi i \vec{h} \cdot \vec{r}_0), \quad |E_{\vec{h}}'| = |E_{\vec{h}}|, \qquad (2.5)$$

$$\phi_{\vec{h}}' = \phi_{\vec{h}} - 2\pi \vec{h} \cdot \vec{r}_0 \qquad (2.6)$$

where $\phi_{\vec{h}}'$ is the phase of the structure factor $E_{\vec{h}}'$ with respect to the new origin. Consider any finite linear combination of the both sides of (2.6) having integer coefficients $A_{\vec{h}}$ which depend upon $\vec{h}$:

$$\sum_{\vec{h}} A_{\vec{h}}\ \phi_{\vec{h}}' = \sum_{\vec{h}} A_{\vec{h}}\ \phi_{\vec{h}} - 2\pi \left(\sum_{\vec{h}} A_{\vec{h}}\ \vec{h} \right) \cdot \vec{r}_0. \qquad (2.7)$$

Clearly, if

$$\sum_{\vec{h}} A_{\vec{h}}\ \vec{h} = 0, \qquad (2.8)$$

then

$$\sum_{\vec{h}} A_{\vec{h}}\ \phi_{\vec{h}}' = \sum_{\vec{h}} A_{\vec{h}}\ \phi_{\vec{h}}, \qquad (2.9)$$

no matter what the vector $\vec{r}_0$ may be, and the linear combination of the phases (2.9) is a structure invariant since it has the same value for every choice of origin. This completes the proof of the fundamental theorems:

Theorem 2.1. If the origin of coordinates is shifted to a new point having position vector $\vec{r}_0$ with respect to the old origin, then the phase $\phi_{\vec{h}}$ of the normalized structure factor $E_{\vec{h}}$ with respect to the old origin is replaced by the new phase $\phi'_{\vec{h}}$ with respect to the new origin given by

$$\phi'_{\vec{h}} = \phi_{\vec{h}} - 2\pi \vec{h} \cdot \vec{r}_0. \tag{2.6}$$

Theorem 2.2. The linear combination of the phases

$$\sum_{\vec{h}} A_{\vec{h}} \, \phi_{\vec{h}}, \tag{2.10}$$

where the $A_{\vec{h}}$ are integers satisfying

$$\sum_{\vec{h}} A_{\vec{h}} \, \vec{h} = 0, \tag{2.11}$$

is a structure invariant for every space group.

Corollary 2.2.1. If

$$\vec{h}_1 + \vec{h}_2 + \vec{h}_3 = 0 \tag{2.12}$$

then

$$\phi_{\vec{h}_1} + \phi_{\vec{h}_2} + \phi_{\vec{h}_3} \tag{2.13}$$

is a structure invariant for every space group.

Since the linear combinations (2.10), where (2.11) also holds, are structure invariants for all the space groups, they will be called the universal structure invariants. For all space groups other than P1 the origin may not be chosen arbitrarily if the simplification permitted by the space group symmetries is to be realized. For example, if a crystal has a center of symmetry it is natural to place the origin at such a center while if a two-fold screw axis, but no other symmetry element is present, the origin would normally be situated on this symmetry axis. In such cases the permissible origins are greatly restricted and it is therefore plausible to assume that many linear combinations of the phases, in addition to those permitted by Theorem 2.2, will remain unchanged in value when the origin is shifted only in the restricted ways allowed by the space group symmetries. It is therefore to be anticipated that in the space groups of higher symmetry there is a great variety of structure invariants and it is our next goal to show that this is in fact the case and to describe how these structure invariants may be derived.

It is necessary now to make precise the notion of permissible origin and structure invariant and to this end the concept of equivalence must be introduced. First, for any space group, the coordinates of equivalent positions depend upon the choice of origin. Hence the functional form for the geometric stucture factor also depends on the choice of origin. Two origins will be said to be equivalent if they give rise to the same functional form for the geometric structure factor. Alternatively, two points are equivalent if they are geometrically situated in the same way with respect to the symmetry elements. For the primitive centrosymmetric space groups the permissible origins are defined to be the eight centers of symmetry:

$$\varepsilon_1, \ \varepsilon_2, \ \varepsilon_3; \quad \varepsilon_i = 0 \text{ or } \frac{1}{2}, \quad i = 1, 2, 3. \qquad (2.14)$$

For the primitive noncentrosymmetric space groups the permissible origins are defined to be those points equivalent to one of the eight points (2.14). Finally, for any primitive space group, the structure invariants are defined to be those linear combinations of the phases with integer coefficients which have the same value (determined by the crystal structure) for all choices of permissible origin.

3.   Some Examples of Space Group Dependent
Structure Invariants

The universal structure invariants obtained in §2 lead, via the space group dependent relationships among the phases, to a variety of additional structure invariants which depend upon the space group. These have important application in the specification of origin and enantiomorph which is an essential feature in any procedure for direct phase determination.

3.1. Space group P$\bar{1}$. In view of Theorem 2.2 the linear combination

$$2\phi_{-\vec{h}} + \phi_{2\vec{h}} \qquad (3.1)$$

in a structure invariant. In the space group P$\bar{1}$ every phase has the value 0 or π if the origin is chosen at a center of symmetry. Hence $2\phi_{-\vec{h}} = 0$ and, in view of (3.1), the single phase

$$\phi_{2\vec{h}} \qquad (3.2)$$

is a structure invariant for every reciprocal vector $\vec{h}$. In

other words the value of $\phi_{2\vec{h}}$ is uniquely determined by the crystal
structure no matter which of the eight permissible origins
(centers of symmetry) is selected.

Next, suppose that $\vec{h}_1$ and $\vec{h}_2$ are two reciprocal vectors the
corresponding components of which have the same parity. In
symbols,

$$\vec{h}_1 \equiv \vec{h}_2 \ (\text{mod } \vec{\omega}) \qquad\qquad (3.3)$$

where the modulus (vector) $\vec{\omega}$ is defined by

$$\vec{\omega} = (2,2,2). \qquad\qquad (3.4)$$

Then the components of $\vec{h}_1 + \vec{h}_2$ are even integers,

$$\vec{h}_1 + \vec{h}_2 = 2\vec{h}, \qquad\qquad (3.5)$$

the linear combination

$$\phi_{\vec{h}_1} + \phi_{\vec{h}_2} + \phi_{-2\vec{h}} \qquad\qquad (3.6)$$

is (in view of Theorem 2.2) a structure invariant and, since $\phi_{-2\vec{h}}$
is itself a structure invariant, the linear combination

$$\phi_{\vec{h}_1} + \phi_{\vec{h}_2} \qquad\qquad (3.7)$$

is also a structure invariant. In other words, if corresponding
components of $\vec{h}_1$ and $\vec{h}_2$ have the same parity, then the value of
the linear combination (3.7) depends only on the structure and
is independent of the choice of permissible origin.

In a similar way it may be shown that if $\vec{h}_1$, $\vec{h}_2$ and $\vec{h}_3$ are

reciprocal vectors the components of whose sum are even integers,
then the linear combination

$$\phi_{\vec{h}_1} + \phi_{\vec{h}_2} + \phi_{\vec{h}_3} \tag{3.8}$$

is a structure invariant. In fact it is now clear that the
following analogue of Theorem 2.2 is valid:

Theorem 3.1. The linear combination of the phases

$$\sum_{\vec{h}} A_{\vec{h}} \, \phi_{\vec{h}}, \tag{3.9}$$

where the $A_{\vec{h}}$ are integers satisfying

$$\sum_{\vec{h}} A_{\vec{h}} \, \vec{h} \equiv 0 \pmod{\vec{\omega}} \tag{3.10}$$

and

$$\vec{\omega} = (2,2,2), \tag{3.11}$$

is a structure invariant for every centrosymmetric space group
(provided that the origin is restricted to lie on a center of
symmetry).

Corollary 3.1.1. If

$$\vec{h}_1 + \vec{h}_2 + \vec{h}_3 \equiv 0 \pmod{\vec{\omega}}, \tag{3.12}$$

then

$$\phi_{\vec{h}_1} + \phi_{\vec{h}_2} + \phi_{\vec{h}_3} \qquad\qquad (3.13)$$

is a structure invariant for every centrosymmetric space group
(again provided that the origin is at a center of symmetry).

   3.2. Space group P2. If b is the unique axis and the origin
is on a two-fold axis, then

$$\phi_{h\ 0\ \ell} = 0 \text{ or } \pi. \qquad\qquad (3.14)$$

From Theorem 2.2, the linear combination

$$2\phi_{\bar{h}\ 0\ \bar{\ell}} + \phi_{2h\ 0\ 2\ell} \qquad\qquad (3.15)$$

is a structure invariant. In view of (3.14), $2\phi_{\bar{h}\ 0\ \bar{\ell}} = 0$ and the
single phase

$$\phi_{2h\ 0\ 2\ell} \qquad\qquad (3.16)$$

is a structure invariant. In other words the value of $\phi_{2h\ 0\ 2\ell}$ is
uniquely determined by the crystal structure no matter which point
on a two-fold axis is chosen as the origin of coordinates.

   Next, define

$$\vec{h}_1 = h_1\ k\ \ell_1 \qquad\qquad (3.17)$$

$$\vec{h}_2 = h_2\ \bar{k}\ \ell_2 \qquad\qquad (3.18)$$

where it is assumed that h_1 and h_2 are integers having the same
parity, ℓ_1 and ℓ_2 are integers having the same parity, and k is an

arbitrary integer. In symbols

$$\vec{h}_1 \equiv -\vec{h}_2 \pmod{\vec{\omega}} \tag{3.19}$$

where the modulus $\vec{\omega}$, in contrast to (3.4), is now defined by

$$\vec{\omega} = (2,0,2). \tag{3.20}$$

Then the first and third components of $\vec{h}_1 + \vec{h}_2$ are even while the middle component is zero:

$$\vec{h}_1 + \vec{h}_2 = (2h\ 0\ 2\ell). \tag{3.21}$$

The linear combination

$$\phi_{\vec{h}_1} + \phi_{\vec{h}_2} + \phi_{2\bar{h}\ 0\ 2\bar{\ell}} \tag{3.22}$$

is, in view of Theorem 2.2, a structure invariant and, since $\phi_{2\bar{h}\ 0\ 2\bar{\ell}}$ is a structure invariant, it follows that the linear combination

$$\phi_{\vec{h}_1} + \phi_{\vec{h}_2} \tag{3.23}$$

is a structure invariant. In a similar way it may be shown, more generally, that the following analogue of Theorem 2.2 holds:

Theorem 3.2. The linear combination of the phases

$$\sum_{\vec{h}} A_{\vec{h}}\ \phi_{\vec{h}}, \tag{3.24}$$

where the $A_{\vec{h}}$ are integers satisfying

$$\sum_{\vec{h}} A_{\vec{h}} \; \vec{h} \equiv 0 \pmod{\vec{\omega}} \tag{3.25}$$

and

$$\vec{\omega} = (2,0,2), \tag{3.26}$$

is a structure invariant for the space groups P2 and $P2_1$ provided that the origin is chosen to be on a two-fold axis.

    <u>Corollary 3.2.1.</u> If

$$\vec{h}_1 + \vec{h}_2 + \vec{h}_3 \equiv 0 \pmod{\vec{\omega}} \tag{3.27}$$

then

$$\phi_{\vec{h}_1} + \phi_{\vec{h}_2} + \phi_{\vec{h}_3} \tag{3.28}$$

is a structure invariant for P2 and $P2_1$ provided that the origin is on a two-fold axis.

Exercises I. 1.

    1. If the origin is chosen at 222 in the space group P222, find all points equivalent to the origin. What are the structure invariants for this space group?

    2. If the origin is chosen midway between three pairs of non-intersecting screw axes in $P2_12_12_1$, find the equivalence class which contains the origin. What are the structure invariants

for this space group?

3. If the origin is chosen on mm2 in the space group Pmm2, find the equivalence class which contains the origin. What are the structure invariants for this space group?

4. In each of P222, $P2_12_12_1$, Pmm2, how many equivalence classes are there?

4. The Structure Seminvariants

In the examples considered thus far and, in fact, in all the primitive space groups of the triclinic, monoclinic and orthorhombic systems, the origins permitted by the space group symmetries are all geometrically situated in the same way with respect to the symmetry elements. For example, in the space group $P\bar{1}$ the origin is one of the eight centers of symmetry, in the space group $P2_1$ the origin lies anywhere on one of the four twofold screw axes, and in the space group $P2_12_12_1$ the eight permissible origins are located midway between three pairs of nonintersecting screw axes. Thus, in each of these cases, the origins are geometrically indistinguishable from one another and they are equivalent. They have the property that the functional (trigonometric) form for the geometric structure factor is the same no matter which one of the equivalent points is actually chosen as the origin. In the space groups of tetragonal or higher symmetry, however, not all permissible origins are geometrically situated in the same way with respect to the symmetry elements. In these space groups the permissible origins fall into two or more classes, those origins in any one class being indistinguishable from one another in the sense that they bear the same geometric relationship to the symmetry elements. The origins in any one class are equivalent to one another and they have the property that the functional form for the geometric structure

factor is the same for all of them. Origins in different equivalence classes are geometrically distinguishable since they bear
different relationships to the symmetry elements. For non-
equivalent origins the functional forms for the geometric structure factor are distinct.

Next, those linear combinations of the phases which are
unchanged in value when the origin is replaced by an equivalent
origin are said to be structure seminvariants. Alternatively,
if the functional form for the structure factor is fixed, then
the structure seminvariants consist of those linear combinations
of the phases which have the same value for every choice of origin
permitted by the fixed functional form for the geometric structure
factor. It turns out that the identity of the structure seminvariants is independent of the chosen functional form for the
structure factor but that the value of any structure seminvariant
does depend, in general, on the functional form. Clearly any
structure invariant is also a structure seminvariant since the
former must be unchanged in value for all permissible origins not
merely those in a given equivalence class. In the space groups of
tetragonal or higher symmetry, however, there exist structure
seminvariants which are not structure invariants. In the triclinic, monoclinic and orthorhombic systems the two concepts
coincide.

4.1. Space group P4. In order to illustrate the principles
discussed so far, complete details for the space group P4 will be
worked out. Employing the notation and terminology of "International Tables of X-Ray Crystallography", Vol. I, pages 167 and
415, in which an arbitrary point on either of the two four-fold
axes is chosen as origin, equivalent positions are found to be

$$x,y,z; \quad \bar{x},\bar{y},z; \quad \bar{y},x,z; \quad y,\bar{x},z. \qquad (4.1)$$

Hence

$$A = \cos\ 2\pi(hx+ky+\ell z)\ +\ \cos\ 2\pi(-hx-ky+\ell z)\ +$$

$$\cos\ 2\pi(-hy+kx+\ell z)\ +\ \cos\ 2\pi(hy-kx+\ell z),$$

$$A = 2\cos\ 2\pi(hx+ky)\ \cos\ 2\pi\ell z\ +\ 2\cos\ 2\pi(hy-kx)\ \cos\ 2\pi\ell z,$$

$$A = 4\cos\ \pi[(h-k)x+(h+k)y]\cos\pi[(h+k)x-(h-k)y]\ \cos\ 2\pi\ell z. \qquad (4.2)$$

Again

$$B = \sin\ 2\pi(hx+ky+\ell z)\ +\ \sin\ 2\pi(-hx-ky+\ell z)\ +$$

$$\sin\ 2\pi(-hy+kx+\ell z)\ +\ \sin\ 2\pi(hy-kx+\ell z),$$

$$B = 2\cos\ 2\pi(hx+ky)\ \sin\ 2\pi\ell z\ +\ 2\cos\ 2\pi(hy-kx)\ \sin\ 2\pi\ell z,$$

$$B = 4\cos\ \pi[(h-k)x+(h+k)y]\cos\pi[(h+k)x-(h-k)y]\ \sin\ 2\pi\ell z. \qquad (4.3)$$

Next, if the origin is shifted to a point on the other four-fold axis, i.e. the axis passing through the point with $x = 1/2$, $y = 1/2$, the equivalent positions (4.1) are replaced by $x + 1/2$, $y + 1/2, z$; $\bar{x} + 1/2, \bar{y} + 1/2, z$; $\bar{y} + 1/2, x + 1/2, z$; $y + 1/2, \bar{x} + 1/2, z$ or, substituting x for $x + 1/2$ and y for $y + 1/2$, by

$$x,y,z; \quad \bar{x},\bar{y},z; \quad \bar{y},x,z; \quad y,\bar{x},z. \qquad (4.4)$$

Since (4.4) is identical with (4.1) the functional forms for the real and imaginary parts of the geometric structure factor are identical with A and B as given by (4.2) and (4.3) respectively. In short, all points on the two four-fold axes are equivalent to each other as was to have been anticipated since they are geometrically indistiguishable.

Next, if the origin is shifted to a point on a two-fold axis, e.g. the axis passing through the point with $x = 0$, $y = 1/2$, the equivalent positions (4.1) are replaced by $x,y + 1/2,z$; $\bar{x},\bar{y} + 1/2$, z; $\bar{y},x + 1/2,z$; $y,\bar{x} + 1/2,z$, or, substituting y for $y + 1/2$, by

$$x,y,z; \quad \bar{x},\bar{y},z; \quad \bar{y} + \frac{1}{2},x + \frac{1}{2},z; \quad y + \frac{1}{2},\bar{x} + \frac{1}{2},z. \quad (4.5)$$

Since the equivalent positions (4.5) are now distinct from (4.1), it is to be anticipated that the functional form for the real and imaginary parts of the geometric structure factor will be different from (4.2) and (4.3). In fact, one now finds

$$A = \cos 2\pi(hx+ky+\ell z) + \cos 2\pi(-hx-ky+\ell z) +$$

$$\cos 2\pi\left[h(-y + \frac{1}{2})+k(x + \frac{1}{2})+\ell z\right] + \cos 2\pi\left[h(y + \frac{1}{2})+k(-x - \frac{1}{2})+\ell z\right] =$$

$$2\cos 2\pi(hx+ky)\cos 2\pi\ell z + 2(-1)^{h+k}\cos 2\pi(hy-kx)\cos 2\pi\ell z$$

$$A = \left\{ \begin{array}{ll} 4\cos\pi[(h-k)x+(h+k)y]\cos\pi[(h+k)x-(h-k)y]\cos 2\pi\ell z, & (h+k)\text{even} \\ \\ -4\sin\pi[(h-k)x+(h+k)y]\sin\pi[(h+k)x-(h-k)y]\cos 2\pi\ell z, & (h+k)\text{odd} \end{array} \right\}.$$

$$(4.6)$$

$$B = \sin2\pi(hx+ky+\ell z) + \sin2\pi(-hx-ky+\ell z) +$$

$$\sin2\pi\left[h(-y + \frac{1}{2})+k(x + \frac{1}{2})+\ell z\right] + \sin2\pi\left[h(y + \frac{1}{2})+k(-x - \frac{1}{2})+\ell z\right] =$$

$$2\cos2\pi(hx+ky)\sin2\pi\ell z+2(-1)^{h+k}\cos2\pi(hy-kx)\sin2\pi\ell z$$

$$B = \begin{cases} 4\cos\pi[(h-k)x+(h+k)y]\cos\pi[(h+k)x-(h-k)y]\sin2\pi\ell z, & (h+k)\text{even} \\ \\ \\ -4\sin\pi[(h-k)x+(h+k)y]\sin\pi[(h+k)x-(h-k)y]\sin2\pi\ell z, & (h+k)\text{odd} \end{cases}.$$

$$(4.7)$$

Comparison of (4.6) and (4.7) with (4.2) and (4.3) respectively shows that the functional form for the geometric structure factor is now different, as expected.

Finally, if the origin is shifted to a point on the other two-fold axis, with $x = 1/2$, $y = 0$, then (4.1) is replaced by $x + 1/2,y,z;$ $\bar{x} + 1/2,\bar{y},z;$ $\bar{y} + 1/2,x,z;$ $y + 1/2,\bar{x},z$, or, substituting x for $x + 1/2$, by $x,y,z;$ $\bar{x},\bar{y},z;$ $\bar{y} + 1/2,x + 1/2,z;$ $y + 1/2, \bar{x} + 1/2,z$, i.e. the same as (4.5). The real and imaginary parts of the geometric structure factor therefore have the same form as (4.6) and (4.7) respectively.

In summary then, the permissible origins for the space group P4 fall into two equivalence classes. In the one class are the origins which lie on either of the two four-fold axes and in the other are the origins which lie on either of the two two-fold axes. The functional form for the geometric structure factor when the origin lies on a four-fold axis is given by (4.2) and (4.3); the functional form for the geometric structure factor when the origin lies on a two-fold axis is given by (4.6) and (4.7). For each fixed functional form for the geometric structure

factor, those linear combinations of the phases which remain
unchanged in value for every choice of origin permitted by the
functional form for the structure factor are the structure
seminvariants. Clearly every structure invariant is also a
structure seminvariant for the former has the same value for all
permissible origins, not merely those belonging to a fixed
equivalence class. While the structure seminvariants are the
same for the two equivalence classes (or two functional forms for
the structure factor) the value of any structure seminvariant may
change in shifting from one equivalence class to the other.

In order to determine those single phases which are structure
seminvariants, one refers to Theorem 2.1. It is clear that now
$\vec{r}_0$ must be of the form

$$0,0,z \quad \text{or} \quad \frac{1}{2},\frac{1}{2},z \ , \tag{4.8}$$

where z is arbitrary, since these are the origins equivalent to
0,0,0. (Note that this is true independently of whether the
initial origin is chosen to lie on a four-fold axis or on a two-
fold axis, i.e. for either of the two permitted functional forms
for the structure factor.) In order to determine the vectors $\vec{h}$
for which $\phi_{\vec{h}}$ is a structure seminvariant, one sets (refer to
(2.6))

$$2\pi\vec{h}\cdot\vec{r}_0 = 2\pi \text{ times an integer} \tag{4.9}$$

where $\vec{r}_0$ may take any of the values (4.8). Writing $\vec{h} = (h,k,\ell)$,
(4.9) becomes

$$\left.\begin{array}{l} h\cdot 0 + k\cdot 0 + \ell z = \text{an integer} \\[2em] h\cdot\frac{1}{2} + k\cdot\frac{1}{2} + \ell z = \text{an integer} \end{array}\right\} . \tag{4.10}$$

Clearly (4.10) is satisfied for all z if and only if h+k is an
even integer and ℓ is zero. In symbols, the single phase $\phi_{hk\ell}$
is a structure seminvariant if and only if

$$
\left.
\begin{array}{l}
h \equiv k \pmod 2 \\[2em]
\text{and} \\[2em]
\ell = 0.
\end{array}
\right\}
\tag{4.11}
$$

Alternatively, if one defines the seminvariant modulus $\vec{\omega}_s$ by
means of

$$
\vec{\omega}_s = (2,0)
\tag{4.12}
$$

and the vector $\vec{h}_s$ seminvariantly associated with $\vec{h} = (h,k,\ell)$ by
means of

$$
\vec{h}_s = (h+k,\ell),
\tag{4.13}
$$

then $\phi_{\vec{h}}$ is a structure seminvariant if and only if

$$
\vec{h}_s \equiv 0 \pmod{\vec{\omega}_s},
\tag{4.14}
$$

i.e. if and only if there exist intergers m and n such that

$$
h + k = 2m, \quad \ell = 0 \cdot n,
\tag{4.15}
$$

or, if and only if h+k is even and ℓ is zero. Stated still a
different way, the phases ϕ_{gg0} and ϕ_{uu0}, where g means even
(gerade) and u means odd (ungerade), are the structure seminvariants. In summary then, once the functional form for the
geometric structure factor has been fixed then the phases ϕ_{gg0}
and ϕ_{uu0} have values which are determined by the structure alone

and are independent of the choice of origin consistent with the chosen functional form for the structure factor.

Next, in order to determine the linear combinations $\phi_{\vec{h}_1} + \phi_{\vec{h}_2}$ which are structure seminvariants, where

$$\vec{h}_1 = (h_1 \ k_1 \ \ell_1), \quad \vec{h}_2 = (h_2 \ k_2 \ \ell_2), \tag{4.16}$$

one notes that Eq. (2.6) implies

$$(\vec{h}_1 + \vec{h}_2) \cdot \vec{r}_0 = \text{an integer} \tag{4.17}$$

where $\vec{r}_0$ is an arbitrary one of the equivalent origins (4.8). In other words, from (4.16) and (4.8),

$$\left.\begin{array}{l} (h_1+h_2)\cdot 0 + (k_1+k_2)\cdot 0 + (\ell_1+\ell_2)z = \text{an integer} \\[3em] (h_1+h_2)\cdot\frac{1}{2} + (k_1+k_2)\cdot\frac{1}{2} + (\ell_1+\ell_2)z = \text{an integer} \end{array}\right\} \tag{4.18}$$

for all values of z. Clearly (4.18) holds if and only if h_1+h_2 and k_1+k_2 are even integers and $\ell_1+\ell_2$ is zero. In short $\phi_{\vec{h}_1} + \phi_{\vec{h}_2}$ is a structure seminvariant if and only if

$$\vec{h}_{1s} + \vec{h}_{2s} \equiv 0 \pmod{\vec{\omega}_s} \tag{4.19}$$

where the seminvariant modulus $\vec{\omega}_s$ is given by (4.12) and the vectors $\vec{h}_{1s}$ and $\vec{h}_{2s}$, seminvariantly associated with $\vec{h}_1$ and $\vec{h}_2$ respectively, are defined by (4.13). The results obtained so far are readily generalized and summarized in

<u>Theorem 4.1.</u> The linear combination of the phases

$$\sum_{\vec{h}} A_{\vec{h}} \; \phi_{\vec{h}}, \tag{4.20}$$

where the $A_{\vec{h}}$ are integers satisfying

$$\sum_{\vec{h}} A_{\vec{h}} \; \vec{h}_s \equiv 0 \pmod{\vec{\omega}_s}, \tag{4.21}$$

and $\vec{\omega}_s$ and $\vec{h}_s$ are defined by (4.12) and (4.13) respectively, is a structure seminvariant in the space group P4.

It is instructive next to derive the structure invariants for this space group. The permissible origins consist of all vectors $\vec{r}_0$ of the form

$$0,0,z; \quad \frac{1}{2},\frac{1}{2},z; \quad 0,\frac{1}{2},z; \quad \frac{1}{2},0,z; \tag{4.22}$$

where z is arbitrary. Referring to Theorem 2.1, it is clear that the single phase $\phi_{\vec{h}} = \phi_{hk\ell}$ is a structure invariant if and only if $\vec{h} \cdot \vec{r}_0$ is an integer for all permissible origins $\vec{r}_o$, i.e., in view of (4.22), if and only if, for all values of z,

$$\left. \begin{array}{l} h \cdot 0 + k \cdot 0 + \ell z \text{ is an integer} \\[2mm] h \cdot \dfrac{1}{2} + k \cdot \dfrac{1}{2} + \ell z \text{ is an integer} \\[2mm] h \cdot 0 + k \cdot \dfrac{1}{2} + \ell z \text{ is an integer} \\[2mm] h \cdot \dfrac{1}{2} + k \cdot 0 + \ell z \text{ is an integer} \end{array} \right\}. \tag{4.23}$$

It follows that $\phi_{hk\ell}$ is a structure invariant if and only if h and k are even integers and ℓ is zero. Alternatively, if the invariant modulus $\vec{\omega}_i$ is defined by

$$\vec{\omega}_i = (2,2,0), \qquad\qquad (4.24)$$

then $\phi_{\vec{h}}$ is a structure invariant if and only if

$$\vec{h} \equiv 0 \pmod{\vec{\omega}_i}. \qquad\qquad (4.25)$$

The same argument which led to Theorem 4.1 now yields

 Theorem 4.2.   The linear combination of the phases

$$\sum_{\vec{h}} A_{\vec{h}}\, \phi_{\vec{h}}, \qquad\qquad (4.26)$$

where the $A_{\vec{h}}$ are integers satisfying

$$\sum_{\vec{h}} A_{\vec{h}}\, \vec{h} \equiv 0 \pmod{\vec{\omega}_i}, \qquad\qquad (4.27)$$

and $\vec{\omega}_i$ is defined by (4.24), is a structure invariant in the space group P4.

 Comparison of Theorems 4.1 and 4.2 is instructive. Clearly every structure invariant is also a structure seminvariant. In particular, the phases ϕ_{gg0} and ϕ_{uu0} are structure seminvariants but of these only the phases ϕ_{gg0} are structure invariants.

Exercises I.2.

1. There are two equivalence classes in P4. The first
consists of all points of the form (0,0,z) or (1/2,1/2,z) (z
arbitrary) and, if the primary origin is on a four-fold axis,
contains just those points lying on either four-fold axis. The
second consists of all points (0,1/2,z) or (1/2,0,z) and comprises
the points which lie on either two-fold axis, provided of course
that the primary origin is again on a four-fold axis.

2. Determine the equivalence classes and identify the
structure seminvariants, in particular the single phases which
are structure seminvariants, in each of the space groups:

   a. $P4_1$.
   b. $P\bar{4}$ and P422.
   c. P3 and $P3_1$.
   d. P312 and $P3_1 12$.
   e. P321 and $P3_1 21$.
   f. R3.
   g. R32.

5. The General Theory of Seminvariance

The results obtained so far suggest the concepts and notation
which will be useful and the proofs already given may be generalized
to all the space groups without essential change. Although the
theories of both the structure invariants and seminvariants were
initially developed simultaneously, it is the concept of the structure seminvariant which plays the crucial role in the actual applications. For this reason only the latter theory will be elaborated
here and the reader interested in the parallel theory of the structure invariants is referred to the literature. It will therefore
be assumed that the origin has been restricted in accordance with
the selection made in Vol. I of "International Tables for X-Ray

Crystallography". In the monoclinic system b is chosen as the unique axis. Only the definitions and final results are given here. The reader may refer to the literature for the proofs or may himself extend those given in the previous sections. These earlier sections contain also the notions of equivalence and the definitions of the structure invariants and seminvariants.

5.1. The two prototypes, P1 and P$\bar{1}$. The motivation for the concepts introduced in describing the general theory will be better understood if the details are first worked out for the two simplest cases, P1 and P$\bar{1}$. It is assumed throughout this discussion that the crystal structure is given.

5.1.1. The space group P1. The origin may be chosen arbitrarily, i.e. there is just one equivalence class and it coincides with the totality of points in the unit cell. Assume that the origin has been fixed and denote by $\phi_{\vec{h}}$ the phase of $E_{\vec{h}}$ where $\vec{h}$ is an arbitrary reciprocal vector not $(0,0,0)$. First we seek an answer to the question whether, by suitable choice of origin, the value of the phase of $E_{\vec{h}}$ may be arbitrarily specified, e.g. $\phi'_{\vec{h}}$. In view of (2.6) this is equivalent to asking whether there exists a vector $\vec{r}_0$, the position vector of the desired new origin with respect to the old one, which satisfies

$$2m\pi + \phi'_{\vec{h}} = \phi_{\vec{h}} - 2\pi \vec{h} \cdot \vec{r}_0$$

$$= \phi_{\vec{h}} - 2\pi(h_1 r_{01} + h_2 r_{02} + h_3 r_{03}) \tag{5.1}$$

where m is an arbitrary integer and h_1, h_2, h_3 and r_{01}, r_{02}, r_{03} are the components of $\vec{h}$ and $\vec{r}_0$ respectively. Clearly this question has an affirmative answer provided only that h_1, h_2, and h_3 are not all zero (as is postulated) since in this case it is certainly possible to find values for r_{01}, r_{02}, and r_{03} which satisfy (5.1). In other words the phase of $E_{\vec{h}}$ may be arbitrarily specified and the origin will then lie anywhere on any one of the finite number

of parallel planes defined by (5.1) and bounded, of course, by
the unit cell. (The reader should verify that, as m ranges over
all the integers, (5.1) defines a finite number of parallel planes
within the unit cell.)

Suppose next that the phase of $E_{\vec{h}}$ has been specified to be
$\phi'_{\vec{h}}$ so that the origin is restricted to lie on the two-dimensional
subset (5.1) of the unit cell. What restriction, if any, does
this impose on the value of the phase ($\phi_{\vec{k}}$ with respect to the
old origin) of any other structure factor $E_{\vec{k}}$ where $\vec{k} \neq (0,0,0)$?
Clearly the permissible origins must satisfy, in addition to (5.1),
an equation

$$2n\pi + \phi'_{\vec{k}} = \phi_{\vec{k}} - 2\pi(k_1 r_{01} + k_2 r_{02} + k_3 r_{03}) \tag{5.2}$$

where n is an arbitrary integer, k_1, k_2, k_3 are the components of
$\vec{k}$ and $\phi'_{\vec{k}}$ is an arbitrary value for the phase of $E_{\vec{k}}$ with respect
to some presumed new origin permitted by the chosen value $\phi'_{\vec{h}}$.
The question now is what conditions must be imposed on $\vec{k}$ in order
that (5.1) and (5.2) be solvable for r_{01}, r_{02}, r_{03} for arbitrary
choice of $\phi'_{\vec{k}}$. A sufficient condition certainly is that $\vec{h}$ and $\vec{k}$
be linearly independent, i.e. that there exist no integers a and
b, other than a = b = 0, which satisfy

$$a\,\vec{h} + b\,\vec{k} = 0. \tag{5.3}$$

In fact, assume that $\vec{h}$ and $\vec{k}$ are linearly independent. Then,
eliminate r_{03} from (5.1) and (5.2):

$$(2m\pi + \phi'_{\vec{h}} - \phi_{\vec{h}})k_3 - (2n\pi + \phi'_{\vec{k}} - \phi_{\vec{k}})h_3 =$$

$$-2\pi[\,(h_1 k_3 - k_1 h_3)r_{01} + (h_2 k_3 - k_2 h_3)r_{02}]. \tag{5.4}$$

If

$$h_1k_3 - k_1h_3 = h_2k_3 - k_2h_3 = 0, \qquad (5.5)$$

then

$$k_3\vec{h} - h_3\vec{k} = k_3(h_1,h_2,h_3) - h_3(k_1,k_2,k_3) = 0 \qquad (5.6)$$

and the assumed linear independence of $\vec{h}$ and $\vec{k}$ implies that $h_3 = k_3 = 0$. If $h_1k_2 - k_1h_2 = 0$, then $k_2\vec{h} - h_2\vec{k} = 0$ and the assumed independence of $\vec{h}$ and $\vec{k}$ implies that $h_2 = k_2 = 0$. Hence $k_1\vec{h} - h_1\vec{k} = 0$, with h_1 and k_1 not both zero, contradicting the assumed independence of $\vec{h}$ and $\vec{k}$. It follows that $h_1k_2 - k_1h_2 \neq 0$ and (5.1) and (5.2) are solvable for r_{01} and r_{02} for arbitrary choice of $\phi_{\vec{h}}$ and $\phi_{\vec{k}}$ and r_{03} may be arbitrarily chosen.

Next, if not both equations (5.5) hold, then (5.4) is clearly solvable for r_{01} and r_{02}, (5.1) is then solvable for r_{03} and (5.2) is satisfied by these values for r_{01}, r_{02}, r_{03}. In any event then, if $\vec{h}$ and $\vec{k}$ are linearly independent, the values of the phases of $E_{\vec{h}}$ and $E_{\vec{k}}$ may be arbitrarily specified and the permissible origins will then lie on the intersection of two finite families of parallel planes, (defined by (5.1) and (5.2)) i.e. may be placed anywhere on any one of a finite set of parallel lines bounded by the unit cell.

Suppose next that $\vec{h}$ and $\vec{k}$ are not linearly independent. Since $\vec{h} \neq (0,0,0)$ and $\vec{k} \neq (0,0,0)$ it follows that there exist integers a and b, both different from 0, such that (5.3) is satisfied. Without loss of generality, a and b may be taken to be relatively prime (i.e. having no common factor >1). Then

$$\vec{k} = \frac{a}{b}\vec{h} \qquad (5.7)$$

where a and b are integers, and $\vec{k}$ is said to be linearly dependent
on $\vec{h}$ or rationally dependent on $\vec{h}$ according as $|b| = 1$ or $|b| \geq 1$.
Equation (5.2) now becomes

$$2n\pi + \phi'_{\vec{k}} = \phi_{\vec{k}} - 2\pi \frac{a}{b}(h_1 r_{01} + h_2 r_{02} + h_3 r_{03})$$

$$= \phi_{\vec{k}} + (2m\pi + \phi'_{\vec{h}} - \phi_{\vec{h}}) \frac{a}{b} \tag{5.8}$$

in view of (5.1). If $\vec{k}$ is linearly dependent on $\vec{h}$, i.e. $b = \pm 1$,
then, except for an irrelevant additive integer multiple of 2π,

$$\phi'_{\vec{k}} = \phi_{\vec{k}} \pm (\phi'_{\vec{h}} - \phi_{\vec{h}})\, a, \tag{5.9}$$

i.e. the arbitrarily chosen new value $\phi'_{\vec{h}}$ for the phase of $E_{\vec{h}}$
restricts the origin in such a way that the corresponding new value
$\phi'_{\vec{k}}$ for the phase of $E_{\vec{k}}$ is uniquely determined. In other words,
once the phase $\phi'_{\vec{h}}$ for $E_{\vec{h}}$ is arbitrarily specified then the value
of the phase of any structure factor $E_{\vec{k}}$, where $\vec{k}$ is linearly
dependent on $\vec{h}$, is uniquely determined by the crystal structure.

If $\vec{k}$ is rationally dependent on $\vec{h}$, i.e. $|b| \geq 1$, then the
integer m in (5.8) may take on the $|b|$ possible values $0,1,2,\ldots,$
$|b| - 1$, so that there are $|b|$ possible values for the phase of
$E_{\vec{k}}$ forming an arithmetic progression with common difference
$2\pi/|b|$. In this case then, once the phase of $E_{\vec{h}}$ has been arbitrar-
ily specified, there are $|b|$ possible values for the phase of any
other structure factor $E_{\vec{k}}$ where $\vec{k}$ is rationally dependent on $\vec{h}$,
i.e. $\vec{k} = (a/b) \vec{h}$, a and b relatively prime.

Suppose finally that $\vec{h}$ and $\vec{k}$ are linearly independent so that
the phases for $E_{\vec{h}}$ and $E_{\vec{k}}$ may be arbitrarily specified, thus
restricting the origin to lie on one of a finite number of
parallel lines in the unit cell. We seek to learn the effect of
this restriction on the value of a third phase $\phi'_{\vec{l}}$, where $\vec{l} \neq$

$(0,0,0)$ is an arbitrary reciprocal vector with components ℓ_1, ℓ_2, ℓ_3. The argument is similar to that just given for $\phi_{\vec{k}}'$. In addition to (5.1) and (5.2), r_{01}, r_{02}, and r_{03} must now also satisfy an equation of the form

$$2p\pi + \phi_{\vec{\ell}}' = \phi_{\vec{\ell}} - 2\pi(\ell_1 r_{01} + \ell_2 r_{02} + \ell_3 r_{03}) \qquad (5.10)$$

where p is an arbitrary integer and $\phi_{\vec{\ell}}'$ is the value for the phase of $E_{\vec{\ell}}$ with respect to some presumed new origin permitted by the (arbitrarily) chosen values $\phi_{\vec{h}}'$, $\phi_{\vec{k}}'$ for the phases of $E_{\vec{h}}$, $E_{\vec{k}}$ respectively. Hence now we seek an answer to the question: What conditions must be imposed on $\vec{\ell}$ in order that (5.1), (5.2), and (5.10) be solvable for r_{01}, r_{02}, r_{03} for arbitrary choice of $\phi_{\vec{\ell}}'$ (as well, of course, for arbitrary choice of $\phi_{\vec{h}}'$ and $\phi_{\vec{k}}'$)? As before, it may be shown that if $\vec{h}$, $\vec{k}$, $\vec{\ell}$ are linearly independent, i.e. if there exist no integers a, b, c not all zero such that

$$a\vec{h} + b\vec{k} + c\vec{\ell} = 0, \qquad (5.11)$$

or, equivalently, if

$$\begin{vmatrix} h_1 & h_2 & h_3 \\ k_1 & k_2 & k_3 \\ \ell_1 & \ell_2 & \ell_3 \end{vmatrix} = \Delta \neq 0, \qquad (5.12)$$

then the values for the phases of $E_{\vec{h}}$, $E_{\vec{k}}$, $E_{\vec{\ell}}$ may be arbitrarily chosen. The number of origins in the unit cell permitted by these choices for $\phi_{\vec{h}}'$, $\phi_{\vec{k}}'$, $\phi_{\vec{\ell}}'$ is equal to $|\Delta|$. Hence the origin will in this way be uniquely chosen if and only if $\Delta = \pm 1$.

If, on the other hand, $\Delta = 0$ or, equivalently, if $\vec{h}$, $\vec{k}$, and $\vec{\ell}$ are linearly dependent, then two possibilities arise: Either $\vec{\ell}$ is linearly dependent on $\vec{h}$ and $\vec{k}$, i.e.

$$\vec{\ell} = a_1\vec{h} + a_2\vec{k} \tag{5.13}$$

where a_1 and a_2 are integers; or $\vec{\ell}$ is rationally dependent on $\vec{h}$ and $\vec{k}$, i.e.

$$\vec{\ell} = b_1\vec{h} + b_2\vec{k} \tag{5.14}$$

where b_1 and b_2 are rational. In the first case the value for $\phi_{\vec{\ell}}'$ is uniquely determined by the chosen values for $\phi_{\vec{h}}'$ and $\phi_{\vec{k}}'$; in the second case the value for $\phi_{\vec{\ell}}'$ may be arbitrarily chosen from a finite set of values determined by the chosen values for $\phi_{\vec{h}}'$ and $\phi_{\vec{k}}'$. In the latter case (5.14) may be written $\vec{\ell} = 1/d(c_1\vec{h} + c_2\vec{k})$ where c_1, c_2, and d are integers having no common factor >1. Then $\phi_{\vec{\ell}}'$ may be arbitrarily chosen from a set of d values in arithmetic progression with common difference equal to $2\pi/d$.

Exercises I. 3.

1. Carry out the proofs of the results quoted in the last two paragraphs.

2. If $\vec{k} = (k_1, k_2, k_3)$ is rationally dependent on $\vec{h} = (h_1, h_2, h_3)$, show that $\vec{k}$ is linearly dependent on $\vec{h}$ provided that the greatest common divisor (g.c.d.) of h_1, h_2, h_3 is unity (i.e., by definition, provided that $\vec{h}$ is primitive). Illustrate with $\vec{h} = (6, 10, 15)$.

3. If $\vec{\ell} = (\ell_1, \ell_2, \ell_3)$ is rationally dependent on the linearly independent pair $\vec{h}$, $\vec{k}$, show that $\vec{\ell}$ is linearly dependent on $\vec{h}$ and $\vec{k}$ provided that the g.c.d. of the three determinants

$$\begin{vmatrix} h_1 & h_2 \\ k_1 & k_2 \end{vmatrix}, \quad \begin{vmatrix} h_1 & h_3 \\ k_1 & k_3 \end{vmatrix}, \quad \begin{vmatrix} h_2 & h_3 \\ k_2 & k_3 \end{vmatrix},$$

not all of which are zero, is unity (i.e., by definition, provided that the pair $\vec{h}$, $\vec{k}$ is primitive). Illustrate with $\vec{h}$ = (6,10,15), $\vec{k}$ = (1,2,4).

4. If $\Delta \neq 0$ is the determinant of the three vectors $\vec{h}$, $\vec{k}$, $\vec{\ell}$ defined by (5.12), show that any vector $\vec{m}$ (with 3 components) is rationally dependent on $\vec{h}$, $\vec{k}$, $\vec{\ell}$. Show, further, that $\vec{m}$ is linearly dependent on $\vec{h}$, $\vec{k}$, $\vec{\ell}$ (for arbitrary $\vec{m}$) if and only if $\Delta = \pm 1$ (i.e., by definition, if and only if the triple $\vec{h}$, $\vec{k}$, $\vec{\ell}$ is primitive). Illustrate with $\vec{h}$ = (6,10,15), $\vec{k}$ = (1,2,4), $\vec{\ell}$ = (2,3,4).

5. If $\vec{h}$ is linearly independent (i.e. $\vec{h} \neq (0,0,0)$), then there exists $\vec{k}$ such that the pair $\vec{h}$, $\vec{k}$ is linearly independent. If $\vec{h}$ is primitive, then there exists $\vec{k}$ such that the pair $\vec{h}$, $\vec{k}$ is primitive. See Ex. 2 and 3 for illustration.

6. If the pair $\vec{h}$, $\vec{k}$ is linearly independent (primitive), then there exists $\vec{\ell}$ such that the triple $\vec{h}$, $\vec{k}$, $\vec{\ell}$ is linearly independent (primitive). See Ex. 3 and 4 for illustration.

Under the hypothesis that the crystal structure is given, the results obtained so far may be conveniently summarized in the following:

Theorem 5.1. The value of any phase $\phi_{\vec{h}}$ which is linearly independent (i.e. $\vec{h} \neq (0,0,0)$) may be specified arbitrarily. Once this is done then the value of any phase $\phi_{\vec{k}}$ which is linearly dependent on $\phi_{\vec{h}}$ (i.e. $\vec{k}$ is linearly dependent on $\vec{h}$) is uniquely determined. Any phase $\phi_{\vec{k}}$ which is rationally dependent on $\phi_{\vec{h}}$ (i.e. $\vec{k}$ is rationally dependent on $\vec{h}$) is also linearly dependent on $\phi_{\vec{h}}$, whence its value is uniquely determined, provided that $\phi_{\vec{h}}$ is primitive (i.e. provided that $\vec{h}$ is primitive).

Theorem 5.2. The values of any two phases $\phi_{\vec{h}}$, $\phi_{\vec{k}}$, constituting a linearly independent set (i.e. the pair $\{\vec{h},\vec{k}\}$ is linearly independent), may be specified arbitrarily. Once this is done then the value of any phase $\phi_{\vec{\ell}}$, which is linearly dependent on the pair $\phi_{\vec{h}}$, $\phi_{\vec{k}}$, is uniquely determined. Any phase $\phi_{\vec{\ell}}$ which is rationally dependent on the pair $\phi_{\vec{h}}$, $\phi_{\vec{k}}$ is also linearly dependent

on this pair, whence its value is uniquely determined, provided
that the pair $\phi_{\vec{h}}$, $\phi_{\vec{k}}$ is primitive (i.e. provided that the pair
$\{\vec{h},\vec{k}\}$ is primitive).

   Theorem 5.3. The values of any three phases $\phi_{\vec{h}}$, $\phi_{\vec{k}}$, $\phi_{\vec{\ell}}$,
constituting a linearly independent set (i.e. the triple $\{\vec{h},\vec{k},\vec{\ell}\}$
is linearly independent), may be specified arbitrarily. Once this
is done then the value of any phase $\phi_{\vec{m}}$ which is linearly dependent
on the triple $\phi_{\vec{h}}$, $\phi_{\vec{k}}$, $\phi_{\vec{\ell}}$ is uniquely determined. Any phase $\phi_{\vec{m}}$, of
necessity rationally dependent on the triple $\phi_{\vec{h}}$, $\phi_{\vec{k}}$, $\phi_{\vec{\ell}}$, is also
linearly dependent on this triple, whence its value is uniquely
determined, provided that the triple $\phi_{\vec{h}}$, $\phi_{\vec{k}}$, $\phi_{\vec{\ell}}$ is primitive
($\Delta = \pm 1$), i.e. provided that the origin has been uniquely fixed
by the chosen values for $\phi_{\vec{h}}$, $\phi_{\vec{k}}$, $\phi_{\vec{\ell}}$.

   If it is not assumed that the crystal structure is given,
but it is assumed instead that a sufficiently large number of
normalized structure factor magnitudes $|E|$ are known, then, in
general, two possible (enantiomorphous) structures, related to
each other by reflection through a point, are determined. (The
complications presented by the possible existence of other homo-
metric structures, i.e. a multiple solution, greater than two-fold,
of the phase problem, are avoided by simply assuming that such
multiplicity of solutions does not occur.) Since the two values
for any structure invariant corresponding to the two enantiomorphs
have the same magnitude but opposite signs, Theorems 5.1-5.3 are
still valid under the assumption that a sufficiently large number
of normalized structure factor magnitudes $|E|$ are given and the
enantiomorph has been selected by specifying arbitrarily the
sign of any structure invariant whose magnitude is different from
0 or π.

   5.1.1.1. Primitivity and accessibility. Suppose that $\Delta \neq 0$
where Δ is defined by (5.12). Then the values for $\phi_{\vec{h}}$, $\phi_{\vec{k}}$, $\phi_{\vec{\ell}}$ may
be arbitrarily specified. If $|\Delta| = 1$ then every reciprocal
vector $\vec{m}$ is linearly dependent on the triple $\{\vec{h},\vec{k},\vec{\ell}\}$ and the value

of $\phi_{\vec{m}}$ is uniquely determined by the specified values for $\phi_{\vec{h}}$, $\phi_{\vec{k}}$, $\phi_{\vec{\ell}}$. We shall say that the triple $\{\vec{h},\vec{k},\vec{\ell}\}$ spans reciprocal space and every reciprocal vector $\vec{m}$ is linearly accessible modulo $(\vec{h},\vec{k},\vec{\ell})$. Also, every phase $\phi_{\vec{m}}$ is said to be linearly accessible modulo $(\phi_{\vec{h}},\phi_{\vec{k}},\phi_{\vec{\ell}})$. Alternatively, the totality of reciprocal space is linearly accessible modulo $(\vec{h},\vec{k},\vec{\ell})$.

Suppose next that $|\Delta|>1$ so that there are $|\Delta|$ choices for the origin permitted by the chosen values for $\phi_{\vec{h}}$, $\phi_{\vec{k}}$, $\phi_{\vec{\ell}}$. The totality of linear combinations of $\vec{h}$, $\vec{k}$, $\vec{\ell}$ with integer coefficients now spans a proper subspace of reciprocal space and not every reciprocal vector $\vec{m}$ is linearly accessible modulo $(\vec{h},\vec{k},\vec{\ell})$ nor is every phase linearly accessible modulo $(\phi_{\vec{h}},\phi_{\vec{k}},\phi_{\vec{\ell}})$. Those vectors $\vec{m}$ which are expressible as linear combinations of $\vec{h}$, $\vec{k}$, $\vec{\ell}$ with integer coefficients are said to be linearly accessible modulo $(\vec{h},\vec{k},\vec{\ell})$ and the corresponding phases $\phi_{\vec{m}}$ are said to be linearly accessible modulo $(\phi_{\vec{h}},\phi_{\vec{k}},\phi_{\vec{\ell}})$. The arbitrary reciprocal vector $\vec{m}$ is certainly rationally dependent on $\vec{h}$, $\vec{k}$, $\vec{\ell}$ (since $\Delta \neq 0$) but now a finite number $D \geq 1$ of values for $\phi_{\vec{m}}$ are possible and these form an arithmetic progression with common difference $2\pi/D$. Further, D is a divisor of Δ so that $1 \leq D \leq |\Delta|$ and $D = 1$ or $D > 1$ according as $\vec{m}$ is or is not linearly dependent on $\vec{h}$, $\vec{k}$, $\vec{\ell}$. Any of these D values for $\phi_{\vec{m}}$ may be chosen and the origin is either not further restricted by adjoining $\phi_{\vec{m}}$ to the basic set $\{\phi_{\vec{h}},\phi_{\vec{k}},\phi_{\vec{\ell}}\}$ or the origin is further restricted (and possible uniquely fixed) depending on whether $D = 1$ or $D > 1$. In fact once $\phi_{\vec{m}}$ is adjoined to the basic set $\{\phi_{\vec{h}},\phi_{\vec{k}},\phi_{\vec{\ell}}\}$ of known phases, then the number of permitted origins is equal to the g.c.d., G, of all 3×3 subdeterminants of the matrix,

$$\begin{pmatrix} h_1 & h_2 & h_3 \\ k_1 & k_2 & k_3 \\ \ell_1 & \ell_2 & \ell_3 \\ m_1 & m_2 & m_3 \end{pmatrix},$$

not all of whose 3 X 3 subdeterminants vanish. Further $G = |\Delta|/D$
so that G divides $|\Delta|$ and $1 \leq G \leq |\Delta|$. The chosen values for
$\phi_{\vec{h}}$, $\phi_{\vec{k}}$, $\phi_{\vec{\ell}}$, $\phi_{\vec{m}}$ fix the origin uniquely if and only if $G = 1$, i.e.
if and only if $D = |\Delta|$. In this case an arbitrary reciprocal
vector $\vec{n}$ is linearly dependent on the quadruple $\{\vec{h},\vec{k},\vec{\ell},\vec{m}\}$ and the
phase $\phi_{\vec{n}}$ is uniquely determined by the specified values for $\phi_{\vec{h}}$,
$\phi_{\vec{k}}$, $\phi_{\vec{\ell}}$, $\phi_{\vec{m}}$. Now we say that the quadruple $\{\vec{h},\vec{k},\vec{\ell},\vec{m}\}$ spans
reciprocal space and every reciprocal vector $\vec{n}$ is linearly
accessible modulo $(\vec{h},\vec{k},\vec{\ell},\vec{m})$. Alternatively, the totality of
reciprocal space is linearly accessible modulo $(\vec{h},\vec{k},\vec{\ell},\vec{m})$ and every
phase $\phi_{\vec{n}}$ is said to be linearly accessible modulo $(\phi_{\vec{h}},\phi_{\vec{k}},\phi_{\vec{\ell}},\phi_{\vec{m}})$.

Suppose on the other hand that $G > 1$ so that there are G
choices for the origin permitted by the chosen values for $\phi_{\vec{h}}$, $\phi_{\vec{k}}$,
$\phi_{\vec{\ell}}$, $\phi_{\vec{m}}$. The totality of linear combinations of $\vec{h}$, $\vec{k}$, $\vec{\ell}$, $\vec{m}$ with
integer coefficients now spans a proper subspace of reciprocal
space so that not every reciprocal vector $\vec{n}$ is linearly accessible
modulo $(\vec{h},\vec{k},\vec{\ell},\vec{m})$ nor is every phase $\phi_{\vec{n}}$ linearly accessible
modulo $(\phi_{\vec{h}},\phi_{\vec{k}},\phi_{\vec{\ell}},\phi_{\vec{m}})$. Those vectors $\vec{n}$ which happen to be expressible as linear combinations of $\vec{h}$, $\vec{k}$, $\vec{\ell}$, $\vec{m}$ with integer coefficients
are said to be linearly accessible modulo $(\vec{h},\vec{k},\vec{\ell},\vec{m})$, the corre-
sponding phases $\phi_{\vec{n}}$ are linearly accessible modulo $(\phi_{\vec{h}},\phi_{\vec{k}},\phi_{\vec{\ell}},\phi_{\vec{m}})$,
and the value of $\phi_{\vec{n}}$ is uniquely determined by the chosen values
for $\phi_{\vec{h}}$, $\phi_{\vec{k}}$, $\phi_{\vec{\ell}}$, $\phi_{\vec{m}}$. The arbitrary reciprocal vector $\vec{n}$ is clearly
rationally dependent on $\vec{h}$, $\vec{k}$, $\vec{\ell}$, $\vec{m}$ (since $\vec{n}$ is rationally
dependent on $\vec{h}$, $\vec{k}$, $\vec{\ell}$) but now a finite number $D' \geq 1$ of values
for $\phi_{\vec{n}}$ are possible and these form an arithmetic progression with
common difference $2\pi/D'$. Further, D' is a divisor of G so that
$1 \leq D' \leq G$ and $D' = 1$ or $D' > 1$ according as $\vec{n}$ is or is not
accessible modulo $(\vec{h},\vec{k},\vec{\ell},\vec{m})$. Any of these D' values for $\phi_{\vec{n}}$ may be
chosen and the origin is either not further restricted by adjoining
$\phi_{\vec{n}}$ to the set $\{\phi_{\vec{h}},\phi_{\vec{k}},\phi_{\vec{\ell}},\phi_{\vec{m}}\}$ or the origin is further restricted
(and possibly uniquely fixed) depending on whether $D' = 1$ or
$D' > 1$. In fact once $\phi_{\vec{n}}$ is adjoined to the set $\{\phi_{\vec{h}},\phi_{\vec{k}},\phi_{\vec{\ell}},\phi_{\vec{m}}\}$ of

known phases, then the number of permitted origins is equal to
the g.c.d., G', of all 3 X 3 subdeterminants of the matrix,

$$\begin{pmatrix} h_1 & h_2 & h_3 \\ k_1 & k_2 & k_3 \\ \ell_1 & \ell_2 & \ell_3 \\ m_1 & m_2 & m_3 \\ n_1 & n_2 & n_3 \end{pmatrix} ,$$

not all of whose 3 X 3 subdeterminants vanish. Further $G' = G/D'$
so that G' divides G and $1 \leq G' \leq G$. The chosen values for $\phi_{\vec{h}}$,
$\phi_{\vec{k}}$, $\phi_{\vec{\ell}}$, $\phi_{\vec{m}}$, $\phi_{\vec{n}}$ fix the origin uniquely if and only if $G' = 1$, i.e.
if and only if $D' = G$. In this case an arbitrary reciprocal
vector $\vec{p}$ is linearly dependent on the quintuple $\{\vec{h},\vec{k},\vec{\ell},\vec{m},\vec{n}\}$ and
the phase $\phi_{\vec{p}}$ is uniquely determined by the specified values for
$\phi_{\vec{h}}$, $\phi_{\vec{k}}$, $\phi_{\vec{\ell}}$, $\phi_{\vec{m}}$, $\phi_{\vec{n}}$. The quintuple $\{\vec{h},\vec{k},\vec{\ell},\vec{m},\vec{n}\}$ spans reciprocal
space, every reciprocal vector $\vec{p}$ is linearly accessible modulo
$(\vec{h},\vec{k},\vec{\ell},\vec{m},\vec{n})$, and every phase $\phi_{\vec{p}}$ is linearly accessible modulo
$(\phi_{\vec{h}},\phi_{\vec{k}},\phi_{\vec{\ell}},\phi_{\vec{m}},\phi_{\vec{n}})$.

If $G' > 1$ the process may be continued until finally one
reaches a primitive matrix,

$$\begin{pmatrix} h_1 & h_2 & h_3 \\ k_1 & k_2 & k_3 \\ \ell_1 & \ell_2 & \ell_3 \\ \cdots\cdots \end{pmatrix} ,$$

i.e. a matrix the g.c.d. of all of whose 3 X 3 subdeterminants is
unity. The origin is then finally uniquely fixed and every phase
is accessible. This discussion may be partially summarized in
the important **theorem:**

Theorem 5.4. In the space group P1, the concepts of total
linear accessibility and primitivity coincide.

Exercise I. 4.

1. Confirm Theorem 5.4 and the previous discussion by means
of the example

$$\vec{h} = (1\ 3\ 1),$$
$$\vec{k} = (0\ 2\ 3),$$
$$\vec{\ell} = (1\ 1\ 4),$$
$$\vec{m} = (1\ 5\ 2),$$
$$\vec{n} = (3\ 6\ 6).$$

Verify that $\Delta = 12$ so that arbitrarily chosen values for $\phi_{\vec{h}}$, $\phi_{\vec{k}}$,
$\phi_{\vec{\ell}}$ do not serve to fix the origin uniquely; nor does adjoining $\phi_{\vec{m}}$
to the basic set restrict the origin to a unique value. In fact,
since $G = 4$, $D = |\Delta|/G = 12/4 = 3$, there are three possible
values for $\phi_{\vec{m}}$ (confirm by expressing $\vec{m}$ as a rational combination
of $\vec{h}$, $\vec{k}$, $\vec{\ell}$) and, when one of these is chosen, there still remains
a four-fold ambiguity in the position of the origin. Clearly,
not every phase is accessible modulo $\{\phi_{\vec{h}}, \phi_{\vec{k}}, \phi_{\vec{\ell}}\}$ nor is every
phase accessible even modulo $\{\phi_{\vec{h}}, \phi_{\vec{k}}, \phi_{\vec{\ell}}, \phi_{\vec{m}}\}$. However, express
$\vec{n}$ as a rational combination of $\vec{h}$, $\vec{k}$, $\vec{\ell}$, $\vec{m}$ in order to verify that,
once $\phi_{\vec{h}}$, $\phi_{\vec{k}}$, $\phi_{\vec{\ell}}$, $\phi_{\vec{m}}$ have been specified, there are four possible
values for $\phi_{\vec{n}}$. Hence $D' = 4$. Confirm that $G' = 1$ so that
$G' = G/D' = 4/4$ and the specification of the values of $\phi_{\vec{h}}$, $\phi_{\vec{k}}$,
$\phi_{\vec{\ell}}$, $\phi_{\vec{m}}$, $\phi_{\vec{n}}$ finally serves to fix the origin uniquely. Verify,
finally, that every reciprocal vector $\vec{p}$ is in fact a linear (not
merely rational) combination of $\vec{h}$, $\vec{k}$, $\vec{\ell}$, $\vec{m}$, $\vec{n}$ so that every
phase $\phi_{\vec{p}}$ is linearly accessible modulo $\{\phi_{\vec{h}}, \phi_{\vec{k}}, \phi_{\vec{\ell}}, \phi_{\vec{m}}, \phi_{\vec{n}}\}$ and the
value of $\phi_{\vec{p}}$ is uniquely determined by the chosen values for
$\{\phi_{\vec{h}}, \phi_{\vec{k}}, \phi_{\vec{\ell}}, \phi_{\vec{m}}, \phi_{\vec{n}}\}$.

   5.1.2. The space group $P\bar{1}$. The origin may be any one of
the eight centers of symmetry

$$(\varepsilon_1, \varepsilon_2, \varepsilon_3), \quad \varepsilon_i = 0 \text{ or } \frac{1}{2}, \quad i = 1, 2, 3, \qquad (5.15)$$

which clearly constitute the unique equivalence class. Assume
that the origin has been fixed and denote by $\phi_{\vec{h}}$ (= 0 or π) the
phase of $E_{\vec{h}}$ where $\vec{h}$ is an arbitrary reciprocal vector different
from (0,0,0). It is already known (Theorem 3.1) that $\phi_{\vec{h}}$ is a
structure invariant if all components of $\vec{h}$ are even integers.
It is not difficult to show that the converse also holds, i.e.
if the components of $\vec{h}$ are not all even then, by suitable choice
of origin, the phase of $E_{\vec{h}}$ may take on the arbitrary value $\phi_{\vec{h}}'$
(= 0 or π). In fact, in view of (2.6), it is sufficient to
determine values of r_{01}, r_{02}, r_{03} (0 or 1/2) such that

$$\phi_{\vec{h}}' = \phi_{\vec{h}} - 2\pi(h_1 r_{01} + h_2 r_{02} + h_3 r_{03}) \qquad (5.16)$$

in which at least one of h_1, h_2, h_3 is odd. For example, if h_1
is odd, one merely sets $r_{02} = r_{03} = 0$ and $r_{01} = 0$ or 1/2
according as $\phi_{\vec{h}}' = \phi_{\vec{h}}$ or $\phi_{\vec{h}}' \neq \phi_{\vec{h}}$. It is possible to go even further
since it is easy to show that no matter what the values (0 or π)
for $\phi_{\vec{h}}'$ and $\phi_{\vec{h}}$ and no matter what the parities of h_1, h_2, h_3
(provided of course that h_1, h_2, h_3 are not all even), then
precisely four of the eight possible sets of values (0 or 1/2)
for r_{01}, r_{02}, r_{03} actually satisfy (5.16). In short, if the
components of $\vec{h}$ are not all even, then the phase of $E_{\vec{h}}$ may be
specified arbitrarily and the origin will then be restricted to
four of the eight points (5.15).

Suppose next that the phase of $E_{\vec{h}}$ has been specified to be
$\phi_{\vec{h}}'$ so that the origin is restricted to a subset of (5.15)
consisting of four points. What restriction, if any, does this
impose on the value of the phase ($\phi_{\vec{k}}$ with respect to the old
origin) of any other structure factor $E_{\vec{k}}$ where not all the components of $\vec{k}$ are even? Clearly the permissible origins must
satisfy, in addition to (5.16), an equation

$$\phi'_{\vec{k}} = \phi_{\vec{k}} - 2\pi(k_1 r_{01} + k_2 r_{02} + k_3 r_{03}) \qquad (5.17)$$

where k_1, k_2, k_3 are the components of $\vec{k}$ and $\phi'_{\vec{k}}$ is an arbitrary value (0 or π) for the phase of $E_{\vec{k}}$ with respect to some presumed new origin permitted by the chosen value $\phi'_{\vec{h}}$. The question now is what conditions must be imposed on $\vec{k}$ in order that (5.15) and (5.16) be solvable for r_{01}, r_{02}, r_{03} (0 or 1/2) for arbitrary choice of $\phi'_{\vec{k}}$. A sufficient condition certainly is that $\vec{h}$ and $\vec{k}$ be linearly independent modulo (2,2,2), i.e. that there exist no integers a and b, not both even, such that the three components of

$$a\vec{h} + b\vec{k} \qquad (5.17)$$

are even. (An equivalent statement is that the components of $\vec{h} + \vec{k}$ be not all even.) In fact, assume that $\vec{h}$ and $\vec{k}$ are linearly independent modulo (2,2,2). Then, eliminate r_{03} from (5.15) and (5.16):

$$(\phi'_{\vec{h}} - \phi_{\vec{h}})k_3 - (\phi'_{\vec{k}} - \phi_{\vec{k}})h_3 =$$

$$-2\pi[(h_1 k_3 - k_1 h_3)r_{01} + (h_2 k_3 - k_2 h_3)r_{02}]. \qquad (5.18)$$

If

$$h_1 k_3 - k_1 h_3 \equiv h_2 k_3 - k_2 h_3 \equiv 0 \ (\text{mod } 2), \qquad (5.19)$$

i.e. if $h_1 k_3 - k_1 h_3$ and $h_2 k_3 - k_2 h_3$ are both even, then

$$k_3\vec{h} - h_3\vec{k} = k_3(h_1,h_2,h_3) - h_3(k_1,k_2,k_3) \equiv 0 \mod (2,2,2), \qquad (5.20)$$

i.e. the three components of $k_3\vec{h} - h_3\vec{k}$ are all even, and the assumed linear independence modulo $(2,2,2)$ of $\vec{h}$ and $\vec{k}$ implies that h_3 and k_3 are both even. If $h_1k_2 - k_1h_2$ is even, then the components of $k_2\vec{h} - h_2\vec{k}$ are all even and the assumed linear independence modulo $(2,2,2)$ of $\vec{h}$ and $\vec{k}$ implies that h_2 and k_2 are both even. Hence the components of $k_1\vec{h} - h_1\vec{k}$ are all even, with h_1 and k_1 not both even, contradicting the assumed linear independence modulo $(2,2,2)$ of $\vec{h}$ and $\vec{k}$. It follows that $h_1k_2 - k_1h_2$ is odd and (5.16) and (5.17) are solvable for r_{01} and r_{02}, for arbitrary choice of $\phi_{\vec{h}}'$ and $\phi_{\vec{k}}'$, and r_{03} is arbitrary (0 or 1/2). In fact, (5.16) and (5.17) imply

$$(h_1k_2 - k_1h_2)r_{01} = \frac{1}{2\pi}[k_2(\phi_{\vec{h}} - \phi_{\vec{h}}') - h_2(\phi_{\vec{k}} - \phi_{\vec{k}}')] = 0 \text{ or } \frac{1}{2} \qquad (5.21)$$

and, since $h_1k_2 - k_1h_2$ is odd, except for an irrelevant additive integer, $r_{01} = 0$ or 1/2 respectively. One solves for r_{02} in a similar way and r_{03} is arbitrary since h_3 and k_3 are both even.

Next, if $h_1k_3 - k_1h_3$ and $h_2k_3 - k_2h_3$ are not both even, i.e. if (5.19) does not hold, then (5.18) is clearly solvable for r_{01} and r_{02}, (5.16) is solvable for r_{03}, and (5.17) is satisfied by these values for r_{01}, r_{02}, r_{03}. In any event then, if $\vec{h}$ and $\vec{k}$ are linearly independent modulo $(2,2,2)$, the values (0 or π) of the phases of $E_{\vec{h}}$ and $E_{\vec{k}}$ may be arbitrarily specified and, as a closer analysis easily shows, the permissible origins will then lie on a subset of (5.15) consisting of just two points.

Suppose next that $\vec{h}$ and $\vec{k}$ are linearly dependent modulo $(2,2,2)$, i.e. that the components of $\vec{h} + \vec{k}$ are all even (assuming of course that the components of $\vec{h}$ are not all even and that the components of $\vec{k}$ are not all even). Then k_1 has the same parity as h_1, k_2 the same parity as h_2, k_3 the same parity as h_3 and the difference $(k_1 r_{01} + k_2 r_{02} + k_3 r_{03}) - (h_1 r_{01} + h_2 r_{02} + h_3 r_{03})$ is an integer (since $r_{01} = 0$ or $1/2$, etc.). Hence, from (5.16) and (5.17),

$$\phi'_{\vec{k}} = \phi'_{\vec{h}} - \phi'_{\vec{h}} - \phi'_{\vec{k}} \qquad (5.22)$$

except for an irrelevant additive integer multiple of 2π. In other words, in the case that $\vec{h}$ and $\vec{k}$ are linearly dependent modulo (2,2,2), the chosen value for $\phi'_{\vec{h}}$ restricts the origin in such a way that the value for $\phi'_{\vec{k}}$ is determined and may not (as in the previous case) be arbitrarily specified.

Suppose finally that $\vec{h}$ and $\vec{k}$ are linearly independent modulo (2,2,2), i.e. not all components of $\vec{h}$ are even, not all components of $\vec{k}$ are even, and not all components of $\vec{h} + \vec{k}$ are even. Then the phases $\phi'_{\vec{h}}$, $\phi'_{\vec{k}}$ (0 or π) for $E_{\vec{h}}$ and $E_{\vec{k}}$ may be arbitrarily specified, thus restricting the origin to two points of the set (5.15). We seek to learn the effect of this restriction on the value of a third phase $\phi'_{\vec{\ell}}$ where $\vec{\ell}$ is an arbitrary reciprocal vector with components ℓ_1, ℓ_2, ℓ_3 not all even. The argument is similar to that just given for $\phi'_{\vec{k}}$ so that it will suffice to quote the final results. If the components of $\vec{h} + \vec{\ell}$ are all even, the value of $\phi'_{\vec{\ell}}$ is determined by that of $\phi'_{\vec{h}}$. Again, if the components of $\vec{k} + \vec{\ell}$ are all even, then the value of $\phi'_{\vec{\ell}}$ is uniquely determined by that of $\phi'_{\vec{k}}$. Next, if the components of $\vec{h} + \vec{k} + \vec{\ell}$ are all even (i.e. $\vec{\ell}$ is linearly dependent modulo (2,2,2) on the pair $\{\vec{h},\vec{k}\}$) then the value of $\phi'_{\vec{\ell}}$ is uniquely determined by the chosen values for $\phi'_{\vec{h}}$ and $\phi'_{\vec{k}}$. Finally, if not all components of $\vec{h} + \vec{\ell}$ are even, not all components of $\vec{k} + \vec{\ell}$ are even, and not all components of $\vec{h} + \vec{k} + \vec{\ell}$ are even (i.e. $\vec{\ell}$ is linearly independent modulo (2,2,2) of the pair $\{\vec{h},\vec{k}\}$), the value (0 or π) for $\phi'_{\vec{\ell}}$ may be chosen arbitrarily and the origin will then be uniquely fixed. If the triple $\{\vec{h},\vec{k},\vec{\ell}\}$ is linearly independent modulo (2,2,2), i.e. at least one component of each of $\vec{h}$, $\vec{k}$, $\vec{\ell}$, $\vec{h} + \vec{k}$, $\vec{h} + \vec{\ell}$, $\vec{k} + \vec{\ell}$, $\vec{h} + \vec{k} + \vec{\ell}$ is odd, so that the values for $\phi_{\vec{h}}$, $\phi_{\vec{k}}$, $\phi_{\vec{\ell}}$ may be arbitrarily specified, thus uniquely fixing the origin, the value of an arbitrary phase $\phi_{\vec{m}}$ is then uniquely determined by the crystal structure. In this case too, $\vec{m}$ is of necessity linearly dependent modulo (2,2,2) on the triple $\{\vec{h},\vec{k},\vec{\ell}\}$, i.e. all three components of one of $\vec{m}$, $\vec{h} + \vec{m}$, $\vec{k} + \vec{m}$, $\vec{\ell} + \vec{m}$, $\vec{h} + \vec{k} + \vec{m}$, $\vec{h} + \vec{\ell} + \vec{m}$, $\vec{k} + \vec{\ell} + \vec{m}$, $\vec{h} + \vec{k} + \vec{\ell} + \vec{m}$ are even integers.

   5.1.2.1. <u>Accessibility</u>. Let $\vec{h}$, $\vec{k}$, $\vec{\ell}$, ... be a set of reciprocal vectors and let $\vec{n}$ be an arbitrary reciprocal vector. In accordance with the concept of accessibility as defined in P1, $\vec{n}$ (or $\phi_{\vec{n}}$) will be said to be linearly accessible modulo $(\vec{h},\vec{k},\vec{\ell},...)$ (or modulo $(\phi_{\vec{h}},\phi_{\vec{k}},\phi_{\vec{\ell}},...)$) if there exist integers a, b, c, ... such that

$$\vec{n} = a\vec{h} + b\vec{k} + c\vec{\ell} + \ldots . \tag{5.23}$$

Since, in P1, condition (5.23) is synonymous with linear dependence, $\phi_{\vec{n}}$ is accessible modulo $(\phi_{\vec{h}},\phi_{\vec{k}},\phi_{\vec{\ell}},...)$ if and only if $\phi_{\vec{n}}$ is linearly dependent on $\phi_{\vec{h}}$, $\phi_{\vec{k}}$, $\phi_{\vec{\ell}}$, ... However, in the space group P$\bar{1}$, these concepts no longer coincide. The phase $\phi_{\vec{n}}$ may be linearly dependent modulo (2,2,2) on the set $\phi_{\vec{h}}$, $\phi_{\vec{k}}$, $\phi_{\vec{\ell}}$, ... but $\phi_{\vec{n}}$ is not necessarily linearly accessible modulo $(\phi_{\vec{h}},\phi_{\vec{k}},\phi_{\vec{\ell}},...)$. In contrast to space group P1, even primitivity is not sufficient to secure accessibility in P$\bar{1}$ (Ex. I.5.). However, if the basis set $\phi_{\vec{h}}$, $\phi_{\vec{k}}$, $\phi_{\vec{\ell}}$, ... contains at least three phases, then primitivity is sufficient. In fact, it is quite easy to confirm the following theorem:

   <u>Theorem 5.5.</u> If the set of three or more phases $\phi_{\vec{h}}$, $\phi_{\vec{k}}$, $\phi_{\vec{\ell}}$, ... is primitive, i.e. if the g.c.d. of all 3 X 3 sub-determinants of

$$\begin{pmatrix} h_1 & h_2 & h_3 \\ k_1 & k_2 & k_3 \\ \ell_1 & \ell_2 & \ell_3 \\ \cdots & \cdots & \cdots \end{pmatrix}$$

is unity, then any phase $\phi_{\vec{n}}$ is linearly accessible modulo $(\phi_{\vec{h}}, \phi_{\vec{k}}, \phi_{\vec{\ell}}, \ldots)$.

Exercise I. 5.

1. Since the g.c.d. of all 2 X 2 subdeterminants of the matrix

$$\begin{pmatrix} 1 & 2 & 2 \\ 3 & 5 & 2 \end{pmatrix}$$

is unity, the pair $\{\phi_{1\ 2\ 2}, \phi_{3\ 5\ 2}\}$ is primitive. Evidently also $\phi_{4\ 5\ 4}$ is linearly dependent modulo (2,2,2) on $\phi_{1\ 2\ 2}$ and $\phi_{3\ 5\ 2}$. Show however that $\phi_{4\ 5\ 4}$ is not linearly accessible modulo $(\phi_{1\ 2\ 2}, \phi_{3\ 5\ 2})$.

5.2. Linear dependence and independence. The detailed exposition of the theory of seminvariance (invariance) for the space groups P1 and P$\bar{1}$ suggests the notation and terminology which will be useful for the remaining space groups. In general, the treatment for any space group is a kind of hybridization of those for P1 and P$\bar{1}$ except that the dimension may change and, in the trigonal and hexagonal systems, the modulus 3 may arise.

It is convenient to introduce the notion of linear dependence and independence modulo $\vec{\omega}$ where $\vec{\omega} = (\omega_1, \ldots, \omega_p)$ is a p-dimensional vector (the modulus) with integer components $\omega_1, \ldots, \omega_p$ some or all of which may be zero. Let $\vec{h} = (h_1, \ldots, h_p)$ be a p-dimensional vector with integer components $h_1, \ldots, h_p$. Then $\vec{h}$ is said to be divisible by $\vec{\omega}$ if the following two conditions are satisfied:

1. $\omega_i = 0$ implies $h_i = 0$,

2. $\omega_i \neq 0$ implies h_i is divisible by ω_i.

In other words $\vec{h}$ is divisible by $\vec{\omega}$ if there exist p integers $q_1, \ldots, q_p$ such that $h_i = q_i \omega_i$, $i = 1, \ldots, p$. If $\vec{h}$ is divisible by $\vec{\omega}$ then the notation

$$\vec{h} \equiv 0 \ (\text{mod } \vec{\omega}) \qquad (5.24)$$

is used and $\vec{h}$ is said to be congruent to zero modulo $\vec{\omega}$. If, for example, p = 1, then the integer h is congruent to zero modulo 0 if and only if h = 0; h is congruent to zero modulo $\omega \neq 0$ if and only if h is divisible by ω. Two p-dimensional vectors $\vec{h}_1$ and $\vec{h}_2$ are said to be congruent modulo $\vec{\omega}$ if the difference $\vec{h}_1 - \vec{h}_2$ is divisible by $\vec{\omega}$ and the notation

$$\vec{h}_1 \equiv \vec{h}_2 \ (\text{mod } \vec{\omega}) \qquad (5.25)$$

is used. If $\vec{h}_1$ and $\vec{h}_2$ are not congruent to each other modulo $\vec{\omega}$ they are said to be incongruent modulo $\vec{\omega}$ and the corresponding notation is

$$\vec{h}_1 \not\equiv \vec{h}_2 \ (\text{mod } \vec{\omega}).$$

Examples are

$(5 \ 1 \ 2) \equiv (3 \ 1 \ \bar{4}) \equiv (1 \ 1 \ 0) \ (\text{mod } (2 \ 0 \ 2))$,

$(1 \ 1 \ 3) \equiv (1 \ 5 \ 3) \equiv (1 \ \bar{3} \ 3) \ (\text{mod } (0 \ 2 \ 0))$,

$(1 \ 1 \ 2) \equiv (3 \ 5 \ 0) \equiv (7 \ 3 \ 4) \ (\text{mod } (2 \ 2 \ 2))$,

$(4 \ 1) \equiv (1 \ 3) \equiv (7 \ \bar{1}) \ (\text{mod } (3 \ 2))$,

but

$(1 \ 3) \not\equiv (3 \ 2) \ (\text{mod } (2 \ 2))$,

$(4 \ 3) \not\equiv (1 \ 1) \ (\text{mod } (3 \ 0))$.

Next, a set of n vectors $\{\vec{h}_j\}$, $j = 1,\ldots,n$, $(n \geq 1)$, is said
to be linearly dependent modulo $\vec{\omega}$ if there exists a set of n
integers $a_1,\ldots,a_n$ at least one of which is incongruent to zero
modulo ω_i for every i $(i = 1,\ldots,p)$, such that

$$\sum_{j=1}^{n} a_j \vec{h}_j \equiv 0 \ (\text{mod } \vec{\omega}). \qquad (5.26)$$

The set $\{\vec{h}_j\}$ is said to be linearly independent modulo $\vec{\omega}$ if it is
not linearly dependent.

The vector $\vec{h}$ is linearly dependent modulo $\vec{\omega}$ on, or linearly
independent modulo $\vec{\omega}$ of, the set $\{\vec{h}_j\}$, $j = 1,2,\ldots,n$, according
as there exist or there do not exist n integers a_j, $j = 1,2,\ldots,n$,
some or all of which may be zero, such that

$$\vec{h} \equiv \sum_{j=1}^{n} a_j \vec{h}_j \ (\text{mod } \vec{\omega}). \qquad (5.27)$$

In particular, any vector $\vec{h}$ which is divisible by $\vec{\omega}$ is linearly
dependent modulo $\vec{\omega}$ on any set of vectors since every a_j in (5.27)
may then be taken to be equal to zero.

In case $\omega_i = 0$ for every $i = 1,2,\ldots,p$, our concepts reduce
to ordinary linear dependence and independence, and the term
'modulo $\vec{\omega}$' will usually be omitted. For this case the following
two theorems are well known:

Theorem 5.6. If $n > p$, the set of n vectors $\vec{h}_j$ =
$(h_{j1}, h_{j2}, \ldots, h_{jp})$, $j = 1,2,\ldots,p$, is linearly dependent.

Theorem 5.7. If $n \leq p$, the set of n vectors $\vec{h}_j$ is linearly
dependent if and only if every nxn sub-determinant of the nxp
matrix

$$H = \begin{pmatrix} h_{11} & h_{12} & \cdots & h_{1p} \\ h_{21} & h_{22} & \cdots & h_{2p} \\ \cdots\cdots\cdots\cdots\cdots \\ h_{n1} & h_{n2} & \cdots & h_{np} \end{pmatrix} \qquad (5.28)$$

vanishes, i.e. if and only if the rank of H is less than n.

5.3. Rational dependence and independence. The vector $\vec{h}$ is rationally dependent modulo $\vec{\omega}$ on, or rationally independent modulo $\vec{\omega}$ of, the set $\{\vec{h}_j\}$, j = 1,2,...,n, according as there exist or there do not exist n rational numbers a_j, j = 1,2,...,n, some or all of which may be zero, such that

$$\vec{h} \equiv \sum_{j=1}^{n} a_j \vec{h}_j \pmod{\vec{\omega}}. \qquad (5.29)$$

The following theorem is an immediate consequence of the previous definitions:

Theorem 5.8. If the vector $\vec{h}$ is linearly dependent modulo $\vec{\omega}$ on the set $\{\vec{h}_j\}$, then $\vec{h}$ is rationally dependent modulo $\vec{\omega}$ on the set $\{\vec{h}_j\}$.

The converse of Theorem 5.8 is not true. However, when $\vec{\omega} = 0$, we obtain a partial converse of this theorem by means of the important concept of the primitive set.

5.4. Primitive sets. Let the set M of n vectors $\{\vec{h}_j\}$, where $\vec{h}_j = (h_{j1}, h_{j2}, ..., h_{jp})$, j = 1,2,...,n, be given. Denote by r the rank of the matrix H (5.28). Then r≤n, r≤p and not all the rxr sub-determinants of H vanish. (However, by definition of rank, all sxs sub-determinants of H, where s>r, do vanish). If the g.c.d. of all rxr sub-determinants of H is unity, then the set M is said to be primitive. In particular, if n = p and the set M is linearly independent (whence r = n = p), then M is

primitive if and only if the determinant of H is ±1. The
importance of the concept of the primitive set is due to the
following fundamental theorem, a partial converse of Theorem 5.8
when every ω_i = 0:

Theorem 5.9. If the vector $\vec{h}$ is rationally dependent on the
primitive set M, then $\vec{h}$ is linearly dependent on the set M.

Exercises I.6.

(All vectors are assumed to have integer components.)

1. One dimension. Define the 1-D (one-dimensional) vectors
$\vec{h}_1$, $\vec{h}_2$, $\vec{h}_3$ by means of

$$\vec{h}_1 = (1), \ \vec{h}_2 = (2), \ \vec{h}_3 = (3),$$

and the sets M_1, M_2, M_3 by

$$M_1 = \{\vec{h}_1\}, \ M_2 = \{\vec{h}_2\}, \ M_3 = \{\vec{h}_3\}.$$

Verify that each of M_1, M_2, M_3 is linearly independent but only M_1
is primitive. Infer that every 1-D vector $\vec{h}$, of necessity
rationally dependent on each of M_1, M_2, M_3, is also linearly
dependent on M_1 but may or may not be linearly dependent on M_2 or
M_3. Illustrate. Next, define M_4, M_5, M_6 by

$$M_4 = \{\vec{h}_1, \vec{h}_2\}, \ M_5 = \{\vec{h}_1, \vec{h}_3\}, \ M_6 = \{\vec{h}_2, \vec{h}_3\}$$

and confirm that each of M_4, M_5, M_6 is linearly dependent and
primitive. Infer that any 1-D vector $\vec{h}$, of necessity rationally
dependent on each of M_4, M_5, M_6, is also linearly dependent on
each of these sets. Verify by examples.

2. Two dimensions. Define the 2-D vectors $\vec{h}_1$, $\vec{h}_2$ by

$$\vec{h}_1 = (2,5), \ \vec{h}_2 = (1,3)$$

and the sets M_1, M_2, M_3 by

$$M_1 = \{\vec{h}_1\}, \ M_2 = \{\vec{h}_2\}, \ M_3 = \{\vec{h}_1, \vec{h}_2\}.$$

Then, each of M_1, M_2, and M_3 is linearly independent and primitive. A 2-D vector $\vec{h}$ which happens to be rationally dependent on M_1 or M_2 is also linearly dependent on M_1 or M_2 respectively. Any 2-D vector $\vec{h}$ is of necessity rationally dependent on M_3 and is therefore linearly dependent on M_3. Illustrate.

3. Two dimensions. Define the 2-D vectors $\vec{h}_1$, $\vec{h}_2$, $\vec{h}_3$, $\vec{h}_4$ by means of $\vec{h}_1 = (1,5)$, $\vec{h}_2 = (1,2)$, $\vec{h}_3 = (3,2)$, $\vec{h}_4 = (2,4)$ and the sets M_j by

$$M_j = \{\vec{h}_j\}, \quad j = 1,2,3,4.$$

Verify that each of M_1, M_2, M_3 is linearly independent and primitive while M_4 is linearly independent but not primitive. Infer that any 2-D vector $\vec{h}$ which happens to be rationally dependent on any one of M_1, M_2, M_3 is also linearly dependent on M_1, M_2 or M_3 respectively. However if $\vec{h}$ is rationally dependent on M_4, $\vec{h}$ is not necessarily linearly dependent on M_4. Illustrate.

Next, define the sets M_j, $j = 5,6,\ldots,10$ by

$$M_5 = \{\vec{h}_1,\vec{h}_2\}, \quad M_6 = \{\vec{h}_1,\vec{h}_3\}, \quad M_7 = \{\vec{h}_1,\vec{h}_4\},$$
$$M_8 = \{\vec{h}_2,\vec{h}_3\}, \quad M_9 = \{\vec{h}_2,\vec{h}_4\}, \quad M_{10} = \{\vec{h}_3,\vec{h}_4\}.$$

Verify that each of M_5, M_6, M_7, M_8, M_{10} is linearly independent and not primitive while M_9 is linearly dependent and primitive. Conclude that any 2-D vector $\vec{h}$, of necessity rationally dependent on each of M_5, M_6, M_7, M_8, M_{10}, is not necessarily linearly dependent on any of these sets. However, if $\vec{h}$ happens to be rationally dependent on M_9, then $\vec{h}$ is linearly dependent on M_9.

Finally, define M_{11} by

$$M_{11} = \{\vec{h}_1,\vec{h}_2,\vec{h}_3\}.$$

Then M_{11} is linearly dependent and primitive. It follows that any 2-D vector $\vec{h}$, of necessity rationally dependent on M_{11}, is also linearly dependent on M_{11}.

Illustrate with $\vec{h} = (1,3)$. Clearly $\vec{h}$ is rationally dependent on each of M_5, M_6, M_7, M_8, M_{10} but is linearly dependent on none of these. However, in view of

$$\vec{h} = 3\vec{h}_1 - 8\vec{h}_2 + 2\vec{h}_3,$$

$\vec{h}$ is linearly dependent on M_{11}.

4. Three dimensions. Define 3-D vectors $\vec{h}_1$, $\vec{h}_2$, $\vec{h}_3$, $\vec{h}_4$ and suitable sets M_1, M_2,... to illustrate Theorem 5.9 in three dimensions.

Let the set of n vectors $\{\vec{h}_j\}$, $j = 1,2,\ldots,n$, be linearly independent, and let n<p. It is known that there exists a vector $\vec{h}_{n+1}$ such that the set of n+1 vectors $\{\vec{h}_j\}$, $j = 1,2,\ldots,n+1$, is also linearly independent. The following extension of this result will find important application in the sequel:

Theorem 5.10. There exists a primitive, linearly independent set $\{\vec{h}\}$ consisting of a single vector. Let the linearly independent set of n vectors $\{\vec{h}_j\}$, $j = 1,2,\ldots,n$, where n>p, be primitive. Then there exists a vector $\vec{h}_{n+1}$ such that the set of n+1 vectors $\{\vec{h}_j\}$, $j = 1,2,\ldots,n+1$, is also primitive and linearly independent.

5.4.1. Primitive set modulo $\vec{\omega}$. Let there be given a set of n vectors $\{\vec{h}_j\}$ and let at least one component of $\vec{\omega} = (\omega_1,\omega_2,\ldots,\omega_p)$ be equal to zero. Suppress those components (if any) of each of the n vectors $\vec{h}_j$ the corresponding components of which in $\vec{\omega}$ are different from zero. We obtain the derived set of n vectors $\vec{h}'_j = (h'_{j_1},h'_{j_2},\ldots,h'_{j_{p'}})$, $j = 1,2,\ldots,n$, where p'≥1 is the number of those components of $\vec{\omega}$ which are equal to zero. The given set $\{\vec{h}_j\}$ is said to be primitive modulo $\vec{\omega}$ if the derived set $\{h'_{j_k}\}$ is primitive. Evidently, if every $\omega_i = 0$, $i = 1,2,\ldots,p$, then the motion of primitive set modulo $\vec{\omega}$ reduces to that of primitive set.

5.5. The three categories of primitive space groups. Complete details of the theory have now been worked out for space groups P1 (§5.1.1), P$\bar{1}$ (§5.1.2) and P4 (§4.1) and the reader is urged to solve some of the problems in Exercises I.2. Only the final results are summarized here, but the reader is referred to the literature for further details.

In order to avoid the unnecessary complexities resulting
from the choice of non-primitive unit cells, the present discussion
will be restricted to the primitive space groups. In "International Tables for X-Ray Crystallography", Vol. 1, 1952, the
unit cell is chosen to be primitive for 94 of the 138 non-centrosymmetric space groups and for 62 of the 92 centrosymmetric space
groups. In the remaining 44 non-centrosymmetric and 30 centrosymmetric space groups, structure factors appropriate to the
choice of a primitive unit cell are readily obtained and the
present methods then are applicable to these space groups also.
Alternatively, since the transformations from primitive unit cells
to the conventional non-primitive unit cells are well known, the
results obtained by the methods described here are readily
interpreted in terms of the latter choice of unit cell.

The primitive space groups fall into three different
categories depending upon the number of equivalence classes.
Category 1 consists of those space groups having one equivalence
class, Category 2 of those having two equivalence classes, and
Category 3 of those having four equivalence classes. As shown
in Tables 1 and 2, each category is further subdivided into
several types depending upon the nature of the equivalence classes.
The latter have been worked out in great detail earlier (§4) for
P4 and the same methods are applicable to all the space groups.
No further description of the equivalence classes is given here
but the reader is referred to Exercises I.2 and the literature
for additional details. The following definitions are found to
be convenient:

Def. 1. There is associated with each type a vector $\vec{\omega}_s$,
called the seminvariant vector of the type, defined by row 5 of
Tables 1 and 2.

Def. 2. For each type the vector $\vec{h}_s$, seminvariantly
associated with the phase $\phi_{\vec{h}}$, is defined by row 6 of the Tables.

Table 1. The thirteen types of primitive non-centrosymmetric space groups

Category	No. of equiv. classes	Type	Space groups	Seminvariant modulus $\vec{\omega}_s$	Vector $\vec{h}_s$ seminvariantly associated with $\phi_{\vec{h}}$ or with $\vec{h}=(h,k,l)$	For fixed form for structure factor, number of phases linearly semi-independent to be specified
1	1	1P000	$P1$	$(0,0,0)$	(h,k,l)	3
		1P202	$P2$, $P2_1$	$(2,0,2)$	(h,k,l)	3
		1P020	Pm, Pc	$(0,2,0)$	(h,k,l)	3
		1P222	$P222$, $P222_1$, $P2_12_12$, $P2_12_12_1$	$(2,2,2)$	(h,k,l)	3
		1P220	$Pmm2$, $Pmc2_1$, $Pcc2$, $Pma2$, $Pca2_1$, $Pnc2$, $Pmn2_1$, $Pba2$, $Pna2_1$, $Pnn2$	$(2,2,0)$	(h,k,l)	3
2	2	2P20	$P4$, $P4_1$, $P4_2$, $P4_3$, $P4mm$, $P4bm$, $P4_2cm$, $P4_2nm$, $P4cc$, $P4nc$, $P4_2mc$, $P4_2bc$	$(2,0)$	$(h+k,l)$	2
		2P22	$P\bar{4}$, $P422$, $P42_12$, $P4_122$, $P4_12_12$, $P4_222$, $P4_22_12$, $P4_322$, $P4_32_12$, $P\bar{4}2m$, $P\bar{4}2c$, $P\bar{4}2_1m$, $P\bar{4}2_1c$, $P\bar{4}m2$, $P\bar{4}c2$, $P\bar{4}b2$, $P\bar{4}n2$	$(2,2)$	$(h+k,l)$	2
3	4	3P30	$P3$, $P3_1$, $P3_2$, $P3m1$, $P3c1$	$(3,0)$	$(h-k,l)$	2
		3P32	$P312$, $P3_112$, $P3_212$, $P\bar{6}$, $P\bar{6}m2$, $P\bar{6}c2$	$(3,2)$	$(h-k,l)$	2
		$3P_10$	$P31m$, $P31c$, $P6$, $P6_1$, $P6_5$, $P6_2$, $P6_4$, $P6_3$, $P6mm$, $P6cc$, $P6_3cm$, $P6_3mc$	(0)	(l)	1
		$3P_12$	$P321$, $P3_121$, $P3_221$, $P622$, $P6_122$, $P6_522$, $P6_222$, $P6_422$, $P6_322$, $P\bar{6}2m$, $P\bar{6}2c$	(2)	(l)	1
		$3P_20$	$R3$, $R3m$, $R3c$	(0)	$(h+k+l)$	1
		$3P_22$	$R32$, $P23$, $P2_13$, $P432$, $P4_232$, $P4_332$, $P4_132$, $P\bar{4}3m$, $P\bar{4}3n$	(2)	$(h+k+l)$	1

Table 2. The four types of primitive centrosymmetric space groups

	1P	2P	3P$_1$	3P$_2$
Category	1	2	3	3
No. of equivalence classes	1	2	4	4
Type	1P	2P	3P$_1$	3P$_2$
Space groups	$P\bar{1}$, $P2/m$, $P2_1/m$, $P2/c$, $P2_1/c$, $Pmmm$, $Pnnn$, $Pccm$, $Pban$, $Pmma$, $Pnna$, $Pmna$, $Pcca$, $Pbam$, $Pccn$, $Pbcm$, $Pnnm$, $Pmmn$, $Pbcn$, $Pbca$, $Pnma$	$P4/m$, $P4_2/m$, $P4/n$, $P4_2/n$, $P4/mmm$, $P4/mcc$, $P4/nbm$, $P4/nnc$, $P4/mbm$, $P4/mnc$, $P4/nmm$, $P4/ncc$, $P4_2/mmc$, $P4_2/mcm$, $P4_2/nbc$, $P4_2/nnm$, $P4_2/mbc$, $P4_2/mnm$, $P4_2/nmc$, $P4_2/ncm$	$P\bar{3}$, $P\bar{3}1m$, $P\bar{3}1c$, $P\bar{3}m1$, $P\bar{3}c1$, $P6/m$, $P6_3/m$, $P6/mmm$, $P6/mcc$, $P6_3/mcm$, $P6_3/mmc$	$R\bar{3}$, $R\bar{3}m$, $R\bar{3}c$, $Pm\bar{3}$, $Pn\bar{3}$, $Pa\bar{3}$, $Pm\bar{3}m$, $Pn\bar{3}n$, $Pm\bar{3}n$, $Pn\bar{3}m$
Seminvariant modulus $\vec{\omega}_s$	$(2,2,2)$	$(2,2)$	(2)	(2)
Vector $\vec{h}_s$ seminvariantly associated with $\phi_{\vec{h}}$ or with $\vec{h} = (h,k,\ell)$	(h,k,ℓ)	$(h+k,\ell)$	(ℓ)	$(h+k+\ell)$
For fixed form for structure factor, number of phases linearly semi-independent to be specified	3	2	1	1

Def. 3.   For each type, a set of phases $\{\phi_{\vec{h}_j}\}$ is said to be
linearly semi-dependent or linearly semi-independent according as
the set of seminvariantly associated vectors is linearly dependent
or independent modulo $\vec{\omega}_s$, where $\vec{\omega}_s$ is the seminvariant modulus
of the type.   The phase $\phi_{\vec{h}}$ is linearly (rationally) dependent on
or linearly (rationally) independent of, the set of phases $\{\phi_{\vec{h}_j}\}$
according as the vector $\vec{h}_s$ seminvariantly associated with $\phi_{\vec{h}}$ is
linearly (rationally) dependent modulo $\vec{\omega}_s$ on, or linearly
(rationally) independent modulo $\vec{\omega}_s$ of, the set of vectors
seminvariantly associated with the set $\{\phi_{\vec{h}_j}\}$.

Def. 4.   For each type, a set of phases is said to be semi-
primitive if the set of seminvariantly associated vectors is
primitive modulo $\vec{\omega}_s$, where $\vec{\omega}_s$ is the seminvariant modulus of the
type.

The major results of the general theory may be briefly
summarized in the following:

Theorem 5.11.   For each type, the structure seminvariants
are the linear combinations

$$\sum_{\vec{h}} A_{\vec{h}}\, \phi_{\vec{h}} \qquad\qquad (5.30)$$

where the $A_{\vec{h}}$ are integers satisfying

$$\sum_{\vec{h}} A_{\vec{h}}\, \vec{h}_s \equiv 0 \ (\mathrm{mod}\ \vec{\omega}_s), \qquad\qquad (5.31)$$

$\vec{h}_s$ is the vector seminvariantly associated with the vector $\vec{h}$ and
$\vec{\omega}_s$ is the modulus of the type.

Theorem 5.12.   If the functional form for the geometric
structure factor is fixed, the values of the magnitudes of the

normalized structure factors determine, in general, the values of
the cosine seminvariants. Thus two possible values for the
structure seminvariants, differing only in sign, are determined
and these values coincide if they happen to be 0 or π. In
general, one of the two possible values for a structure
seminvariant corresponds to one of the enantiomorphous structures
permitted by the structure factor magnitudes and the second value
corresponds to the other enantiomorph. When the two enantiomorphs
are distinct, one may be selected by specifying arbitrarily the
sign of any structure seminvariant whose value is different from
0 and π. (The latter, of course, have the same value for both
enantiomorphs, even when the enantiomorphs are distinct, and are
therefore not useful in enantiomorph selection).

Theorems 5.11 and 5.12, together with Tables 1 and 2, lead
in an obvious way to recipes for origin selection in all the
space groups. Details for Type 2P20 are given in the following
Theorems 5.13 - 5.16. For detailed procedures in the other space
groups the interested reader is referred to the literature. It
should be emphasized that in any given case there are usually
several different modes of origin selection and, depending on the
particular problem at hand, one method may well be preferable to
the others. For this reason a proper understanding of the under-
lying rationale is essential. Finally, the role of the concept
of primitivity and its relationship to accessibility are
particularly noteworthy and should be emphasized.

In the following Theorems 5.13 - 5.16, it is assumed that the
type is 2P20, the magnitudes of the normalized structure factors
$|E|$ are given, the functional form for the geometric structure
factor has been specified, and the enantiomorph has been selected
by specifying arbitrarily the sign of a particular structure
seminvariant the value of which is different from 0 and π.

Theorem 5.13. A single phase $\phi_{hk\ell}$ is a structure
seminvariant, i.e. its value is uniquely determined, if and only

if h + k is even and $\ell = 0$.

    <u>Theorem 5.14.</u> A phase $\phi_{h_1 k_1 0}$ which is linearly semi-independent, i.e. $h_1 + k_1$ is odd, has just two possible values and these differ from each other by π. Either one of these two values may be chosen. Once this is done, then the value of any phase $\phi_{hk\ell}$ which is linearly semi-dependent on $\phi_{h_1 k_1 0}$, i.e. $\ell = 0$, is uniquely determined.

    <u>Theorem 5.15.</u> Let $h_2 + k_2$ be even. Then the value of any phase $\phi_{h_2 k_2 \ell_2}$ which is linearly semi-independent, i.e. $\ell_2 \neq 0$, may be specified arbitrarily. Once this is done, then the value of any phase $\phi_{hk\ell}$ which is linearly semi-dependent on $\phi_{h_2 k_2 \ell_2}$, i.e. h + k is even and ℓ is divisible by ℓ_2, is uniquely determined. In view of Theorem 5.9, any phase $\phi_{hk\ell}$ which is rationally semi-dependent on $\phi_{h_2 k_2 \ell_2}$ is also linearly semi-dependent on $\phi_{h_2 k_2 \ell_2}$, whence its value is uniquely determined, provided that $\phi_{h_2 k_2 \ell_2}$ is semi-primitive, i.e. provided that $\ell_2 = \pm 1$.

    <u>Theorem 5.16.</u> Let $\ell_1 = 0$ and $h_2 + k_2$ be even. Let $\phi_{h_1 k_1 \ell_1}$ and $\phi_{h_2 k_2 \ell_2}$ be any two phases which constitute a linearly semi-independent set, i.e. $h_1 + k_1$ is odd and $\ell_2 \neq 0$. In accordance with Theorems 5.14 and 5.15, either one of the two permitted values for $\phi_{h_1 k_1 \ell_1}$ may be chosen while the value for $\phi_{h_2 k_2 \ell_2}$ may be specified arbitrarily. Once this is done, then the value of any phase $\phi_{hk\ell}$ which is linearly semi-dependent on the pair $\phi_{h_1 k_1 \ell_1}$, $\phi_{h_2 k_2 \ell_2}$ is uniquely determined. In view of Theorem 5.9, any phase $\phi_{hk\ell}$, of necessity rationally semi-dependent on the pair $\phi_{h_1 k_1 \ell_1}$, $\phi_{h_2 k_2 \ell_2}$, is also linearly semi-dependent on this pair, whence its value is uniquely determined (and the origin uniquely specified), provided that

the pair $\phi_{h_1 k_1 \ell_1}$, $\phi_{h_2 k_2 \ell_2}$ is semi-primitive, i.e. provided
that $\ell_2 = \pm 1$.

Exercises I.7.

1. For space groups of Type 2P20, what is the seminvariant
modulus? What is the vector seminvariantly associated with
$\phi_{2\ 1\ 0}$? with $\phi_{1\ 1\ 5}$? Show that $\phi_{2\ 1\ 0}$, $\phi_{1\ 1\ 5}$ constitute a
linearly semi-independent pair so that their values may be
specified in accordance with Theorem 5.16. Show that an arbitrary
phase (e.g. $\phi_{1\ 3\ 2}$ or $\phi_{1\ 2\ 2}$) is rationally semi-dependent on the
pair $\phi_{2\ 1\ 0}$, $\phi_{1\ 1\ 5}$. Are all phases (e.g. $\phi_{1\ 3\ 2}$) linearly
semi-dependent on this pair? Is the value of the phase $\phi_{1\ 3\ 2}$
uniquely determined by the specified values for $\phi_{2\ 1\ 0}$, $\phi_{1\ 1\ 5}$?
Is the value of $\phi_{1\ 3\ 5}$ uniquely determined? Is the pair $\phi_{2\ 1\ 0}$,
$\phi_{1\ 1\ 5}$ suitable for origin specification?
2. The same as Ex. 1, but replace the pair $\phi_{2\ 1\ 0}$, $\phi_{1\ 1\ 5}$
by the pair $\phi_{2\ 1\ 0}$, $\phi_{1\ 1\ 1}$.

6. The Values of the Cosine Seminvariants Determine
the Values of the Individual Phases

In order to show that the values of the cosine seminvariants
lead directly and unambiguously to the values of the individual
phases, assume first that the space group is P1 so that the
cosine seminvariants (in this case cosine invariants) in question
are all of the form

$$\cos (\phi_1 + \phi_2 + \phi_3) \hspace{3cm} (6.1)$$

in which the abbreviations

$$\phi_i = \phi_{\vec{h}_i}, \quad i = 1,2,3 \tag{6.2}$$

have been introduced and it is assumed that

$$\vec{h}_1 + \vec{h}_2 + \vec{h}_3 = 0. \tag{6.3}$$

Denote the origin fixing triple by $\vec{h}$, $\vec{k}$, $\vec{\ell}$ so that the triple of phases $\{\phi_{\vec{h}}, \phi_{\vec{k}}, \phi_{\vec{\ell}}\}$ is primitive, i.e.

$$\begin{vmatrix} h_1 & h_2 & h_3 \\ k_1 & k_2 & k_3 \\ \ell_1 & \ell_2 & \ell_3 \end{vmatrix} = \pm 1 \tag{6.4}$$

where

$$\vec{h} = (h_1, h_2, h_3) \tag{6.5}$$

$$\vec{k} = (k_1, k_2, k_3) \tag{6.6}$$

$$\vec{\ell} = (\ell_1, \ell_2, \ell_3). \tag{6.7}$$

For simplicity fix the origin by setting

$$\phi_{\vec{h}} = \phi_{\vec{k}} = \phi_{\vec{\ell}} = 0. \tag{6.8}$$

Suppose further that the origin fixing triple is chosen in such a way that the value of the cosine invariant

$$\cos (\phi_{\vec{h}} + \phi_{\vec{k}} + \phi_{-\vec{h}-\vec{k}}) \approx +1. \qquad (6.9)$$

Assume finally that the values of all cosine invariants (6.1) which will be needed in the iterative process to be described are known, at least approximately. Then the values of individual phases may be determined, for example, in accordance with the following sequence:

    1. $\phi_{\vec{h}+\vec{k}}$. In view of (6.8) and (6.9), it is clear that $\phi_{\vec{h}+\vec{k}} \approx 0$.

    2. $\phi_{\vec{k}+\vec{l}}$. Since the value of the cosine invariant

$$\cos (\phi_{-\vec{k}} + \phi_{-\vec{l}} + \phi_{\vec{k}+\vec{l}}) = c_1 \qquad (6.10)$$

is presumed to be known and since $\phi_{-\vec{k}} = \phi_{-\vec{l}} = 0$, there are two possible values for $\phi_{\vec{k}+\vec{l}}$ differing only in sign (assuming that $c_1 \neq \pm 1$). Either of these two values for $\phi_{\vec{k}+\vec{l}}$ may be chosen, thus specifying the enantiomorph and leading to a unique value for $\phi_{\vec{k}+\vec{l}}$. If it should happen that $c_1 = \pm 1$, then the value of $\phi_{\vec{k}+\vec{l}}$ is 0 or π and the decision to select the enantiomorph must be deferred.

    3. $\phi_{\vec{h}-\vec{l}}$. The presumed known values of the cosine invariants

$$\cos (\phi_{-\vec{h}} + \phi_{\vec{l}} + \phi_{\vec{h}-\vec{l}}) = c_2 \qquad (6.11)$$

and

$$\cos (\phi_{-\vec{h}-\vec{k}} + \phi_{\vec{k}+\vec{l}} + \phi_{\vec{h}-\vec{l}}) = c_3, \qquad (6.12)$$

together with the known values 0 for $\phi_{-\vec{h}}$, $\phi_{\vec{\ell}}$, $\phi_{-\vec{h}-\vec{k}}$ and arc cos c_1 for $\phi_{\vec{k}+\vec{\ell}}$, are clearly more than sufficient to determine the value of $\phi_{\vec{h}-\vec{\ell}}$. In fact, $\phi_{\vec{h}-\vec{\ell}}$ can be determined, in accordance with the principle of least-squares, as that (in general, unique) value which makes the function

$$
\Phi = \left(\cos\,(\phi_{-\vec{h}} + \phi_{\vec{\ell}} + \phi_{\vec{h}-\vec{\ell}}) - c_2 \right)^2 +
$$

$$
\left(\cos\,(\phi_{-\vec{h}-\vec{k}} + \phi_{\vec{k}+\vec{\ell}} + \phi_{\vec{h}-\vec{\ell}}) - c_3 \right)^2 \tag{6.13}
$$

a minimum. Since the values c_2, c_3 are assumed to be only approximations to the true values of the respective cosines, it is expected that $\Phi_{min} > 0$, but the value of Φ_{min} may be taken as a measure of the accuracy of the phase determination, the smaller the value of Φ_{min} the more reliable would be the calculated value of $\phi_{\vec{h}-\vec{\ell}}$. If it should happen that the values of c_1, c_2, c_3 are all ± 1 then $\phi_{\vec{h}-\vec{\ell}} = 0$ or π and the decision to choose the enantiomorph would again have to be deferred.

    4. $\phi_{\vec{h}+\vec{k}+\vec{\ell}}$. This phase would also be determined by least-squares using

$$
\cos\,(\phi_{-h-k} + \phi_{-\ell} + \phi_{h+k+\ell}) = c_4 \tag{6.14}
$$

and

$$
\cos\,(\phi_{-k-\ell} + \phi_{-h} + \phi_{h+k+\ell}) = c_5 \tag{6.15}
$$

as well as the previously determined values for $\phi_{-\vec{h}-\vec{k}}$, $\phi_{-\vec{\ell}}$, $\phi_{-\vec{k}-\vec{\ell}}$, and $\phi_{-\vec{h}}$.

    The process may clearly be continued and phases determined, for example, in the following order:

$$\phi_{-\vec{h}+\vec{k}+\vec{\ell}}, \quad \phi_{\vec{h}+\vec{k}-\vec{\ell}}, \quad \phi_{2\vec{k}}, \quad \phi_{\vec{h}-\vec{k}}, \quad \phi_{\vec{k}-\vec{\ell}}, \quad \phi_{\vec{h}-\vec{k}+\vec{\ell}}, \quad \phi_{\vec{h}+\vec{\ell}}, \quad \cdots$$

employing respectively the cosine invariants

$$\left.\begin{array}{c} \cos\ (\phi_{\vec{h}-\vec{\ell}} + \phi_{-\vec{k}} + \phi_{-\vec{h}+\vec{k}+\vec{\ell}}) \\[20pt] \cos\ (\phi_{\vec{h}} + \phi_{-\vec{k}-\vec{\ell}} + \phi_{-\vec{h}+\vec{k}+\vec{\ell}}) \end{array}\right\} \qquad (6.16)$$

$$\left.\begin{array}{c} \cos\ (\phi_{-\vec{h}+\vec{\ell}} + \phi_{-\vec{k}} + \phi_{\vec{h}+\vec{k}-\vec{\ell}}) \\[20pt] \cos\ (\phi_{-\vec{h}-\vec{k}} + \phi_{\vec{\ell}} + \phi_{\vec{h}+\vec{k}-\vec{\ell}}) \end{array}\right\} \qquad (6.17)$$

$$\left.\begin{array}{c} \cos\ (\phi_{-\vec{k}} + \phi_{-\vec{k}} + \phi_{2\vec{k}}) \\[20pt] \cos\ (\phi_{\vec{h}-\vec{k}-\vec{\ell}} + \phi_{-\vec{h}-\vec{k}+\vec{\ell}} + \phi_{2\vec{k}}) \end{array}\right\} \qquad (6.18)$$

$$\left.\begin{array}{c} \cos\ (\phi_{-\vec{h}} + \phi_{\vec{k}} + \phi_{\vec{h}-\vec{k}}) \\[20pt] \cos\ (\phi_{-\vec{h}-\vec{k}} + \phi_{2\vec{k}} + \phi_{\vec{h}-\vec{k}}) \\[20pt] \cos\ (\phi_{-\vec{h}+\vec{k}+\vec{\ell}} + \phi_{-\vec{\ell}} + \phi_{\vec{h}-\vec{k}}) \end{array}\right\} \qquad (6.19)$$

etc. Although, in these early cycles, only a two or three-fold redundancy is present, it is not unusual to find redundancies of twenty-fold or more in the later cycles.

A number of variations and improvements on this simple theme suggest themselves immediately. For example, at intervals, one may redetermine the values of phases determined earlier by employing subsequently determined phases. Thus, once the values of $\phi_{\vec{k}+\vec{\ell}}$, $\phi_{\vec{h}-\vec{\ell}}$, $\phi_{\vec{h}+\vec{k}+\vec{\ell}}$ have been determined, one may use the cosine invariants

$$
\left\{
\begin{array}{l}
\cos\ (\phi_{-\vec{h}} + \phi_{-\vec{k}} + \phi_{\vec{h}+\vec{k}}) \\[2ex]
\cos\ (\phi_{-\vec{h}+\vec{\ell}} + \phi_{-\vec{k}-\vec{\ell}} + \phi_{\vec{h}+\vec{k}}) \\[2ex]
\cos\ (\phi_{-\vec{h}-\vec{k}-\vec{\ell}} + \phi_{\vec{\ell}} + \phi_{\vec{h}+\vec{k}})
\end{array}
\right\}
\qquad (6.20)
$$

to redetermine, with three-fold redundancy, the value of $\phi_{\vec{h}+\vec{k}}$
which had been initially determined, probably not very accurately,
from the single cosine invariant (6.9). This feed-back feature,
which increases the overdeterminacy and improves the accuracy of
the phase determination, particularly in the early stages, is a
useful addendum to the basic procedure.

Another variation on the basic theme, which may reduce
considerably the computing time required, is to vary the rate of
acquisition of new phases. Thus, instead of determining only
one phase at a time, one may, after the first two or three cycles,
determine new phases at an accelerating rate, for example
according to the scheme 1,3,5,... in successive cycles.

Finally, one may use the value of Φ_{min} as a figure of merit
in order to decide whether the value of any phase may not be
determined with sufficient accuracy to be used in subsequent
cycles.

If the space group is not P1 then one usually adds several
phases, determined by the Σ_1 formula to be described in Chapter
II, to the origin determining set. Also, for each phase which
has been determined, a number of other phases, related to the
first by the space group symmetries, are also known. For example,
in the space group $P2_1$, if the value of any phase $\phi_{hk\ell}$ has been
found, then the values of $\phi_{h\bar{k}\ell}$, $\phi_{\bar{h}k\bar{\ell}}$ and $\phi_{\bar{h}\bar{k}\bar{\ell}}$ are also known.
Finally, in all space groups other than P1 there exist cosine
seminvariants which involve only two phases whose values may

often be found with sufficient accuracy that they play a major
role in the phase determination. Thus the space group symmetries
may be used in several ways to facilitate the process of evaluating
the individual phases once the values of the cosine seminvariants
are assumed to be known.

Now that it has been shown how the values of the cosine
seminvariants may be used to determine the values of individual
phases, at least in principle, it is necessary to emphasize the
limitations of this procedure which arise in practice. These
stem exclusively from our restricted ability to calculate the
values of the cosine seminvariants with sufficient accuracy,
especially when $|E|$ values are relatively small. Thus it may
happen, in the process just described, that the values of cosine
seminvariants are required with associated low values of $|E|$.
Since such cosine seminvariants cannot ordinarily be calculated
with great accuracy, they are usually deleted, with resulting
loss of redundancy, for, if included, they may introduce
unacceptably large inaccuracies in the phase determination. Thus,
it is vital, in choosing the origin determining set, that this be
done in such a way that the phases reached in the early cycles
have, for the most part, large associated $|E|$ values.

The remainder of Part A is devoted, almost exclusively, to
an account of the basic theory underpinning the methods which
have thus far been devised for calculating the values of the
cosine seminvariants with sufficient accuracy to be useful in
phase determination.

CHAPTER II

EXPLICIT FORMULAS FOR THE COSINE SEMINVARIANTS:

THE ALGEBRAIC APPROACH

1. Introduction

In all space groups other than P1 there exist single phases
which are structure seminvariants and there are linear combinations of two phases which are structure seminvariants. In every
space group there exist structure seminvariants which are linear
combinations of three phases. Finally, there exist space group
dependent structure seminvariants of special type, e.g. $2\phi_{\vec{h}}$ in
the space group $P2_12_12_1$. The cosine of a structure invariant
(seminvariant) is said to be a cosine invariant (cosine seminvariant) and, in general, the values of the cosine invariants are
determined by the observed magnitudes of the normalized structure
factors. For each fixed functional form for the structure factor,
the magnitudes of the structure factors also determine the values
of the cosine seminvariants. It is the purpose in the present
chapter to derive explicit formulas which exhibit these relationships. In general, the formulas have exact validity if all atoms
are identical and if certain rational dependencies among the
interatomic vectors (or their projections, depending on the space
group) do not exist. Otherwise, the formulas possess only
approximate validity. They are further limited by the need, in
practice, to estimate certain averages by means of a finite
sample, thus introducing errors of finite sampling. Despite these

67

limitations the formulas to be derived here have found important
application in the solution of many crystal structures and, as
further refinements and improvements are made, may be expected
to be even more useful in future applications. In order to
simplify the analysis, it will be assumed, as usual, that the
unit cell contains N identical, non-vibrating, point atoms.

2. Formulas for the Cosine Seminvariants, Cos ϕ

The first formulas of this type were found in 1953 and have
come to be known as Σ_1 formulas. Although the methodology was
described at that time in a general sort of way, only a few of
the Σ_1 formulas have actually been derived explicitly in the
intervening years. It is noteworthy that, in the space groups
of higher symmetry, the derivations are decidedly non-trivial
and, at least in some cases, the form of final results quite
unexpected. A number of cases of increasing complexity will be
worked out in detail in order to illustrate the principles
involved.

2.1. Space group P2, b as unique axis. Reference to
Chapter I shows that the phases which are structure seminvariants
(or invariants) are of the form $\phi_{2h\ 0\ 2\ell}$. From "International
Tables for X-Ray Crystallography", Vol. I, page 374, one finds

$$\left.\begin{array}{l} A = 2\cos 2\pi(hx + \ell z)\cos 2\pi ky \\ B = 2\cos 2\pi(hx + \ell z)\sin 2\pi ky \end{array}\right\} . \qquad (2.1)$$

Hence

$$E_{hk\ell} = \frac{2}{N^{1/2}} \sum_{\mu=1}^{N/2} \left\{ \cos 2\pi(hx_\mu + \ell z_\mu)\cos 2\pi k y_\mu + \right.$$

$$\left. i \cos 2\pi(hx_\mu + \ell z_\mu)\sin 2\pi k y_\mu \right\}. \tag{2.2}$$

$$|E_{hk\ell}|^2 = \frac{4}{N} \sum_{\mu,\nu}^{N/2} \left\{ \cos 2\pi(hx_\mu + \ell z_\mu)\cos 2\pi k y_\mu + \right.$$

$$\left. i \cos 2\pi(hx_\mu + \ell z_\mu)\sin 2\pi k y_\mu \right\} \times$$

$$\left\{ \cos 2\pi(hx_\nu + \ell z_\nu)\cos 2\pi k y_\nu - \right.$$

$$\left. i \cos 2\pi(hx_\nu + \ell z_\nu)\sin 2\pi k y_\nu \right\}. \tag{2.3}$$

Splitting the double sum into two parts, the first corresponding to the case that $\mu=\nu$ and the second to the case that $\mu\neq\nu$, one obtains

$$|E_{hk\ell}|^2 = \frac{4}{N} \sum_{\mu=1}^{N/2} \left\{ \cos^2 2\pi(hx_\mu + \ell z_\mu)\cos^2 2\pi k y_\mu + \right.$$

$$\left. \cos^2 2\pi(hx_\mu + \ell z_\mu)\sin^2 2\pi k y_\mu \right\} +$$

$$\frac{4}{N} \sum_{\substack{\mu \neq \nu \\ 1}}^{N/2} \left\{ \cos 2\pi(hx_\mu + \ell z_\mu)\cos 2\pi(hx_\nu + \ell z_\nu)\cos 2\pi k y_\mu \cos 2\pi k y_\nu + \right.$$

$$\left. \cos 2\pi(hx_\mu + \ell z_\mu)\cos 2\pi(hx_\nu + \ell z_\nu)\sin 2\pi k y_\mu \sin 2\pi k y_\nu \right\}. \qquad (2.4)$$

The trigonometric formulas $\cos^2 x = 1/2 + 1/2 \cos 2x$ and $\cos x$ $\cos y + \sin x \sin y = \cos (x-y)$ then lead to

$$|E_{hk\ell}|^2 = \frac{4}{N} \sum_{\mu=1}^{N/2} \left\{ \frac{1}{2} + \frac{1}{2} \cos 4\pi(hx_\mu + \ell z_\mu) \right\} +$$

$$\frac{4}{N} \sum_{\substack{\mu \neq \nu \\ 1}}^{N/2} \cos 2\pi(hx_\mu + \ell z_\mu)\cos 2\pi(hx_\nu + \ell z_\nu)\cos 2\pi k(y_\mu - y_\nu). \qquad (2.5)$$

Employing (2.2) with h replaced by 2h, ℓ by 2ℓ, k by 0, one finds

$$|E_{hk\ell}|^2 = 1 + \frac{1}{N^{1/2}} E_{2h\ 0\ 2\ell} +$$

$$\frac{4}{N} \sum_{\substack{\mu \neq \nu \\ 1}}^{N/2} \cos 2\pi(hx_\mu + \ell z_\mu) \cos 2\pi(hx_\nu + \ell z_\nu) \cos 2\pi k(y_\mu - y_\nu). \qquad (2.6)$$

Next, average both sides of (2.6) over all integers k. If

$$y_\mu - y_\nu \neq 0 \text{ whenever } \mu \neq \nu, \qquad (2.7)$$

then

$$\left\langle \cos 2\pi k(y_\mu - y_\nu) \right\rangle_k = 0 \qquad (2.8)$$

and (2.6) becomes

$$E_{2h\ 0\ 2\ell} = |E_{2h\ 0\ 2\ell}| \cos \phi_{2h\ 0\ 2\ell} = N^{1/2} \left\langle |E_{hk\ell}|^2 - 1 \right\rangle_k$$

$$(2.9)$$

the desired formula. Hence $\phi_{2h\ 0\ 2\ell} = 0$ or π according as the average on the right side of Eq. (2.9) is positive or negative.

Exercise II. 1.

1. Show that in the space group $P2_1$, b as unique axis, the Σ_1 formula is

$$E_{2h\ 0\ 2\ell} = \left|E_{2h\ 0\ 2\ell}\right| \cos \phi_{2h\ 0\ 2\ell} = N^{1/2} \left\langle (-1)^k (\left|E_{hk\ell}\right|^2 - 1) \right\rangle_k ,$$

$$(2.10)$$

so that $\phi_{2h\ 0\ 2\ell}$ is 0 or π according as the right side of (2.10) is positive or negative.

2.2. Space group $P2_1 2_1 2_1$. From Chapter I, it follows that the single phases which are structure seminvariants (or invariants) are of the form $\phi_{2h\ 2k\ 2\ell}$. The Σ_1 formulas are explicit expressions for $\cos \phi_{2h\ 2k\ 2\ell}$ where any one or any two of h, k, ℓ are zero. Only the case for $\cos \phi_{0\ 2k\ 2\ell}$ is considered here. The others follow by symmetry. Reference to "International Tables for Crystallography", Vol. I, page 385 shows that

$$\left. \begin{aligned} A &= 4\cos 2\pi\left(hx - \frac{h-k}{4}\right)\cos 2\pi\left(ky - \frac{k-\ell}{4}\right)\cos 2\pi\left(\ell z - \frac{\ell-h}{4}\right) \\ B &= -4\sin 2\pi\left(hx - \frac{h-k}{4}\right)\sin 2\pi\left(ky - \frac{k-\ell}{4}\right)\sin 2\pi\left(\ell z - \frac{\ell-h}{4}\right) \end{aligned} \right\}.$$

$$(2.11)$$

Thus

$$E_{hk\ell} = \frac{4}{N^{1/2}} \sum_{\mu=1}^{N/4} \left\{ \cos 2\pi\left(hx_\mu - \frac{h-k}{4}\right)\cos 2\pi\left(ky_\mu - \right. \right.$$

$$\left. \frac{k-\ell}{4}\right)\cos 2\pi\left(\ell z_\mu - \frac{\ell-h}{4}\right) - i \sin 2\pi\left(hx_\mu - \right.$$

$$\left. \left. \frac{h-k}{4}\right)\sin 2\pi\left(ky_\mu - \frac{k-\ell}{4}\right)\sin 2\pi\left(\ell z_\mu - \frac{\ell-h}{4}\right) \right\} ,$$

$$(2.12)$$

and

$$|E_{hk\ell}|^2 = \frac{16}{N} \sum_{\substack{\mu,\nu \\ 1}}^{N/4} \left\{ \cos 2\pi\left(hx_\mu - \frac{h-k}{4}\right)\cos 2\pi\left(ky_\mu - \right. \right.$$

$$\left. \frac{k-\ell}{4}\right)\cos 2\pi\left(\ell z_\mu - \frac{\ell-h}{4}\right) - i \sin 2\pi\left(hx_\mu - \right.$$

$$\left. \left. \frac{h-k}{4}\right)\sin 2\pi\left(ky_\mu - \frac{k-\ell}{4}\right)\sin 2\pi\left(\ell z_\mu - \frac{\ell-h}{4}\right) \right\} \times$$

$$\left\{ \cos 2\pi\left(hx_\nu - \frac{h-k}{4}\right)\cos 2\pi\left(ky_\nu - \frac{k-\ell}{4}\right)\cos 2\pi\left(\ell z_\nu - \frac{\ell-h}{4}\right) + \right.$$

$$\left. i \sin 2\pi\left(hx_\nu - \frac{h-k}{4}\right)\sin 2\pi\left(ky_\nu - \frac{k-\ell}{4}\right)\sin 2\pi\left(\ell z_\nu - \frac{\ell-h}{4}\right) \right\},$$

$$(2.13)$$

or

$$|E_{hk\ell}|^2 = \frac{16}{N} \sum_{\substack{\mu,\nu \\ 1}}^{N/4} \left\{ \cos 2\pi\left(hx_\mu - \frac{h-k}{4}\right)\cos 2\pi\left(hx_\nu - \frac{h-k}{4}\right)\cos 2\pi\left(ky_\mu - \right.\right.$$

$$\left. \frac{k-\ell}{4}\right)\cos 2\pi\left(ky_\nu - \frac{k-\ell}{4}\right)\cos 2\pi\left(\ell z_\mu - \frac{\ell-h}{4}\right)\cos 2\pi\left(\ell z_\nu - \frac{\ell-h}{4}\right) +$$

$$\sin 2\pi\left(hx_\mu - \frac{h-k}{4}\right)\sin 2\pi\left(hx_\nu - \frac{h-k}{4}\right)\sin 2\pi\left(ky_\mu - \frac{k-\ell}{4}\right)\sin 2\pi\left(ky_\nu - \right.$$

$$\left.\left. \frac{k-\ell}{4}\right)\sin 2\pi\left(\ell z_\mu - \frac{\ell-h}{4}\right)\sin 2\pi\left(\ell z_\nu - \frac{\ell-h}{4}\right)\right\}. \qquad (2.14)$$

Splitting the double sum into two parts, the first corresponding to $\mu=\nu$, the second to $\mu\neq\nu$, one finds

$$|E_{hk\ell}|^2 = \frac{16}{N} \sum_{\mu=1}^{N/4} \left\{ \cos^2 2\pi\left(hx_\mu - \frac{h-k}{4}\right)\cos^2\left(ky_\mu - \frac{k-\ell}{4}\right)\cos^2\left(\ell z_\mu - \frac{\ell-h}{4}\right) + \right.$$

$$\left. \sin^2 2\pi\left(hx_\mu - \frac{h-k}{4}\right)\sin^2\left(ky_\mu - \frac{k-\ell}{4}\right)\sin^2\left(\ell z_\mu - \frac{\ell-h}{4}\right)\right\} +$$

$$\frac{16}{N} \sum_{\substack{\mu\neq\nu \\ 1}}^{N/4} \left\{ \cos 2\pi\left(hx_\mu - \frac{h-k}{4}\right)\cos 2\pi\left(hx_\nu - \frac{h-k}{4}\right)\cos 2\pi\left(ky_\mu - \right.\right.$$

$$\left.\frac{k-\ell}{4}\right)\cos2\pi\left(ky_\nu - \frac{k-\ell}{4}\right)\cos2\pi\left(\ell z_\mu - \frac{\ell-h}{4}\right)\cos2\pi\left(\ell z_\nu - \frac{\ell-h}{4}\right) +$$

$$\sin2\pi\left(hx_\mu - \frac{h-k}{4}\right)\sin2\pi\left(hx_\nu - \frac{h-k}{4}\right)\sin2\pi\left(ky_\mu - \right.$$

$$\left.\frac{k-\ell}{4}\right)\sin2\pi\left(ky_\nu - \frac{k-\ell}{4}\right)\sin2\pi\left(\ell z_\mu - \frac{\ell-h}{4}\right)\sin2\pi\left(\ell z_\nu - \frac{\ell-h}{4}\right)\right\}.$$

$$(2.15)$$

Next, employing the trigonometric identities

$$\cos^2 x = \frac{1}{2} + \frac{1}{2}\cos2x, \quad \sin^2 x = \frac{1}{2} - \frac{1}{2}\cos2x, \quad\quad (2.16)$$

$$\cos x \cos y = \frac{1}{2}\cos(x+y) + \frac{1}{2}\cos(x-y),$$

$$\sin x \sin y = -\frac{1}{2}\cos(x+y) + \frac{1}{2}\cos(x-y), \quad\quad (2.17)$$

(2.15) becomes

$$|E_{hk\ell}|^2 = \frac{2}{N}\sum_{\mu=1}^{N/4}\left\{\left[1 + \cos4\pi\left(hx_\mu - \frac{h-k}{4}\right)\right]\left[1 + \cos4\pi\left(ky_\mu - \frac{k-\ell}{4}\right)\right] \times\right.$$

$$\left[1 + \cos4\pi\left(\ell z_\mu - \frac{\ell-h}{4}\right)\right] + \left[1 - \cos4\pi\left(hx_\mu - \frac{h-k}{4}\right)\right] \times$$

$$
\left.\left[1 - \cos4\pi\left(ky_\mu - \frac{k-\ell}{4}\right)\right]\left[1 - \cos4\pi\left(\ell z_\mu - \frac{\ell-h}{4}\right)\right]\right]\right\} +
$$

$$
\frac{2}{N}\sum_{\substack{\mu\neq\nu \\ 1}}^{N/4}\left\{\left[\cos2\pi\left(h(x_\mu+x_\nu) - \frac{h-k}{2}\right) + \cos2\pi h(x_\mu-x_\nu)\right] \times\right.
$$

$$
\left[\cos2\pi\left(k(y_\mu+y_\nu) - \frac{k-\ell}{2}\right) + \cos2\pi k(y_\mu-y_\nu)\right] \times
$$

$$
\left[\cos2\pi\left(\ell(z_\mu+z_\nu) - \frac{\ell-h}{2}\right) + \cos2\pi\ell(z_\mu-z_\nu)\right] +
$$

$$
\left[-\cos2\pi\left(h(x_\mu+x_\nu) - \frac{h-k}{2}\right) + \cos2\pi h(x_\mu-x_\nu)\right] \times
$$

$$
\left[-\cos2\pi\left(k(y_\mu+y_\nu) - \frac{k-\ell}{2}\right) + \cos2\pi k(y_\mu-y_\nu)\right] \times
$$

$$
\left.\left[-\cos2\pi\left(\ell(z_\mu+z_\nu) - \frac{\ell-h}{2}\right) + \cos2\pi\ell(z_\mu-z_\nu)\right]\right\} , \qquad (2.18)
$$

$$
|E_{hk\ell}|^2 - 1 = \frac{4}{N}\sum_{\mu=1}^{N/4}\left\{ (-1)^{\ell+h}\cos4\pi hx_\mu\cos4\pi ky_\mu + \right.
$$

$$(-1)^{h+k}\cos 4\pi k y_\mu \cos 4\pi \ell z_\mu + (-1)^{k+\ell}\cos 4\pi \ell z_\mu \cos 4\pi h x_\mu \Bigg\} +$$

$$\frac{4}{N}\sum_{\substack{\mu \neq \nu \\ 1}}^{N/4}\Bigg\{(-1)^{\ell+h}\cos 2\pi h(x_\mu + x_\nu)\cos 2\pi k(y_\mu +$$

$$y_\nu)\cos 2\pi \ell(z_\mu - z_\nu) + (-1)^{h+k}\cos 2\pi k(y_\mu + y_\nu)\cos 2\pi \ell(z_\mu +$$

$$z_\nu)\cos 2\pi h(x_\mu - x_\nu) + (-1)^{k+\ell}\cos 2\pi \ell(z_\mu + z_\nu)\cos 2\pi h(x_\mu +$$

$$x_\nu)\cos 2\pi k(y_\mu - y_\nu) + \cos 2\pi h(x_\mu - x_\nu)\cos 2\pi k(y_\mu -$$

$$y_\nu)\cos 2\pi \ell(z_\mu - z_\nu)\Bigg\}. \tag{2.19}$$

Next, multiply both sides of (2.19) by $(-1)^{h+k}$ and observe that if h is replaced by 0, k by 2k, and ℓ by 2ℓ, then (2.12) becomes

$$E_{0\ 2k\ 2\ell} = \frac{4}{N^{1/2}}\sum_{\mu=1}^{N/4}\cos 4\pi k y_\mu \cos 4\pi \ell z_\mu. \tag{2.20}$$

Hence, from (2.19),

$$(-1)^{h+k}(|E_{hk\ell}|^2 - 1) = \frac{1}{N^{1/2}}E_{0\ 2k\ 2\ell} +$$

$$\frac{4}{N} \sum_{\mu=1}^{N/4} \left\{ (-1)^{k+\ell} \cos 4\pi h x_\mu \cos 4\pi k y_\mu + \right.$$

$$\left. (-1)^{\ell+h} \cos 4\pi \ell z_\mu \cos 4\pi h x_\mu \right\} +$$

$$\frac{4}{N} \sum_{\substack{\mu \neq \nu \\ 1}}^{N/4} \left\{ (-1)^{k+\ell} \cos 2\pi h (x_\mu + x_\nu) \cos 2\pi k (y_\mu + y_\nu) \cos 2\pi \ell (z_\mu - z_\nu) + \right.$$

$$\cos 2\pi k (y_\mu + y_\nu) \cos 2\pi \ell (z_\mu + z_\nu) \cos 2\pi h (x_\mu - x_\nu) +$$

$$(-1)^{\ell+h} \cos 2\pi \ell (z_\mu + z_\nu) \cos 2\pi h (x_\mu + x_\nu) \cos 2\pi k (y_\mu - y_\nu) +$$

$$\left. (-1)^{h+k} \cos 2\pi h (x_\mu - x_\nu) \cos 2\pi k (y_\mu - y_\nu) \cos 2\pi \ell (z_\mu - z_\nu) \right\} .$$

$$(2.21)$$

Finally, average both sides of (2.21) over all integers h and observe that, under the assumptions that

$$x_\mu \neq 0 \text{ and } x_\mu \pm x_\nu \neq 0 \text{ if } \mu \neq \nu, \qquad (2.22)$$

all averages under the summation signs on the right side of (2.21) vanish. It follows that, subject to (2.22),

$$E_{0\ 2k\ 2\ell} = |E_{0\ 2k\ 2\ell}| \cos \phi_{0\ 2k\ 2\ell} = N^{1/2} \left\langle (-1)^{h+k} (|E_{hk\ell}|^2 - 1) \right\rangle_h ,$$

$$(2.23)$$

and, by symmetry, under conditions like those of (2.22),

$$E_{2h\ 0\ 2\ell} = |E_{2h\ 0\ 2\ell}|\cos\phi_{2h\ 0\ 2\ell} = N^{1/2}\left\langle(-1)^{k+\ell}(|E_{hk\ell}|^2 - 1)\right\rangle_k,$$

$$(2.24)$$

$$E_{2h\ 2k\ 0} = |E_{2h\ 2k\ 0}|\cos\phi_{2h\ 2k\ 0} = N^{1/2}\left\langle(-1)^{\ell+h}(|E_{hk\ell}|^2 - 1)\right\rangle_\ell,$$

$$(2.25)$$

the Σ_1 formulas for the space group $P2_12_12_1$. Observe that k or ℓ may be zero in (2.23), ℓ or h may be zero in (2.24), and h or k may be zero in (2.25).

Exercise II. 2.

1. Derive the Σ_1 formulas for each of the space groups $P222$, $P222_1$, $P2_12_12$.

2.3. __Space group P4.__ It was shown in Chapter I that the single phase $\phi_{hk\ell}$ is a structure seminvariant in the space group P4 if h+k is even and ℓ is zero i.e. if

$$h \equiv k \pmod 2 \text{ and } \ell = 0. \qquad (2.26)$$

Reference to page 415 of Vol. I of "International Tables of Crystallography" shows that, if the origin is on a four-fold axis,

$$\left.\begin{array}{l} A = 4\cos\pi[(h-k)x+(h+k)y]\cos\pi[(h+k)x-(h-k)y]\cos2\pi\ell z, \\[2ex] B = 4\cos\pi[(h-k)x+(h+k)y]\cos\pi[(h+k)x-(h-k)y]\sin2\pi\ell z. \end{array}\right\} \qquad (2.27)$$

Hence

$$
E_{hk\ell} = \frac{4}{N^{1/2}} \sum_{\mu=1}^{N/4} \cos\pi[(h-k)x_\mu+(h+k)y_\mu]\cos\pi[(h+k)x_\mu-(h-k)y_\mu] \; \times
$$

$$
(\cos2\pi\ell z_\mu + i \sin2\pi\ell z_\mu). \tag{2.28}
$$

$$
|E_{hk\ell}|^2 = \frac{16}{N} \sum_{\substack{\mu,\nu \\ 1}}^{N/4} \cos\pi[(h-k)x_\mu+(h+k)y_\mu]\cos\pi[(h+k)x_\mu-(h-k)y_\mu] \; \times
$$

$$
(\cos2\pi\ell z_\mu + i \sin2\pi\ell z_\mu) \; \times
$$

$$
\cos\pi[(h-k)x_\nu+(h+k)y_\nu]\cos\pi[(h+k)x_\nu-(h-k)y_\nu] \; \times
$$

$$
(\cos2\pi\ell z_\nu - i \sin2\pi\ell z_\nu). \tag{2.29}
$$

Decomposing the sum into a simple sum, when $\mu=\nu$, and a double sum, when $\mu\neq\nu$, one obtains, in view of the trigonometric identity

$$
\cos x \cos y + \sin x \sin y = \cos(x-y) \tag{2.30}
$$

$$
|E_{hk\ell}|^2 = \frac{16}{N} \sum_{\mu=1}^{N/4} \cos^2\pi[(h-k)x_\mu+(h+k)y_\mu]\cos^2\pi[(h+k)x_\mu-(h-k)y_\mu] \; +
$$

$$
\frac{16}{N} \sum_{\substack{\mu\neq\nu \\ 1}}^{N/4} \cos\pi[(h-k)x_\mu+(h+k)y_\mu]\cos\pi[(h+k)x_\mu-(h-k)y_\mu] \; \times
$$

$$
\cos\pi[(h-k)x_\nu+(h+k)y_\nu]\cos\pi[(h+k)x_\nu-(h-k)y_\nu]\cos2\pi\ell(z_\mu-z_\nu).
$$
$$
\tag{2.31}
$$

The simple sum on the right side of (2.31) is independent of ℓ whereas the average of each term of the double sum over all integers ℓ vanishes (if $z_\mu \neq z_\nu$) since

$$\left\langle \cos 2\pi\ell(z_\mu - z_\nu) \right\rangle_\ell = 0 \text{ if } z_\mu \neq z_\nu \text{ whenever } \mu \neq \nu. \qquad (2.32)$$

Hence, averaging both sides of (2.31) with respect to ℓ one obtains, in view of (2.16),

$$\left\langle |E_{hk\ell}|^2 \right\rangle_\ell = \frac{4}{N} \sum_{\mu=1}^{N/4} \left\{ 1 + \cos 2\pi[(h-k)x_\mu + (h+k)y_\mu] \right\} \times$$
$$\left\{ 1 + \cos 2\pi[(h+k)x_\mu - (h-k)y_\mu] \right\} \qquad (2.33)$$

or

$$\left\langle |E_{hk\ell}|^2 - 1 \right\rangle_\ell = \frac{4}{N} \sum_{\mu=1}^{N/4} \left\{ \cos 2\pi[(h-k)x_\mu + (h+k)y_\mu] + \right.$$

$$\cos 2\pi[(h+k)x_\mu - (h-k)y_\mu] +$$

$$\left. \cos 2\pi[(h-k)x_\mu + (h+k)y_\mu] \cos 2\pi[(h+k)x_\mu - (h-k)y_\mu] \right\}.$$
$$(2.34)$$

In view of (2.17) the first two terms on the right of (2.34) may be combined so that

$$\left\langle |E_{hk\ell}|^2 - 1 \right\rangle_\ell = \frac{4}{N} \sum_{\mu=1}^{N/4} \left\{ 2\cos2\pi(hx_\mu+ky_\mu)\cos2\pi(kx_\mu-hy_\mu) + \right.$$

$$\left. \cos2\pi[(h-k)x_\mu+(h+k)y_\mu]\cos2\pi[(h+k)x_\mu-(h-k)y_\mu] \right\}.$$

$$(2.35)$$

From (2.28)

$$E_{h+k,-h+k,0} = \frac{4}{N^{1/2}} \sum_{\mu=1}^{N/4} \cos2\pi(hx_\mu+ky_\mu)\cos2\pi(kx_\mu-hy_\mu), \qquad (2.36)$$

$$E_{2h\ 2k\ 0} = \frac{4}{N^{1/2}} \sum_{\mu=1}^{N/4} \cos2\pi[(h-k)x_\mu+(h+k)y_\mu]\cos2\pi[(h+k)x_\mu-(h-k)y_\mu].$$

$$(2.37)$$

Hence (2.35) becomes

$$2E_{h+k,-h+k,0} + E_{2h\ 2k\ 0} = N^{1/2} \left\langle |E_{hk\ell}|^2 - 1 \right\rangle_\ell. \qquad (2.38)$$

In (2.38) replace h+k by h and -h+k by k, where the new h and k are either both odd or both even, in order to obtain

$$E_{h\ k\ 0} = \frac{N^{1/2}}{2} \left\langle \left| E_{\frac{1}{2}(h-k),\ \frac{1}{2}(h+k),\ \ell} \right|^2 - 1 \right\rangle_\ell - \frac{1}{2} E_{h-k,h+k,0}, \qquad (2.39)$$

the basic Σ_1 formula for the space group P4. However, (2.39) expresses the seminvariant $E_{h\,k\,0}$ in terms of the (in general) unknown seminvariant $E_{h-k,h+k,\ell}$. In order to evaluate this term, one employs (2.39) with h replaced by h-k and k replaced by h+k. Thus

$$E_{h-k,h+k,0} = \frac{N^{1/2}}{2}\left\langle |E_{\bar{k}h\ell}|^2 - 1\right\rangle_\ell - \frac{1}{2}E_{2\bar{k}\,2h\,0}, \tag{2.40}$$

or (in view of $E_{\bar{k}h\ell} = E_{hk\ell}$, which is readily verified from (2.28)),

$$E_{h-k,h+k,0} = \frac{N^{1/2}}{2}\left\langle |E_{hk\ell}|^2 - 1\right\rangle_\ell - \frac{1}{2}E_{2h\,2k\,0}. \tag{2.41}$$

Substituting from (2.41) into (2.39), one finds

$$E_{h\,k\,0} = \frac{N^{1/2}}{2}\left\langle (|E_{\frac{1}{2}(h-k),\frac{1}{2}(h+k),\ell}|^2 - 1) - \frac{1}{2}(|E_{hk\ell}|^2 - 1)\right\rangle_\ell +$$

$$\frac{1}{4}E_{2h\,2k\,0}, \tag{2.42}$$

an expression for $E_{h\,k\,0}$ which however depends on the unknown term $E_{2h\,2k\,0}$. In order to evaluate the latter, one again employs (2.39) with h replaced by 2h and k replaced by 2k. Then

$$E_{2h\,2k\,0} = \frac{N^{1/2}}{2}\left\langle |E_{h-k,h+k,\ell}|^2 - 1\right\rangle_\ell -$$

$$\frac{1}{2}E_{2(h-k),2(h+k),0}. \tag{2.43}$$

Substituting from (2.43) into (2.42) one obtains

$$E_{h\ k\ 0} = \frac{N^{1/2}}{2} \left\langle (|E_{\frac{1}{2}(h-k),\frac{1}{2}(h+k),\ell}|^2 - 1) - \frac{1}{2}(|E_{hk\ell}|^2 - 1) + \right.$$

$$\left. \frac{1}{4}(|E_{h-k,h+k,\ell}|^2 - 1) \right\rangle_\ell - \frac{1}{8} E_{2(h-k),2(h+k),0}. \qquad (2.44)$$

One may continue in this way to obtain a sequence of formulas (2.39), (2.42), (2.44), etc. each of which is an explicit expression in closed form for the cosine seminvariant $E_{h\ k\ 0} = |E_{h\ k\ 0}| \cos \phi_{h\ k\ 0}$ with successive remainder terms

$$-\frac{1}{2} E_{h-k,h+k,0}, \quad \frac{1}{4} E_{2h\ 2k\ 0}, \quad -\frac{1}{8} E_{2(h-k),2(h+k),0}, \quad \frac{1}{16} E_{4h\ 4k\ 0}, \cdots$$

$$(2.45)$$

only the magnitudes of which are known. The sequence (2.45) however, clearly approaches zero, although not necessarily mono- tonically, so that one finally obtains the desired Σ_1 formula in the space group P4 in the form of the rapidly converging series,

$$E_{h\ k\ 0} = \frac{N^{1/2}}{2} \left\langle (|E_{\frac{1}{2}(h-k),\frac{1}{2}(h+k),\ell}|^2 - 1) - \frac{1}{2}(|E_{hk\ell}|^2 - 1) + \right.$$

$$\frac{1}{4}(|E_{(h-k),h+k),\ell}|^2 - 1) - \frac{1}{8}(|E_{2h\ 2k\ \ell}|^2 - 1) +$$

$$\left. \frac{1}{16}(|E_{2(h-k),2(h+k),\ell}|^2 - 1) - \cdots \right\rangle_\ell. \qquad (2.46)$$

In practice however, it is likely that one of the finite forms
(2.39), (2.42), ..., with remainder term, will prove to be more
useful than (2.46). For example, if the known magnitude
$1/2|E_{h-k,h+k,0}|$ is very small compared to $|E_{h\ k\ 0}|$ one would use
(2.39). If, on the other hand, $1/2|E_{h-k,h+k,0}|$ is large but
$1/4|E_{2h\ 2k\ 0}|$ is small relative to $|E_{h\ k\ 0}|$, then (2.42) would
be used. Again, if $1/2|E_{h-k,h+k,0}|$ and $1/4|E_{2h\ 2k\ 0}|$ are both
large but $1/8|E_{2(h-k),2(h+k),0}|$ is small compared with $|E_{h\ k\ 0}|$,
then one would use (2.44), etc. It is noteworthy that Eq. (2.46),
the Σ_1 formula for P4, differs markedly from the Σ_1 formula for
$P4_1$ (Ex. 5, below).

Exercises II. 3.

   1. Space group P4, origin on 2. Show that when the origin
is on a two-fold axis the functional form for the geometric
structure factor in the space group P4 ((4.6) and (4.7) of Chapter
I) may be written

$$
\left.
\begin{aligned}
A &= 4\cos\pi\left[(h-k)x+(h+k)y+\frac{h+k}{2}\right]\cos\pi\left[(h+k)x-(h-k)y-\frac{h+k}{2}\right]\cos 2\pi\ell z, \\[2ex]
B &= 4\cos\pi\left[(h-k)x+(h+k)y+\frac{h+k}{2}\right]\cos\pi\left[(h+k)x-(h-k)y-\frac{h+k}{2}\right]\sin 2\pi\ell z.
\end{aligned}
\right\}
$$

$$(2.47)$$

   2. Show that, in the space group P4, if the origin is on a
two-fold axis,

$$
2(-1)^{h+k}E_{h+k,-h+k,0} + E_{2h\ 2k\ 0} = N^{1/2}\left\langle |E_{hk\ell}|^2 - 1 \right\rangle_\ell, \qquad (2.48)
$$

(the analogue of (2.38)) for all h and k.

    3. For the space group P4 with origin on a two-fold axis, the analogues of (2.39), (2.42), ... are

$$(-1)^h E_{h\ k\ 0} = \frac{N^{1/2}}{2} \left\langle \left| E_{\frac{1}{2}(h-k),\frac{1}{2}(h+k),\ell} \right|^2 - 1 \right\rangle_\ell - \frac{1}{2} E_{h-k,h+k,0},$$

$$(2.49)$$

$$(-1)^h E_{h\ k\ 0} = \frac{N^{1/2}}{2} \left\langle \left(\left| E_{\frac{1}{2}(h-k),\frac{1}{2}(h+k),\ell} \right|^2 - 1 \right) - \right.$$

$$\left. \frac{1}{2}\left(\left| E_{hk\ell} \right|^2 - 1 \right) \right\rangle_\ell + \frac{1}{4} E_{2h\ 2k\ 0}, \qquad (2.50)$$

etc., and finally the analogue of (2.46) is

$$(-1)^h E_{h\ k\ 0} = \frac{N^{1/2}}{2} \left\langle \left(\left| E_{\frac{1}{2}(h-k),\frac{1}{2}(h+k),\ell} \right|^2 - 1 \right) - \frac{1}{2}\left(\left| E_{hk\ell} \right|^2 - 1 \right) + \right.$$

$$\left. \frac{1}{4}\left(\left| E_{h-k,h+k,\ell} \right|^2 - 1 \right) - \ldots \right\rangle_\ell \qquad (2.51)$$

where h and k have the same parity.

    4. Infer that, in the space group P4, the value of a cosine seminvariant $\cos \phi_{h\ k\ 0}$ which is also a cosine invariant (i.e. h and k are both even) is independent of the functional form for the geometric structure factor. However, the value of a cosine seminvariant $\cos \phi_{h\ k\ 0}$ which is not a cosine invariant (i.e. h and k are both odd) depends on the functional form for the

geometric structure factor and, in fact, is reversed in sign as
the origin is shifted from one equivalence class to the other.

5. Space group $P4_1$. If the origin is on a four-fold axis
in the space group $P4_1$, show that

$$E_{2h\ 2k\ 0} + 2E_{h+k,-h+k,0} = N^{1/2} \left\langle |E_{hk\ell}|^2 - 1 \right\rangle_{\ell=4n}, \qquad (2.52)$$

$$E_{2h\ 2k\ 0} - 2E_{h+k,-h+k,0} = N^{1/2} \left\langle |E_{hk\ell}|^2 - 1 \right\rangle_{\ell=4n+2}, \qquad (2.53)$$

$$E_{2h\ 2k\ 0} = -N^{1/2} \left\langle |E_{hk\ell}|^2 - 1 \right\rangle_{\ell=2n+1}. \qquad (2.54)$$

Infer that

$$E_{2h\ 2k\ 0} = N^{1/2} \left\langle (-1)^\ell (|E_{hk\ell}|^2 - 1) \right\rangle_\ell, \qquad (2.55)$$

while, if h and k have the same parity,

$$E_{h\ k\ 0} = \frac{N^{1/2}}{2} \left\langle (-1)^{\frac{\ell}{2}} (|E_{\frac{1}{2}(h-k),\frac{1}{2}(h+k),\ell}|^2 - 1) \right\rangle_{\ell=2n}, \qquad (2.56)$$

the Σ_1 formulas in the space group $P4_1$.

6. Space group $P4_1$. If the origin is on a two-fold axis
in space group $P4_1$, confirm (by deriving the appropriate Σ_1
formula) that each cosine seminvariant $\cos \phi_{h\ k\ 0}$ which is also
a cosine invariant (i.e. h and k are both even) has the same
value as that which obtains when the origin is on a four-fold axis;
if, however, $\cos \phi_{h\ k\ 0}$ is a cosine seminvariant which is not
a cosine invariant (i.e. h and k are both odd), then its sign is

opposite that which obtains when the origin is on a four-fold
axis. Could this result have been predicted from seminvariance
theory?

3. Formulas for the Cosine Seminvariants, Cos $(\phi_1 + \phi_2)$

Although the space group dependent linear combinations of
two phases which are structure seminvariants have been known
since 1953, explicit formulas for their cosines in terms of the
magnitudes of the normalized structure factors have been
discovered only recently (1971). Two cases, the space groups
P2 and $P2_1$, which illustrate the methods and results, are
considered here. Although the general approach is similar in
these two space groups, the derivations differ considerably in
detail and the final formulas are surprisingly different.

3.1. Space group P2. In this space group, if the origin
is on a two-fold axis,

$$E_{hk\ell} = \frac{2}{N^{1/2}} \sum_{\mu=1}^{N/2} \cos 2\pi(hx_\mu + \ell z_\mu) \exp(2\pi i k y_\mu) \tag{3.1}$$

$$|E_{hk\ell}|^2 = \frac{4}{N} \sum_{\substack{\mu,\nu \\ 1}}^{N/2} \cos 2\pi(hx_\mu + \ell z_\mu)\cos 2\pi(hx_\nu + \ell z_\nu)\exp\left\{2\pi i k(y_\mu - y_\nu)\right\}.$$

$$\tag{3.2}$$

Employing the usual device of splitting the double sum into a
simple sum, when $\mu=\nu$, and the residual double sum, when $\mu\neq\nu$, one
obtains

$$|E_{hk\ell}|^2 = \frac{4}{N} \sum_{\mu=1}^{N/2} \cos^2 2\pi(hx_\mu + \ell z_\mu) +$$

$$\frac{4}{N} \sum_{\substack{\mu \neq \nu \\ 1}}^{N/2} \cos 2\pi(hx_\mu + \ell z_\mu)\cos 2\pi(hx_\nu + \ell z_\nu)\cos 2\pi k(y_\mu - y_\nu). \quad (3.3)$$

From

$$\cos^2 2\pi(hx_\mu + \ell z_\mu) = \frac{1}{2} + \frac{1}{2}\cos 4\pi(hx_\mu + \ell z_\mu) \quad (3.4)$$

and (from (3.1))

$$E_{2h\ 0\ 2\ell} = \frac{2}{N^{1/2}} \sum_{\mu=1}^{N/2} \cos 4\pi(hx_\mu + \ell z_\mu) \quad (3.5)$$

(3.3) leads directly to

$$|E_{hk\ell}|^2 - 1 = \frac{1}{N^{1/2}} E_{2h\ 0\ 2\ell} +$$

$$\frac{4}{N} \sum_{\substack{\mu \neq \nu \\ 1}}^{N/2} \cos 2\pi(hx_\mu + \ell z_\mu)\cos 2\pi(hx_\nu + \ell z_\nu)\cos 2\pi k(y_\mu - y_\nu)$$

$$(3.6)$$

and, replacing h and ℓ by h' and ℓ' respectively,

$$|E_{h'k\ell'}|^2 - 1 = \frac{1}{N^{1/2}} E_{2h'\ 0\ 2\ell'} +$$

$$\frac{4}{N} \sum_{\substack{\mu \neq \nu \\ 1}}^{N/2} \cos 2\pi(h'x_\mu + \ell'z_\mu)\cos 2\pi(h'x_\nu + \ell'z_\nu)\cos 2\pi k(y_\mu - y_\nu).$$

$$(3.7)$$

Multiplying (3.6) and (3.7), one obtains

$$(|E_{hk\ell}|^2 - 1)(|E_{h'k\ell'}|^2 - 1) = \frac{1}{N} E_{2h\ 0\ 2\ell}\ E_{2h'\ 0\ 2\ell'} +$$

$$\frac{4}{N^{3/2}} E_{2h\ 0\ 2\ell} \sum_{\substack{\mu \neq \nu \\ 1}}^{N/2} \cos 2\pi(h'x_\mu + \ell'z_\mu)\cos 2\pi(h'x_\nu + \ell'z_\nu) \times$$

$$\cos 2\pi k(y_\mu - y_\nu) + \frac{4}{N^{3/2}} E_{2h'\ 0\ 2\ell'} \sum_{\substack{\mu \neq \nu \\ 1}}^{N/2} \cos 2\pi(hx_\mu + \ell z_\mu) \times$$

$$\cos 2\pi(hx_\nu + \ell z_\nu)\cos 2\pi k(y_\mu - y_\nu) + \frac{16}{N^2} \sum_{\substack{\mu \neq \nu \\ \rho \neq \sigma \\ 1}}^{N/2} \cos 2\pi(hx_\mu + \ell z_\mu) \times$$

$$\cos 2\pi(hx_\nu + \ell z_\nu)\cos 2\pi(h'x_\rho + \ell'z_\rho)\cos 2\pi(h'x_\sigma + \ell'z_\sigma) \times$$

$$\cos 2\pi k(y_\mu - y_\nu)\cos 2\pi k(y_\sigma - y_\rho).$$

$$(3.8)$$

The quadruple sum in (3.8) may be decomposed into two (equal) double sums obtained when $\rho=\mu$, $\sigma=\nu$ or $\rho=\nu$, $\sigma=\mu$; into four (equal) triple sums obtained when $\rho=\mu$, $\sigma\neq\nu$ or $\rho=\nu$, $\sigma\neq\mu$ or $\sigma=\mu$, $\rho\neq\nu$ or $\sigma=\nu$, $\rho\neq\mu$; and into the residual quadruple sum when no two of μ, ν, ρ, σ are equal. Thus

$$
(|E_{hk\ell}|^2 - 1)(|E_{h'k\ell'}|^2 - 1) = \frac{1}{N} E_{2h\ 0\ 2\ell}\ E_{2h'\ 0\ 2\ell'} +
$$

$$
\frac{4}{N^{3/2}}\ E_{2h\ 0\ 2\ell} \sum_{\substack{\mu\neq\nu \\ 1}}^{N/2} \cos2\pi(h'x_\mu+\ell'z_\mu)\cos2\pi(h'x_\nu+\ell'z_\nu)\ \times
$$

$$
\cos2\pi k(y_\mu-y_\nu)\ + \frac{4}{N^{3/2}}\ E_{2h'\ 0\ 2\ell'} \sum_{\substack{\mu\neq\nu \\ 1}}^{N/2} \cos2\pi(hx_\mu-\ell z_\mu)\ \times
$$

$$
\cos2\pi(hx_\nu+\ell z_\nu)\cos2\pi k(y_\mu-y_\nu)\ + \frac{32}{N^2} \sum_{\substack{\mu\neq\nu \\ 1}}^{N/2} \cos2\pi(hx_\mu+\ell z_\mu)\ \times
$$

$$
\cos2\pi(h'x_\mu+\ell'z_\mu)\cos2\pi(hx_\nu+\ell z_\nu)\cos2\pi(h'x_\nu+\ell'z_\nu)\cos^2 2\pi k(y_\mu-y_\nu)\ +
$$

$$
\frac{64}{N^2} \sum_{\substack{\mu\neq\nu\neq\rho \\ 1}}^{N/2} \cos2\pi(hx_\mu+\ell z_\mu)\cos2\pi(hx_\nu+\ell z_\nu)\cos2\pi(h'x_\mu+\ell'z_\mu)\ \times
$$

$$
\cos2\pi(h'x_\rho+\ell'z_\rho)\cos2\pi k(y_\mu-y_\nu)\cos2\pi k(y_\mu-y_\rho)\ +
$$

$$\frac{16}{N^2} \sum_{\substack{\mu \neq \nu \neq \rho \neq \sigma \\ 1}}^{N/2} \cos 2\pi(hx_\mu + \ell z_\mu)\cos 2\pi(hx_\nu + \ell z_\nu)\cos 2\pi(h'x_\rho + \ell'z_\rho) \ \times$$

$$\cos 2\pi(h'x_\sigma + \ell'z_\sigma)\cos 2\pi k(y_\mu - y_\nu)\cos 2\pi k(y_\rho - y_\sigma). \qquad (3.9)$$

If both sides of (3.9) are averaged over all integers k then each
contributor to the first two double sums on the right will vanish
if it is assumed that

$$y_\mu - y_\nu \neq 0 \text{ when } \mu \neq \nu \qquad (3.10)$$

since

$$\left\langle \cos 2\pi k(y_\mu - y_\nu) \right\rangle_k = 0 \quad \text{if } y_\mu - y_\nu \neq 0, \qquad (3.11)$$

while if, in addition, it is assumed that

$$2y_\mu - y_\nu - y_\rho \neq 0 \quad \text{when } \mu \neq \nu \neq \rho. \qquad (3.12)$$

then

$$\left\langle \cos 2\pi k(y_\mu - y_\nu)\cos 2\pi k(y_\mu - y_\rho) \right\rangle_k =$$

$$\frac{1}{2} \left\langle \cos 2\pi k(2y_\mu - y_\nu - y_\rho) + \cos 2\pi k(y_\nu - y_\rho) \right\rangle_k = 0 \qquad (3.13)$$

so that each contributor to the triple sum on the right side of
(3.9) will also vanish during the averaging process. Clearly,
all contributors to the fourth order sum on the right side of
(3.9) will also vanish upon averaging over k once (3.10) is

assumed to hold. However

$$\left\langle \cos^2 2\pi k(y_\mu - y_\nu)\right\rangle_k = \frac{1}{2} + \frac{1}{2}\left\langle \cos 4\pi k(y_\mu - y_\nu)\right\rangle_k = \frac{1}{2}, \quad (3.14)$$

so that the third second order sum on the right side of (3.9) does not vanish during the averaging process. In short, if (3.10) and (3.12) are assumed to hold, then (3.9) leads to

$$\left\langle (|E_{hk\ell}|^2 - 1)(|E_{h'k\ell'}|^2 - 1)\right\rangle_k = \frac{1}{N} E_{2h\ 0\ 2\ell} E_{2h'\ 0\ 2\ell'} +$$

$$\frac{4}{N^2} \sum_{\substack{\mu \neq \nu \\ 1}}^{N/2} \left\{ \cos 2\pi[(h+h')x_\mu + (\ell+\ell')z_\mu] + \cos 2\pi[(h-h')x_\mu + \right.$$

$$(\ell-\ell')z_\mu] \right\} \left\{ \cos 2\pi[(h+h')x_\nu + (\ell+\ell')z_\nu] + \cos 2\pi[(h-h')x_\nu + \right.$$

$$(\ell-\ell')z_\nu] \Big\}. \quad (3.15)$$

The double sum in (3.15) consists of four terms each of the form

$$\frac{4}{N^2} \sum_{\substack{\mu \neq \nu \\ 1}}^{N/2} \cos 2\pi(h_1 x_\mu + \ell_1 z_\mu)\cos 2\pi(h_2 x_\nu + \ell_2 z_\nu) \quad (3.16)$$

and it is necessary now to evaluate this double sum. To this end write (from (3.1))

$$E_{h_1\ 0\ \ell_1} = \frac{2}{N^{1/2}} \sum_{\mu=1}^{N/2} \cos 2\pi (h_1 x_\mu + \ell_1 z_\mu) \qquad (3.17)$$

$$E_{h_2\ 0\ \ell_2} = \frac{2}{N^{1/2}} \sum_{\mu=1}^{N/2} \cos 2\pi (h_2 x_\mu + \ell_2 z_\mu). \qquad (3.18)$$

Multiply (3.17) and (3.18). Then

$$E_{h_1\ 0\ \ell_1}\ E_{h_2\ 0\ \ell_2} = \frac{4}{N} \sum_{\substack{\mu,\nu \\ 1}}^{N/2} \cos 2\pi (h_1 x_\mu + \ell_1 z_\mu) \cos 2\pi (h_2 x_\nu + \ell_2 z_\nu) \qquad (3.19)$$

$$E_{h_1\ 0\ \ell_1}\ E_{h_2\ 0\ \ell_2} = \frac{4}{N} \sum_{\mu=1}^{N/2} \cos 2\pi (h_1 x_\mu + \ell_1 z_\mu) \cos 2\pi (h_2 x_\mu + \ell_2 z_\mu) +$$

$$\frac{4}{N} \sum_{\substack{\mu \neq \nu \\ 1}}^{N/2} \cos 2\pi (h_1 x_\mu + \ell_1 z_\mu) \cos 2\pi (h_2 x_\nu + \ell_2 z_\nu) \qquad (3.20)$$

$$E_{h_1\ 0\ \ell_1}\ E_{h_2\ 0\ \ell_2} = \frac{2}{N} \sum_{\mu=1}^{N/2} \left\{ \cos 2\pi [(h_1+h_2)x_\mu + (\ell_1+\ell_2)z_\mu] + \right.$$

$$\left. \cos 2\pi [(h_1-h_2)x_\mu + (\ell_1-\ell_2)z_\mu] \right\} +$$

$$\frac{4}{N} \sum_{\substack{\mu \neq \nu \\ 1}}^{N/2} \cos 2\pi (h_1 x_\mu + \ell_1 z_\mu) \cos 2\pi (h_2 x_\nu + \ell_2 z_\nu). \tag{3.21}$$

In view of (3.1) the two simple sums on the right side of (3.21) may be replaced by

$$\frac{1}{N^{1/2}} E_{h_1+h_2, 0, \ell_1+\ell_2} \quad \text{and} \quad \frac{1}{N^{1/2}} E_{h_1-h_2, 0, \ell_1-\ell_2}$$

respectively, so that (3.21) leads to

$$\frac{4}{N} \sum_{\substack{\mu \neq \nu \\ 1}}^{N/2} \cos 2\pi (h_1 x_\mu + \ell_1 z_\mu) \cos 2\pi (h_2 x_\nu + \ell_2 z_\nu) = E_{h_1 \ 0 \ \ell_1} E_{h_2 \ 0 \ \ell_2} -$$

$$\frac{1}{N^{1/2}} E_{h_1+h_2, 0, \ell_1+\ell_2} - \frac{1}{N^{1/2}} E_{h_1-h_2, 0, \ell_1-\ell_2}, \tag{3.22}$$

the desired expression for the double sum (3.16) which is needed in (3.15). Substituting from (3.22) into (3.15), after replacing first, h_1, ℓ_1, h_2, ℓ_2 by h+h', ℓ+ℓ', h+h', ℓ+ℓ', respectively; second, h_1, ℓ_1, h_2, ℓ_2 by h+h', ℓ+ℓ', h-h', ℓ-ℓ', respectively; third h_1, ℓ_1, h_2, ℓ_2 by h-h', ℓ-ℓ', h+h', ℓ+ℓ', respectively; and fourth, h_1, ℓ_1, h_2, ℓ_2 by h-h', ℓ-ℓ', h-h', ℓ-ℓ', respectively, one finds

$$\left\langle (|E_{hk\ell}|^2 - 1)(|E_{h'k\ell'}|^2 - 1) \right\rangle_k = \frac{1}{N} E_{2h \ 0 \ 2\ell} E_{2h' \ 0 \ 2\ell'} +$$

$$\frac{1}{N} E_{h+h', 0, \ell+\ell'}^2 - \frac{1}{N^{3/2}} E_{2(h+h'), 0, 2(\ell+\ell')} - \frac{1}{N^{3/2}} E_{0 \ 0 \ 0} +$$

$$\frac{1}{N} E^2_{h-h',0,\ell-\ell'} - \frac{1}{N^{3/2}} E_{2(h-h'),0,2(\ell-\ell')} - \frac{1}{N^{3/2}} E_{0\ 0\ 0} +$$

$$\frac{2}{N} E_{h+h',0,\ell-\ell'} E_{h-h',0,\ell-\ell'} - \frac{2}{N^{3/2}} E_{2h\ 0\ 2\ell} - \frac{2}{N^{3/2}} E_{2h'\ 0\ 2\ell'} .$$

$$(3.23)$$

Hence,

$$E_{h+h',0,\ell+\ell'} E_{h-h',0,\ell-\ell'} = \frac{N}{2} \left\langle (|E_{hk\ell}|^2 - 1)(|E_{h'k\ell'}|^2 - 1) \right\rangle_k -$$

$$\frac{1}{2} E_{2h,0,2\ell} E_{2h',0,2\ell'} - \frac{1}{2} E^2_{h+h,0,\ell+\ell'} - \frac{1}{2} E^2_{h-h',0,\ell-\ell'} +$$

$$\frac{1}{N^{1/2}} + \frac{1}{N^{1/2}} E_{2h\ 0\ 2\ell} + \frac{1}{N^{1/2}} E_{2h'\ 0\ 2\ell'} +$$

$$\frac{1}{2N^{1/2}} E_{2(h+h'),0,2(\ell+\ell')} + \frac{1}{2N^{1/2}} E_{2(h-h'),0,2(\ell-\ell')} . \quad (3.24)$$

Making the substitution

$$\left. \begin{array}{ll} h + h' = h_1, & \ell + \ell' = \ell_1, \\[2mm] h - h' = h_2, & \ell - \ell' = \ell_2, \end{array} \right\} \qquad (3.25)$$

where $h_1 \pm h_2$ and $\ell_1 \pm \ell_2$ are all even, and neglecting terms of order $1/N^{1/2}$ in (3.24), one finally obtains the basic formula for the cosine seminvariant $\cos(\phi_{h_1\ 0\ \ell_1} + \phi_{h_2\ 0\ \ell_2})$

$$E_{h_1 0 \ell_1} E_{h_2 0 \ell_2} = |E_{h_1 0 \ell_1} E_{h_2 0 \ell_2}| \cos(\phi_{h_1 0 \ell_1} + \phi_{h_2 0 \ell_2}) \approx$$

$$\frac{N}{2} \left\langle (|E_{\frac{1}{2}(h_1+h_2),k,\frac{1}{2}(\ell_1+\ell_2)}|^2 - 1) \times \right.$$

$$\left. (|E_{\frac{1}{2}(h_1-h_2),k,\frac{1}{2}(\ell_1-\ell_2)}|^2 - 1) \right\rangle_k -$$

$$\frac{1}{2} E_{h_1+h_2,0,\ell_1+\ell_2} E_{h_1-h_2,0,\ell_1-\ell_2} - \frac{1}{2}(E^2_{h_1 0 \ell_1} + E^2_{h_2 0 \ell_2}).$$

$$(3.26)$$

However, the right side of (3.26) contains the term

$$- \frac{1}{2} E_{h_1+h_2,0,\ell_1+\ell_2} E_{h_1-h_2,0,\ell_1-\ell_2}$$

only the magnitude of which is, in general, known. Since

$$\left. \begin{array}{c} (h_1 + h_2) + (h_1 - h_2) = 2h_1 \\[2mm] (\ell_1 + \ell_2) + (\ell_1 - \ell_2) = 2\ell_1 \end{array} \right\} \qquad (3.27)$$

are both even, $E_{h_1+h_2,0,\ell_1+\ell_2} E_{h_1-h_2,0,\ell_1-\ell_2}$ is a structure seminvariant and (3.26) may be employed, with h_1 replaced by (h_1+h_2), h_2 by (h_1-h_2), ℓ_1 by $(\ell_1+\ell_2)$, and ℓ_2 by $(\ell_1-\ell_2)$ to yield

$$E_{h_1+h_2,0,\ell_1+\ell_2} E_{h_1-h_2,0,\ell_1-\ell_2} \approx \frac{N}{2} \left\langle (|E_{h_1 k \ell_1}|^2 - 1) \times \right.$$

$$\left. (|E_{h_2 k \ell_2}|^2 - 1) \right\rangle_k - \frac{1}{2} E_{2h_1 0 2\ell_1} E_{2h_2 0 2\ell_2} -$$

$$\frac{1}{2}(E^2_{h_1+h_2,0,\ell_1+\ell_2} + E^2_{h_1-h_2,0,\ell_1-\ell_2}).$$ (3.28)

Substituting from (3.28) into (3.26), one finds

$$E_{h_1\,0\,\ell_1}\,E_{h_2\,0\,\ell_2} \approx \frac{N}{2}\left\langle (|E_{\frac{1}{2}(h_1+h_2),k,\frac{1}{2}(\ell_1+\ell_2)}|^2 - 1)\times\right.$$

$$(|E_{\frac{1}{2}(h_1-h_2),k,\frac{1}{2}(\ell_1-\ell_2)}|^2 - 1) -$$

$$\left.\frac{1}{2}(|E_{h_1k\ell_1}|^2 - 1)(|E_{h_2k\ell_2}|^2 - 1)\right\rangle_k +$$

$$\frac{1}{4}E_{2h_1,0,2\ell_1}\,E_{2h_2,0,2\ell_2} - \frac{1}{2}(E^2_{h_1\,0\,\ell_1} + E^2_{h_2\,0\,\ell_2}) +$$

$$\frac{1}{4}(E^2_{h_1+h_2,0,\ell_1+\ell_2} + E^2_{h_1-h_2,0,\ell_1-\ell_2}).$$ (3.29)

Again, the right side of (3.29) contains the unknown remainder term $1/4\;E_{2h_1,0,2\ell_1}\,E_{2h_2,0,2\ell_2}$ which is a structure seminvariant and may be found from (3.26) by merely replacing h_1 by $2h_1$, ℓ_1 by $2\ell_1$, h_2 by $2h_2$, and ℓ_2 by $2\ell_2$:

$$E_{2h_1,0,2\ell_1}\,E_{2h_2,0,2\ell_2} \approx \frac{N}{2}\left\langle (|E_{(h_1+h_2),k,(\ell_1+\ell_2)}|^2 - 1)\times\right.$$

$$\left.(|E_{(h_1-h_2),k,(\ell_1-\ell_2)}|^2 - 1)\right\rangle_k -$$

$$\frac{1}{2} \, E_{2(h_1+h_2),0,2(\ell_1+\ell_2)} \, E_{2(h_1-h_2),0,2(\ell_1-\ell_2)} \, -$$

$$\frac{1}{2} \, (E^2_{2h_1,0,2\ell_1} + E^2_{2h_2,0,2\ell_2}). \tag{3.30}$$

Substituting from (3.30) into (3.29) yields

$$E_{h_1 \, 0 \, \ell_1} \, E_{h_2 \, 0 \, \ell_2} \approx \frac{N}{2} \Big\langle (|E_{\frac{1}{2}(h_1+h_2),k,\frac{1}{2}(\ell_1+\ell_2)}|^2 - 1) \times$$

$$(|E_{\frac{1}{2}(h_1-h_2),k,\frac{1}{2}(\ell_1-\ell_2)}|^2 - 1) - \frac{1}{2}(|E_{h_1 k \ell_1}|^2 - 1) \times$$

$$(|E_{h_2 k \ell_2}|^2 - 1) + \frac{1}{4}(|E_{h_1+h_2,k,\ell_1+\ell_2}|^2 - 1) \times$$

$$(|E_{h_1-h_2,k,\ell_1-\ell_2}|^2 - 1)\Big\rangle_k - \frac{1}{8} E_{2(h_1+h_2),0,2(\ell_1+\ell_2)} \times$$

$$E_{2(h_1-h_2),0,2(\ell_1-\ell_2)} - \frac{1}{2}(E^2_{h_1 \, 0 \, \ell_1} + E^2_{h_2 \, 0 \, \ell_2}) +$$

$$\frac{1}{4}(E^2_{h_1+h_2,0,\ell_1+\ell_2} + E^2_{h_1-h_2,0,\ell_1-\ell_2}) -$$

$$\frac{1}{8}(E^2_{2h_1,0,2\ell_1} + E^2_{2h_2,0,2\ell_2}). \tag{3.31}$$

Continuing in this way one obtains a sequence of formulas (3.26), (3.29), (3.31), ..., in which the unknown remainder terms,

$$- \frac{1}{2} E_{h_1+h_2,0,\ell_1+\ell_2} \; E_{h_1-h_2,0,\ell_1-\ell_2}; \; \frac{1}{4} E_{2h_1,0,2\ell_1} \; E_{2h_2,0,2\ell_2};$$

$$- \frac{1}{8} E_{2(h_1+h_2),0,2(\ell_1+\ell_2)} \; E_{2(h_1-h_2),0,2(\ell_1-\ell_2)}; \; \cdot \; \cdot \; \cdot \; , \quad (3.32)$$

obviously approach zero. Hence the desired formula is finally
expressed in the form of two rapidly converging series,

$$E_{h_1 \, 0 \, \ell_1} E_{h_2 \, 0 \, \ell_2} = |E_{h_1 \, 0 \, \ell_1} E_{h_2 \, 0 \, \ell_2}| \cos(\phi_{h_1 \, 0 \, \ell_1} + \phi_{h_2 \, 0 \, \ell_2}) \approx$$

$$\frac{N}{2} \Bigg\langle (|E_{\frac{1}{2}(h_1+h_2),k,\frac{1}{2}(\ell_1+\ell_2)}|^2 - 1)(|E_{\frac{1}{2}(h_1-h_2),k,\frac{1}{2}(\ell_1-\ell_2)}|^2 - 1) -$$

$$\frac{1}{2}(|E_{h_1 k \ell_1}|^2 - 1)(|E_{h_2 k \ell_2}|^2 - 1) +$$

$$\frac{1}{4}(|E_{h_1+h_2,k,\ell_1+\ell_2}|^2 - 1)(|E_{h_1-h_2,k,\ell_1-\ell_2}|^2 - 1) -$$

$$\frac{1}{8}(|E_{2h_1,k,2\ell_1}|^2 - 1)(|E_{2h_2,k,2\ell_2}|^2 - 1) + \ldots \Bigg\rangle_k -$$

$$\frac{1}{2}(E_{h_1 \, 0 \, \ell_1}^2 + E_{h_2 \, 0 \, \ell_2}^2) + \frac{1}{4}(E_{h_1+h_2,0,\ell_1+\ell_2}^2 + E_{h_1-h_2,0,\ell_1-\ell_2}^2) -$$

$$\frac{1}{8}(E_{2h_1,0,2\ell_1}^2 + E_{2h_2,0,2\ell_2}^2) + \frac{1}{16}(E_{2(h_1+h_2),0,2(\ell_1+\ell_2)}^2 +$$

$$E_{2(h_1-h_2),0,2(\ell_1-\ell_2)}^2) - \cdot \; \cdot \; \cdot \; \cdot \; , \quad (3.33)$$

in which $h_1 \pm h_2$ and $\ell_1 \pm \ell_2$ are all even.

   3.2. Space group $P2_1$. In this space group, if the origin is on a two-fold axis,

$$E_{hk\ell} = \frac{2}{N^{1/2}} \sum_{\mu=1}^{N/2} \cos 2\pi \left(hx_\mu + \ell z_\mu + \frac{k}{4} \right) \exp \left\{ 2\pi i k \left(y_\mu - \frac{1}{4} \right) \right\} \qquad (3.34)$$

$$|E_{hk\ell}|^2 = \frac{4}{N} \sum_{\substack{\mu,\nu \\ 1}}^{N/2} \cos 2\pi \left(hx_\mu + \ell z_\mu + \frac{k}{4} \right) \cos 2\pi \left(hx_\nu + \ell z_\nu + \frac{k}{4} \right) \times$$

$$\exp \left\{ 2\pi i k (y_\mu - y_\nu) \right\} . \qquad (3.35)$$

Decomposing the double sum into a simple sum, when $\mu = \nu$, and the residual double sum, when $\mu \neq \nu$, one obtains

$$|E_{hk\ell}|^2 = \frac{4}{N} \sum_{\mu=1}^{N/2} \cos^2 2\pi \left(hx_\mu + \ell z_\mu + \frac{k}{4} \right) +$$

$$\frac{4}{N} \sum_{\substack{\mu \neq \nu \\ 1}}^{N/2} \cos 2\pi \left(hx_\mu + \ell z_\mu + \frac{k}{4} \right) \times$$

$$\cos 2\pi \left(hx_\nu + \ell z_\nu + \frac{k}{4} \right) \exp \left\{ 2\pi i k (y_\mu - y_\nu) \right\} . \qquad (3.36)$$

From

$$\cos^2 2\pi \left(hx_\mu + \ell z_\mu + \frac{k}{4} \right) =$$

$$\frac{1}{2} + \frac{1}{2}(-1)^k \cos 4\pi (hx_\mu + \ell z_\mu) \qquad\qquad (3.37)$$

and

$$E_{2h\ 0\ 2\ell} = \frac{2}{N^{1/2}} \sum_{\mu=1}^{N/2} \cos 4\pi (hx_\mu + \ell z_\mu) \qquad\qquad (3.38)$$

(3.36) leads directly to

$$|E_{hk\ell}|^2 - 1 = \frac{(-1)^k}{N^{1/2}} E_{2h\ 0\ 2\ell} + \frac{4}{N} \sum_{\substack{\mu \neq \nu \\ 1}}^{N/2} \cos 2\pi \left(hx_\mu + \ell z_\mu + \frac{k}{4} \right) \times$$

$$\cos 2\pi \left(hx_\nu + \ell z_\nu + \frac{k}{4} \right) \cos 2\pi k (y_\mu - y_\nu) \ . \qquad\qquad (3.39)$$

Similarly

$$|E_{h'k\ell'}|^2 - 1 = \frac{(-1)^k}{N^{1/2}} E_{2h'\ 0\ 2\ell'} + \frac{4}{N} \sum_{\substack{\mu \neq \nu \\ 1}}^{N/2} \cos 2\pi \left(h'x_\mu + \ell'z_\mu + \frac{k}{4} \right) \times$$

$$\cos 2\pi \left(h'x_\nu + \ell'z_\nu + \frac{k}{4} \right) \cos 2\pi k (y_\mu - y_\nu) \qquad\qquad (3.40)$$

and

$$(-1)^k (|E_{hk\ell}|^2 - 1)(|E_{h'k\ell'}|^2 - 1) = \frac{(-1)^k}{N} E_{2h\ 0\ 2\ell} E_{2h'\ 0\ 2\ell'} +$$

$$\frac{2}{N^{3/2}} E_{2h\ 0\ 2\ell} \sum_{\substack{\mu \neq \nu \\ 1}}^{N/2} \left\{ (-1)^k \Bigl(\cos 2\pi[h'(x_\mu + x_\nu) + \ell'(z_\mu + z_\nu)] \Bigr) + \right.$$

$$\left. \cos 2\pi[h'(x_\mu - x_\nu) + \ell'(z_\mu - z_\nu)] \right\} \cos 2\pi k(y_\mu - y_\nu) +$$

$$\frac{2}{N^{3/2}} E_{2h'\ 0\ 2\ell'} \sum_{\substack{\mu \neq \nu \\ 1}}^{N/2} \left\{ (-1)^k \Bigl(\cos 2\pi[h(x_\mu + x_\nu) + \ell(z_\mu + z_\nu)] \Bigr) + \right.$$

$$\left. \cos 2\pi[h(x_\mu - x_\nu) + \ell(z_\mu - z_\nu)] \right\} \cos 2\pi k(y_\mu - y_\nu) +$$

$$\frac{2(-1)^k}{N^2} \sum_{\substack{\mu \neq \nu \\ \rho \neq \sigma \\ 1}}^{N/2} \left\{ \Bigl(\cos 2\pi[h(x_\mu + x_\nu) + \ell(z_\mu + z_\nu)] \cos 2\pi[h'(x_\rho + x_\sigma) + \right.$$

$$\ell'(z_\rho + z_\sigma)] + \cos 2\pi[h(x_\mu - x_\nu) + \ell(z_\mu - z_\nu)] \cos 2\pi[h'(x_\rho - x_\sigma) +$$

$$\ell'(z_\rho - z_\sigma)] \Bigr) + (-1)^k \Bigl(\cos 2\pi[h(x_\mu + x_\nu) + \ell(z_\mu + z_\nu)] \cos 2\pi[h'(x_\rho - x_\sigma) +$$

$$\ell'(z_\rho - z_\sigma)] + \cos 2\pi[h(x_\mu - x_\nu) + \ell(z_\mu - z_\nu)] \cos 2\pi[h'(x_\rho + x_\sigma) +$$

$$\ell'(z_\rho+z_\sigma)]\Bigg]\Bigg\}\Bigg\{\cos2\pi k(y_\mu-y_\nu+y_\rho-y_\sigma) +$$

$$\cos2\pi k(y_\mu-y_\nu-y_\rho+y_\sigma)\Bigg\} . \tag{3.41}$$

The quadruple sum in (3.41) may be decomposed into two (equal) double sums obtained when $\rho=\mu$, $\sigma=\nu$ or $\rho=\nu$, $\sigma=\mu$; into four (equal) triple sums obtained when $\rho=\mu$, $\sigma\neq\nu$ or $\rho=\nu$, $\sigma\neq\mu$ or $\sigma=\mu$, $\rho\neq\nu$ or $\sigma=\nu$, $\rho\neq\mu$; and into the residual quadruple sum when no two of μ, ν, ρ, σ are equal. Thus

$$(-1)^k(|E_{hk\ell}|^2-1)(|E_{h'k\ell'}|^2-1) = \frac{(-1)^k}{N} E_{2h\ 0\ 2\ell}\ E_{2h'\ 0\ 2\ell'} +$$

$$\frac{2}{N^{3/2}}\ E_{2h\ 0\ 2\ell} \sum_{\substack{\mu\neq\nu\\1}}^{N/2} \Bigg\{(-1)^k\Bigg(\cos2\pi[h'(x_\mu+x_\nu)+\ell'(z_\mu+z_\nu)]\Bigg) +$$

$$\cos2\pi[h'(x_\mu-x_\nu)+\ell'(z_\mu-z_\nu)]\Bigg\}\cos2\pi k(y_\mu-y_\nu) +$$

$$\frac{2}{N^{3/2}}\ E_{2h'\ 0\ 2\ell'} \sum_{\substack{\mu\neq\nu\\1}}^{N/2} \Bigg\{(-1)^k\Bigg(\cos2\pi[h(x_\mu+x_\nu)+\ell(z_\mu+z_\nu)]\Bigg) +$$

$$\cos2\pi[h(x_\mu-x_\nu)+\ell(z_\mu-z_\nu)]\Bigg\} \times$$

$$\cos 2\pi k(y_\mu - y_\nu) + \frac{4(-1)^k}{N^2} \sum_{\substack{\mu \neq \nu \\ 1}}^{N/2} \left\{ \left(\cos 2\pi [h(x_\mu + x_\nu) + \ell(z_\mu + z_\nu)] \times \right. \right.$$

$$\cos 2\pi [h'(x_\mu + x_\nu) + \ell'(z_\mu + z_\nu)] + \cos 2\pi [h(x_\mu - x_\nu) + \ell(z_\mu - z_\nu)] \times$$

$$\left. \cos 2\pi [h'(x_\mu - x_\nu) + \ell'(z_\mu - z_\nu)] \right) + (-1)^k \left(\cos 2\pi [h(x_\mu + x_\nu) + \ell(z_\mu + z_\nu)] \times \right.$$

$$\cos 2\pi [h'(x_\mu - x_\nu) + \ell'(z_\mu - z_\nu)] + \cos 2\pi [h(x_\mu - x_\nu) + \ell(z_\mu - z_\nu)] \times$$

$$\left. \left. \cos 2\pi [h'(x_\mu + x_\nu) + \ell'(z_\mu + z_\nu)] \right) \right\} \left\{ 1 + \cos 4\pi k(y_\mu - y_\nu) \right\} +$$

$$\frac{8(-1)^k}{N} \sum_{\substack{\mu \neq \nu \neq \rho \\ 1}}^{N/2} \left\{ \left(\cos 2\pi [h(x_\mu + x_\nu) + \ell(z_\mu + z_\nu)] \times \right. \right.$$

$$\cos 2\pi [h'(x_\mu + x_\rho) + \ell'(z_\mu + z_\rho)] + \cos 2\pi [h(x_\mu - x_\nu) + \ell(z_\mu - z_\nu)] \times$$

$$\left. \cos 2\pi [h'(x_\mu - x_\rho) + \ell'(z_\mu - z_\rho)] \right) + (-1)^k \left(\cos 2\pi [h(x_\mu + x_\nu) + \ell(z_\mu + z_\nu)] \times \right.$$

$$\cos 2\pi [h'(x_\mu - x_\rho) + \ell'(z_\mu - z_\rho)] + \cos 2\pi [h(x_\mu - x_\nu) + \ell(z_\mu - z_\nu)] \times$$

$$\left. \left. \cos 2\pi [h'(x_\mu + x_\rho) + \ell'(z_\mu + z_\rho)] \right) \right\} \left\{ \cos 2\pi k(y_\mu - y_\nu) + \right.$$

$$\left. \cos 2\pi k(2y_\mu - y_\nu - y_\rho) \right\} +$$

$$\frac{2(-1)^k}{N} \sum_{\substack{\mu \neq \nu \neq \rho \neq \sigma \\ 1}}^{N/2} \left\{ \left(\cos 2\pi[h(x_\mu+x_\nu)+\ell(z_\mu+z_\nu)] \times \right. \right.$$

$$\cos 2\pi[h'(x_\rho+x_\sigma)+\ell'(z_\rho+z_\sigma)] + \cos 2\pi[h(x_\mu-x_\nu)+\ell(z_\mu-z_\nu)] \times$$

$$\left. \cos 2\pi[h'(x_\rho-x_\sigma)+\ell'(z_\rho-z_\sigma)] \right) + (-1)^k \left(\cos 2\pi[h(x_\mu+x_\nu)+\ell(z_\mu+z_\nu)] \times \right.$$

$$\cos 2\pi[h'(x_\rho-x_\sigma)+\ell'(z_\rho-z_\sigma)] + \cos 2\pi[h(x_\mu-x_\nu)+\ell(z_\mu-z_\nu)] \times$$

$$\left. \left. \cos 2\pi[h'(x_\rho+x_\sigma)+\ell'(z_\rho+z_\sigma)] \right) \right\} \left\{ \cos 2\pi k(y_\mu-y_\nu+y_\rho-y_\sigma) + \right.$$

$$\left. \cos 2\pi k(y_\mu-y_\nu-y_\rho+y_\sigma) \right\} . \tag{3.42}$$

If both sides of (3.42) are averaged over all integers k then each contributor to the first two double sums on the right will vanish if it is assumed that

$$y_\mu - y_\nu \neq 0 \text{ or } \frac{1}{2} \text{ when } \mu \neq \nu \tag{3.43}$$

since

$$\left\langle (\pm 1)^k \cos 2\pi k(y_\mu-y_\nu) \right\rangle_k = 0 \text{ if } y_\mu - y_\nu \neq 0 \text{ or } \frac{1}{2} \tag{3.44}$$

while if in addition,

$$2y_\mu-y_\nu-y_\rho \neq 0 \text{ or } \frac{1}{2} \text{ when } \mu \neq \nu \neq \rho, \tag{3.45}$$

then

$$\left\langle (\pm 1)^k \cos 2\pi k (2y_\mu - y_\nu - y_\rho) \right\rangle_k = 0, \tag{3.46}$$

so that each contributor to the triple sum on the right side of
(3.42) will also vanish during the averaging process. Clearly,
all contributors to the fourth order sum on the right side of
(3.42) will also vanish upon averaging over k provided it is
assumed that

$$(y_\mu - y_\nu) - (y_\rho - y_\sigma) \neq 0 \text{ or } \frac{1}{2} \text{ if } \mu \neq \nu \neq \rho \neq \sigma. \tag{3.47}$$

Finally,

$$\left\langle (-1)^k \right\rangle_k = 0. \tag{3.48}$$

However, that portion of the third second order sum on the right
side of (3.42) which is independent of k survives the averaging
process. In short, if (3.43), (3.45) and (3.47) are assumed to
hold, (3.42) leads to

$$\left\langle (-1)^k (|E_{hk\ell}|^2 - 1)(|E_{h'k\ell'}|^2 - 1) \right\rangle_k =$$

$$\frac{4}{N^2} \sum_{\substack{\mu \neq \nu \\ 1}}^{N/2} \left\{ \cos 2\pi [h(x_\mu + x_\nu) + \ell(z_\mu + z_\nu)] \cos 2\pi [h'(x_\mu - x_\nu) + \ell'(z_\mu - z_\nu)] + \right.$$

$$\left. \cos 2\pi [h(x_\mu - x_\nu) + \ell(z_\mu - z_\nu)] \cos 2\pi [h'(x_\mu + x_\nu) + \ell'(z_\mu + z_\nu)] \right\} =$$

$$\frac{2}{N^2} \sum_{\substack{\mu \neq \nu \\ 1}}^{N/2} \left\{ \cos 2\pi[(h+h')x_\mu + (\ell+\ell')z_\mu + (h-h')x_\nu + (\ell-\ell')z_\nu] + \right.$$

$$\cos 2\pi[(h-h')x_\mu + (\ell-\ell')z_\mu + (h+h')x_\nu + (\ell+\ell')z_\nu] +$$

$$\cos 2\pi[(h+h')x_\mu + (\ell+\ell')z_\mu - (h-h')x_\nu - (\ell-\ell')z_\nu] +$$

$$\left. \cos 2\pi[(h-h')x_\mu + (\ell-\ell')z_\mu - (h+h')x_\nu - (\ell+\ell')z_\nu] \right\} =$$

$$\frac{8}{N^2} \sum_{\substack{\mu \neq \nu \\ 1}}^{N/2} \cos 2\pi[(h+h')x_\mu + (\ell+\ell')z_\mu]\cos 2\pi[(h-h')x_\nu + (\ell-\ell')z_\nu].$$

$$(3.49)$$

There remains only the task of evaluating the double sum on the right of (3.49). To this end, (3.34) implies

$$E_{h+h',0,\ell+\ell'} = \frac{2}{N^{1/2}} \sum_{\mu=1}^{N/2} \cos 2\pi[(h+h')x_\mu + (\ell+\ell')z_\mu], \qquad (3.50)$$

$$E_{h-h',0,\ell-\ell'} = \frac{2}{N^{1/2}} \sum_{\mu=1}^{N/2} \cos 2\pi[(h-h')x_\mu + (\ell-\ell')z_\mu] . \qquad (3.51)$$

Multiplying (3.50) and (3.51),

$$E_{h+h',0,\ell+\ell'} \ E_{h-h',0,\ell-\ell'} =$$

$$\frac{4}{N} \sum_{\substack{\mu,\nu \\ 1}}^{N/2} \cos2\pi[(h+h')x_\mu+(\ell+\ell')z_\mu]\cos2\pi[(h-h')x_\nu+(\ell-\ell')z_\nu] =$$

$$\frac{4}{N} \sum_{\mu=1}^{N/2} \cos2\pi[(h+h')x_\mu+(\ell+\ell')z_\mu]\cos2\pi[(h-h')x_\mu+(\ell-\ell')z_\mu] \ +$$

$$\frac{4}{N} \sum_{\substack{\mu\neq\nu \\ 1}}^{N/2} \cos2\pi[(h+h')x_\mu+(\ell+\ell')z_\mu]\cos2\pi[(h-h')x_\nu+(\ell-\ell')z_\nu].$$

$$(3.52)$$

However, the first term on the right of (3.52) may be written

$$\frac{2}{N} \sum_{\mu=1}^{N/2} \left\{ \cos4\pi(hx_\mu+\ell z_\mu) \ + \ \cos4\pi(h'x_\mu+\ell'z_\mu) \right\} =$$

$$\frac{1}{N^{1/2}} \ E_{2h \ 0 \ 2\ell} + \frac{1}{N^{1/2}} \ E_{2h' \ 0 \ 2\ell'} \qquad (3.53)$$

in view of (3.34). Combining (3.52) and (3.53) yields

$$\frac{4}{N} \sum_{\substack{\mu\neq\nu \\ 1}}^{N/2} \cos2\pi[(h+h')x_\mu+(\ell+\ell')z_\mu]\cos2\pi[(h-h')x_\nu+(\ell-\ell')z_\nu] =$$

$$E_{h+h',0,\ell+\ell'} \; E_{h-h',0,\ell-\ell'} - \frac{1}{N^{1/2}} E_{2h\;0\;2\ell} - \frac{1}{N^{1/2}} E_{2h'\;0\;2\ell'}.$$

$$(3.54)$$

Substituting from (3.54) into (3.49), one finds

$$\left\langle (-1)^k (|E_{hk\ell}|^2 - 1)(|E_{h'k\ell'}|^2 - 1) \right\rangle_k =$$

$$\frac{2}{N} E_{h+h',0,\ell+\ell'} \; E_{h-h',0,\ell-\ell'} -$$

$$\frac{2}{N^{3/2}} E_{2h\;0\;2\ell} - \frac{2}{N^{3/2}} E_{2h'\;0\;2\ell'}. \qquad (3.55)$$

Finally, under the transformation

$$\left. \begin{array}{ll} h_1 = h+h', & \ell_1 = \ell+\ell' \\[2mm] h_2 = h-h', & \ell_2 = \ell-\ell' \end{array} \right\} \qquad (3.56)$$

(3.55) becomes the desired formula

$$E_{h_1\;0\;\ell_1} \; E_{h_2\;0\;\ell_2} = E_{h_1\;0\;\ell_1} \; E_{h_2\;0\;\ell_2} \cos(\phi_{h_1\;0\;\ell_1} + \phi_{h_2\;0\;\ell_2}) =$$

$$\frac{N}{2} \left\langle (-1)^k (|E_{\frac{1}{2}(h_1+h_2),k,\frac{1}{2}(\ell_1+\ell_2)}|^2 - 1) \times \right.$$

$$\left. (|E_{\frac{1}{2}(h_1-h_2),k,\frac{1}{2}(\ell_1-\ell_2)}|^2 - 1) \right\rangle_k + \frac{1}{N^{1/2}} E_{h_1+h_2,0,\ell_1+\ell_2} +$$

$$\frac{1}{N^{1/2}} E_{h_1-h_2,0,\ell_1-\ell_2} \qquad (3.57)$$

provided, of course, that $h_1 + h_2$ and $\ell_1 + \ell_2$ are both even (from (3.56)). The unexpected differences between (3.33) and (3.57), the formulas for the cosine seminvariants $\cos(\phi_1 + \phi_2)$ in P2 and P2$_1$ respectively, are noteworthy and reminiscent of similar differences between (2.46) and (2.56), the Σ_1 formulas in P4 and P4$_1$ respectively.

Exercises II. 2.

1. Space group P2$_1$. Show that, in the space group P2$_1$, $\phi_{h_1 \, k_1 \, \ell_1} + \phi_{h_2 \, k_2 \, \ell_2}$ is a structure seminvariant if and only if $k_1 = -k_2$ and $h_1 + h_2$ and $\ell_1 + \ell_2$ are both even integers.

2. Space group P2$_1$. Prove the following generalization of (3.57): In the space group P2$_1$,

$$\left| E_{h_1 \, k_1 \, \ell_1} \, E_{h_2 \, k_1 \, \ell_2} \right| \cos(\phi_{h_1 \, k_1 \, \ell_1} + \phi_{h_2 \, \bar{k}_1 \, \ell_2}) \approx$$

$$\frac{N}{2} \left\langle (-1)^k (\left| E_{\frac{1}{2}(h_1 + h_2), \, k, \, \frac{1}{2}(\ell_1 + \ell_2)} \right|^2 - 1) \times \right.$$

$$\left. (\left| E_{\frac{1}{2}(h_1 - h_2), \, k_1 + \, k, \, \frac{1}{2}(\ell_1 - \ell_2)} \right|^2 - 1) \right\rangle_k \qquad (3.58)$$

provided that $h_1 + h_2$ and $\ell_1 + \ell_2$ are both even.

3. Space group P2. Derive the explicit expression for $\left| E_{h_1 \, k_1 \, \ell_1} \, E_{h_2 \, k_1 \, \ell_2} \right| \cos(\phi_{h_1 \, k_1 \, \ell_1} + \phi_{h_2 \, \bar{k}_1 \, \ell_2})$, where $h_1 + h_2$ and $\ell_1 + \ell_2$ are both even, which generalizes (3.33).

4. Space groups P222 and P2$_1$2$_1$2$_1$. Derive explicit formulas

for the cosine seminvariant $\cos(\phi_{h_1 k_1 \ell_1} + \phi_{h_2 \bar{k}_1 \ell_2})$, where $h_1 + h_2$
and $\ell_1 + \ell_2$ are both even, which are the analogues of (3.58). By
cyclic permutation of the indices, infer the formulas for the
cosine seminvariant $\cos(\phi_{h_1 k_1 \ell_1} + \phi_{h_2 k_2 \bar{\ell}_1})$, where $h_1 + h_2$ and
$k_1 + k_2$ are both even, and for $\cos(\phi_{h_1 k_1 \ell_1} + \phi_{\bar{h}_1 k_2 \ell_2})$ where $k_1 + k_2$
and $\ell_1 + \ell_2$ are both even.

5.   In $P2_1 2_1 2_1$, if $\ell_1 + \ell_2$ is even, derive a third formula
(in addition to the two of Ex. 4) for the special cosine
seminvariant $\cos(\phi_{h_1 0 \ell_1} + \phi_{h_1 0 \ell_2})$ of the form

$$
\left| E_{h_1 0 \ell_1} \, E_{h_1 0 \ell_2} \right| \cos(\phi_{h_1 0 \ell_1} + \phi_{h_1 0 \ell_2}) \approx
$$

$$
K \left\langle (-1)^{\ell_1 + h} \left(\left| E_{h, 0, \frac{1}{2}(\ell_1 + \ell_2)} \right|^2 - 1 \right) \times \right.
$$

$$
\left. \left(\left| E_{h_1 + h, 0, \frac{1}{2}(\ell_1 - \ell_2)} \right|^2 - 1 \right) \right\rangle_h .
$$

4.   Formulas for the Cosine Seminvariants, $\cos(\phi_1 + \phi_2 + \phi_3)$

It has been seen that the values of the cosine seminvariants
$\cos\phi$ and $\cos(\phi_1 + \phi_2)$ depend, in general, not only on the
magnitudes $|E|$ of the normalized structure factors but also on
the functional form for the geometric structure factor. Suppose
next that $\vec{h}_1$, $\vec{h}_2$, and $\vec{h}_3$ are three vectors fixed in reciprocal
space which satisfy

$$
\vec{h}_1 + \vec{h}_2 + \vec{h}_3 = 0. \tag{4.1}
$$

Introduce the usual abbreviations

$$E_i = E_{\vec{h}_i}, \quad |E_i| = |E_{\vec{h}_i}|, \quad \phi_i = \phi_{\vec{h}_i}, \quad i = 1,2,3. \tag{4.2}$$

Then, from Corollary 2.21 of Chapter I,

$$\phi_1 + \phi_2 + \phi_3 \tag{4.3}$$

is a structure invariant, and therefore also a structure
seminvariant, for every space group. Hence the values of the
cosine seminvariants, $\cos(\phi_1 + \phi_2 + \phi_3)$, (in contrast to the
cosine seminvariants $\cos \phi$ and $\cos(\phi_1 + \phi)$) are uniquely determined
by the magnitudes alone of the normalized structure factors E
no matter which functional form is chosen for the geometric
structure factor. It is the purpose in the present section to
derive formulas for these consine seminvariants which are valid
under conditions to be specified.

   4.1. The first preliminary formula. Multiplication of

$$E_{\vec{h}} = \frac{1}{N^{1/2}} \sum_{\mu=1}^{N} \exp(2\pi i \vec{h} \cdot \vec{r}_{\mu}) \tag{4.4}$$

by the complex conjugate

$$\bar{E}_{\vec{h}} = \frac{1}{N^{1/2}} \sum_{\mu=1}^{N} \exp(-2\pi i \vec{h} \cdot \vec{r}_{\mu}) \tag{4.5}$$

gives

$$|E_{\vec{h}}|^2 = \frac{1}{N} \sum_{\substack{\mu,\nu \\ 1}}^{N} \exp\left\{ 2\pi i \vec{h} \cdot (\vec{r}_{\mu} - \vec{r}_{\nu}) \right\}. \tag{4.6}$$

The double sum (4.6) may be decomposed into a simple sum, when $\mu = \nu$, and the residual double sum, when $\mu \neq \nu$:

$$|E_{\vec{h}}|^2 = 1 + \frac{1}{N} \sum_{\substack{\mu \neq \nu \\ 1}}^{N} \exp\left\{ 2\pi i \vec{h} \cdot (\vec{r}_\mu - \vec{r}_\nu) \right\}. \qquad (4.7)$$

Introducing the notation

$$\vec{r}_{\mu\nu} = \vec{r}_\mu - \vec{r}_\nu, \qquad (4.8)$$

so that $\vec{r}_{\mu\nu}$ is the interatomic vector joining the atom labeled ν to the atom labeled μ, (4.7) may be written

$$|E_{\vec{h}}|^2 - 1 = \frac{1}{N} \sum_{\substack{\mu \neq \nu \\ 1}}^{N} \cos 2\pi \vec{h} \cdot \vec{r}_{\mu\nu}, \qquad (4.9)$$

the first preliminary formula.

   4.2 The second preliminary formula. Multiplication of

$$E_1 = \frac{1}{N^{1/2}} \sum_{\mu=1}^{N} \exp(2\pi i \vec{h}_1 \cdot \vec{r}_\mu), \qquad (4.10)$$

$$E_2 = \frac{1}{N^{1/2}} \sum_{\mu=1}^{N} \exp(2\pi i \vec{h}_2 \cdot \vec{r}_\mu), \qquad (4.11)$$

and

$$E_3 = \frac{1}{N^{1/2}} \sum_{\mu=1}^{N} \exp(2\pi i \vec{h}_3 \cdot \vec{r}_\mu) \qquad (4.12)$$

yields

$$E_1 E_2 E_3 = \frac{1}{N^{3/2}} \sum_{\substack{\mu,\nu,\rho \\ 1}}^{N} \exp\left\{ 2\pi i (\vec{h}_1 \cdot \vec{r}_\mu + \vec{h}_2 \cdot \vec{r}_\nu + \vec{h}_3 \cdot \vec{r}_\rho) \right\} . \qquad (4.13)$$

The triple sum (4.13) may be decomposed into a simple sum, when $\mu=\nu=\rho$, three double sums, when $\mu=\nu\neq\rho$ or $\nu=\rho\neq\mu$ or $\rho=\mu\neq\nu$, and a residual triple sum, when no two of μ, ν, ρ are equal:

$$E_1 E_2 E_3 = \frac{1}{N^{3/2}} \sum_{\mu=1}^{N} \exp\left\{ 2\pi i (\vec{h}_1 + \vec{h}_2 + \vec{h}_3) \cdot \vec{r}_\mu \right\} +$$

$$\frac{1}{N^{3/2}} \sum_{\substack{\mu\neq\nu \\ 1}}^{N} \exp\left\{ 2\pi i [(\vec{h}_1 + \vec{h}_2) \cdot \vec{r}_\mu + \vec{h}_3 \cdot \vec{r}_\nu] \right\} +$$

$$\frac{1}{N^{3/2}} \sum_{\substack{\mu\neq\nu \\ 1}}^{N} \exp\left\{ 2\pi i [(\vec{h}_2 + \vec{h}_3) \cdot \vec{r}_\mu + \vec{h}_1 \cdot \vec{r}_\nu] \right\} +$$

$$\frac{1}{N^{3/2}} \sum_{\substack{\mu\neq\nu \\ 1}}^{N} \exp\left\{ 2\pi i [(\vec{h}_3 + \vec{h}_1) \cdot \vec{r}_\mu + \vec{h}_2 \cdot \vec{r}_\nu] \right\} +$$

$$\frac{1}{N^{3/2}} \sum_{\substack{\mu \neq \nu \neq \rho \\ 1}}^{N} \exp \left\{ 2\pi i (\vec{h}_1 \cdot \vec{r}_\mu + \vec{h}_2 \cdot \vec{r}_\nu + \vec{h}_3 \cdot \vec{r}_\rho) \right\} . \qquad (4.14)$$

In view of (4.1), (4.14) may be written

$$E_1 E_2 E_3 = \frac{1}{N^{1/2}} + \frac{1}{N^{3/2}} \sum_{\substack{\mu \neq \nu \\ 1}}^{N} \exp \left\{ 2\pi i \vec{h}_1 \cdot (\vec{r}_\mu - \vec{r}_\nu) \right\} +$$

$$\frac{1}{N^{3/2}} \sum_{\substack{\mu \neq \nu \\ 1}}^{N} \exp \left\{ 2\pi i \vec{h}_2 \cdot (\vec{r}_\mu - \vec{r}_\nu) \right\} +$$

$$\frac{1}{N^{3/2}} \sum_{\substack{\mu \neq \nu \\ 1}}^{N} \exp \left\{ 2\pi i \vec{h}_3 \cdot (\vec{r}_\mu - \vec{r}_\nu) \right\} +$$

$$\frac{1}{N^{3/2}} \sum_{\substack{\mu \neq \nu \neq \rho \\ 1}}^{N} \exp \left\{ 2\pi i (\vec{h}_1 \cdot \vec{r}_\mu - \vec{h}_1 \cdot \vec{r}_\nu - \vec{h}_3 \cdot \vec{r}_\nu + \vec{h}_3 \cdot \vec{r}_\rho) \right\} . \qquad (4.15)$$

Using (4.8),

$$E_1 E_2 E_3 = \frac{1}{N^{1/2}} + \frac{1}{N^{3/2}} \sum_{\substack{\mu \neq \nu \\ 1}}^{N} (\cos 2\pi \vec{h}_1 \cdot \vec{r}_{\mu\nu} + \cos 2\pi \vec{h}_2 \cdot \vec{r}_{\mu\nu} + \cos 2\pi \vec{h}_3 \cdot \vec{r}_{\mu\nu}) +$$

$$\frac{1}{N^{3/2}} \sum_{\substack{\mu \neq \nu \neq \rho \\ 1}}^{N} \exp \left\{ 2\pi i (\vec{h}_1 \cdot \vec{r}_{\mu\nu} - \vec{h}_3 \cdot \vec{r}_{\nu\rho}) \right\}.$$ (4.16)

Finally, equating the real parts of (4.16), one obtains, in view of (4.9), the second preliminary formula,

$$|E_1 E_2 E_3| \cos(\phi_1 + \phi_2 + \phi_3) = \frac{1}{N^{3/2}} \sum_{\substack{\mu \neq \nu \neq \rho \\ 1}}^{N} \cos 2\pi (\vec{h}_1 \cdot \vec{r}_{\mu\nu} - \vec{h}_3 \cdot \vec{r}_{\nu\rho}) +$$

$$\frac{1}{N^{1/2}} (|E_1|^2 + |E_2|^2 + |E_3|^2 - 2).$$ (4.17)

4.3. <u>The third preliminary formula.</u> In view of (4.7),

$$|E_{\vec{k}}|^2 - 1 = \frac{1}{N} \sum_{\substack{\mu \neq \nu \\ 1}}^{N} \exp(2\pi i \vec{k} \cdot \vec{r}_{\mu\nu})$$ (4.18)

$$|E_{\vec{h}_1 + \vec{k}}|^2 - 1 = \frac{1}{N} \sum_{\substack{\mu \neq \nu \\ 1}}^{N} \exp \left\{ 2\pi i (\vec{h}_1 + \vec{k}) \cdot \vec{r}_{\mu\nu} \right\}$$ (4.19)

$$|E_{-\vec{h}_3 + \vec{k}}|^2 - 1 = \frac{1}{N} \sum_{\substack{\mu \neq \nu \\ 1}}^{N} \exp \left\{ 2\pi i (-\vec{h}_3 + \vec{k}) \cdot \vec{r}_{\mu\nu} \right\}.$$ (4.20)

Multiply (4.18), (4.19), and (4.20) to get

$$(|E_{\vec{k}}|^2 - 1)(|E_{\vec{h}_1+\vec{k}}|^2 - 1)(|E_{-\vec{h}_3+\vec{k}}|^2 - 1) =$$

$$\frac{1}{N^3} \sum_{\substack{\mu \neq \nu \\ \rho \neq \sigma \\ \xi \neq \eta \\ 1}}^{N} \exp \left\{ 2\pi i (\vec{h}_1 \cdot \vec{r}_{\mu\nu} - \vec{h}_3 \cdot \vec{r}_{\rho\sigma}) + 2\pi i \vec{k} \cdot (\vec{r}_{\mu\nu} + \vec{r}_{\rho\sigma} + \vec{r}_{\xi\eta}) \right\}. \qquad (4.21)$$

Next,

$$\left\langle \exp \left\{ 2\pi i \vec{k} \cdot (\vec{r}_{\mu\nu} + \vec{r}_{\rho\sigma} + \vec{r}_{\xi\eta}) \right\} \right\rangle_{\vec{k}} = 0 \text{ or } 1 \qquad (4.22)$$

according as

$$\vec{r}_{\mu\nu} + \vec{r}_{\rho\sigma} + \vec{r}_{\xi\eta} \neq 0 \text{ or } = 0. \qquad (4.23)$$

Hence, if both sides of (4.21) are averaged over all vectors $\vec{k}$, the only contributors to the sum on the right side of (4.21) which do not vanish are those for which

$$\vec{r}_{\mu\nu} + \vec{r}_{\rho\sigma} + \vec{r}_{\xi\eta} = 0. \qquad (4.24)$$

A triple of interatomic vectors $(\vec{r}_{\mu\nu}, \vec{r}_{\rho\sigma}, \vec{r}_{\xi\eta})$ which satisfies (4.24) is said to constitute an interaction. In view of (4.8) it is clear that every triple $(\vec{r}_{\mu\nu}, \vec{r}_{\nu\rho}, \vec{r}_{\rho\mu})$ which forms an interatomic triangle is an interaction since

$$\vec{r}_{\mu\nu} + \vec{r}_{\nu\rho} + \vec{r}_{\rho\mu} = 0. \qquad (4.25)$$

Such an interaction will be said to be a valid interaction and it will be assumed that only valid interactions are present.

Under this assumption it follows that, if both sides of (4.21) are averaged over all vectors $\vec{k}$, the non-zero contributors to the right hand side require either $\rho=\nu$, $\xi=\sigma$, $\eta=\mu$, or $\xi=\nu$, $\rho=\eta$, $\sigma=\mu$. Hence, (4.21) leads to

$$\left\langle (|E_{\vec{k}}|^2 - 1)(|E_{\vec{h}_1+\vec{k}}|^2 - 1)(|E_{-\vec{h}_3+\vec{k}}|^2 - 1) \right\rangle_{\vec{k}} =$$

$$\frac{1}{N^3} \sum_{\substack{\mu\neq\nu\neq\sigma \\ 1}}^{N} \exp\left\{ 2\pi i(\vec{h}_1 \cdot \vec{r}_{\mu\nu} - \vec{h}_3 \cdot \vec{r}_{\nu\sigma}) \right\} +$$

$$\frac{1}{N^3} \sum_{\substack{\mu\neq\nu\neq\eta \\ 1}}^{N} \exp\left\{ 2\pi i(\vec{h}_1 \cdot \vec{r}_{\mu\nu} - \vec{h}_3 \cdot \vec{r}_{\eta\mu}) \right\}. \qquad (4.26)$$

or

$$\left\langle (|E_{\vec{k}}|^2 - 1)(|E_{\vec{h}_1+\vec{k}}|^2 - 1)(|E_{-\vec{h}_3+\vec{k}}|^2 - 1) \right\rangle_{\vec{k}} =$$

$$\frac{2}{N^3} \sum_{\substack{\mu\neq\nu\neq\rho \\ 1}}^{N} \cos 2\pi(\vec{h}_1 \cdot \vec{r}_{\mu\nu} - \vec{h}_3 \cdot \vec{r}_{\nu\rho}), \qquad (4.27)$$

the third preliminary formula.

4.4. **The main theorem.** Equations (4.17) and (4.27) imply

Theorem 4.1. If the only interactions are valid ones, then

$$|E_1 E_2 E_3| \cos(\phi_1 + \phi_2 + \phi_3) = \frac{N^{3/2}}{2} \left\langle (|E_{\vec{k}}|^2 - 1)(|E_{\vec{h}_1 + \vec{k}}|^2 - 1) \times \right.$$

$$\left. (|E_{-\vec{h}_3 + \vec{k}}|^2 - 1) \right\rangle_{\vec{k}} + \frac{1}{N^{1/2}}(|E_1|^2 + |E_2|^2 + |E_3|^2 - 2). \quad (4.28)$$

Although (4.28) is an explicit expression for the cosine
invariants $\cos(\phi_1 + \phi_2 + \phi_3)$ in terms of observed magnitudes only,
it has not, in practice, proved to be a particularly good method
for evaluating the cosine invariants accurately. There are two
reasons for this. The first is that the condition for (4.28) to
hold, the existence of valid interactions only, is rarely, if
ever, fulfilled. The second reason is that the average on the
right side of (4.28) can, in practice, only be estimated since
only a finite number of magnitudes $|E|$ are obtainable from
experiment. Hence errors arising from the finite sampling are
always present and these are exaggerated by multiplication
by the large factor $N^{3/2}/2$. These obstacles have to some extent
been overcome and refinements of (4.28) have been useful in a
number of crystal structure determinations even for noncentrosymmetric structures having no heavy atom.

In the first place (4.28) may be replaced by

$$|E_1 E_2 E_3| \cos(\phi_1 + \phi_2 + \phi_3) \approx$$

$$\frac{\sigma_1^3 - 3\sigma_1 \sigma_2 + 2\sigma_3}{\sigma_2^{3/2} \left\langle (|E_{\vec{k}}|^2 - 1)^3 \right\rangle_{\vec{k}}} \left\langle (|E_{\vec{k}}|^2 - 1)(|E_{\vec{h}_1 + \vec{k}}|^2 - 1) \times \right.$$

$$\left. (|E_{-\vec{h}_3 + \vec{k}}|^2 - 1) \right\rangle_{\vec{k}} + \frac{\sigma_3}{\sigma_2^{3/2}} (|E_1|^2 + |E_2|^2 + |E_3|^2 - 2),$$

$$(4.29)$$

in which

$$\sigma_n = \sum_{j=1}^{N} z_j^n,$$ (4.30)

and z_j is the atomic number of the atom labeled j (so that dis-similar atoms are permitted). If interactions other than the valid ones (so-called induced and chance interactions) are present, and if not all the atoms are identical, (4.29) is still valid, at least approximately. In fact (4.29) holds even in the centrosymmetric space groups when there may be a large number of induced and chance interactions. However, the combination of limited data and the occurrence of induced (or chance) inter-actions which are approximately valid ones often conspires to render even the improved (4.29) not good enough, in general, to be useful. Instead, a final modification has been devised:

$$|E_1 E_2 E_3| \cos(\phi_1 + \phi_2 + \phi_3) \approx$$

$$K \left\langle (|E_{\vec{k}}|^{1/2} - \overline{|E|^{1/2}})(|E_{\vec{h}_1 + \vec{k}}|^{1/2} - \overline{|E|^{1/2}}) \times \right.$$

$$\left. (|E_{-\vec{h}_3 + \vec{k}}|^{1/2} - \overline{|E|^{1/2}}) \right\rangle_{\vec{k}} + R_3$$ (4.31)

where
$$R_3 = \frac{\sigma_3}{4\sigma_2^{3/2}} \left\{ \frac{3}{2}(|E_1 E_2|^2 + |E_2 E_3|^2 + |E_3 E_1|^2) + \right.$$

$$\left. |E_1|^2 + |E_2|^2 + |E_3|^2 - \frac{7}{2} \right\}$$ (4.32)

and

$$\overline{|E|^{1/2}} = \left\langle |E_{\vec{k}}|^{1/2} \right\rangle_{\vec{k}}. \qquad (4.33)$$

In (4.31) the average may be taken over the restricted set of
reciprocal vectors $\vec{k}$ for which $|E_{\vec{k}}| > t$, where t is a preassigned
number usually in the neighborhood of 1.5.   This abbreviated
calculation results in a considerable savings in computer time
with no loss in accuracy.   The coefficient K in (4.31) is a
sliding scale factor which is adjusted in such a way that the
resulting distribution of values for the cosine invariants
$\cos(\phi_1 + \phi_2 + \phi_3)$ coincides with the theoretical distribution
(Chapter III).   For further details on the implementation of
(4.31) the reader is referred to Part B and, for a fuller account
of the theoretical basis, to the existing literature.

Equation (4.31) has been found to be particularly useful
in the evaluation of those cosine seminvariants which, on account
of space group symmetries, must have one of the values ±1.   In
many instances, particularly in the important space group
$P2_12_12_1$, the values of these two-dimensional cosine seminvariants
suffice, by means of the several tangent techniques to be
discussed later, to give the values of a sufficient number of
three-dimensional phases to yield the crystal structure.

5.   The Double-Angle Formula in $P2_12_12_1$

Reference to Chapter I, in particular §3, Exercises I.1.,
shows that $2\phi_{\vec{h}}$ is a structure invariant (seminvariant) in the
space group $P2_12_12_1$.   Hence, in this space group the value of
$\cos 2\phi_{\vec{h}}$ is uniquely determined by the magnitudes $|E|$ and a
formula for this cosine seminvariant in terms of known magnitudes

$|E|$ is to be anticipated. In view of the trigonometric identity

$$\cos 2\phi = \frac{1 - \tan^2\phi}{1 + \tan^2\phi} \qquad (5.1)$$

the square of the tangent of any phase should also be expressible in terms of known magnitudes. In fact, such a formula, the so-called squared-tangent formula, has only recently been secured. Since the details of the analysis are too lengthy to be reproduced here, the reader is referred to the literature for this work and only the final formula is quoted here:

$$\tan^2\phi_{\vec{h}} \approx$$

$$\frac{\left\langle (|E_{\vec{k}}|^2-1)(|E_{\vec{h}-\vec{k}}|^2-1)\{(|E_{\vec{h}}|^2-1)\sin^2(\phi_{\vec{k}}+\phi_{\vec{h}-\vec{k}})+\frac{1}{2}\}\right\rangle_{\vec{k}}}{\left\langle (|E_{\vec{k}}|^2-1)(|E_{\vec{h}-\vec{k}}|^2-1)\{(|E_{\vec{h}}|^2-1)\cos^2(\phi_{\vec{k}}+\phi_{\vec{h}-\vec{k}})+\frac{1}{2}\}\right\rangle_{\vec{k}}} \qquad (5.2)$$

in which $\vec{h}$ is a fixed three-dimensional reciprocal vector and the averages are taken over two-dimensional vectors $\vec{k}$ for which the vectors $\vec{h}-\vec{k}$ are also two-dimensional. In view of the space group symmetries the required values (0 or 1) for $\sin^2(\phi_{\vec{k}}+\phi_{\vec{h}-\vec{k}})$ and $\cos^2(\phi_{\vec{k}}+\phi_{\vec{h}-\vec{k}})$ are known. In practice the vectors $\vec{k}$ are restricted so that $|E_{\vec{k}}|>1$ and $|E_{\vec{h}-\vec{k}}|>1$ and the formula yields accurate values for $\tan^2\phi_{\vec{h}}$ only if $|E_{\vec{h}}|>1$. In view of (5.1), (5.2) may be written in the more suggestive form

$$\cos 2\phi_{\vec{h}} = \frac{(|E_{\vec{h}}|^2-1)\left\langle (|E_{\vec{k}}|^2-1)(|E_{\vec{h}-\vec{k}}|^2-1)\cos 2\phi_{\vec{k}}\cos 2\phi_{\vec{h}-\vec{k}}\right\rangle_{\vec{k}}}{|E_{\vec{h}}|^2\left\langle (|E_{\vec{k}}|^2-1)(|E_{\vec{h}-\vec{k}}|^2-1)\right\rangle_{\vec{k}}} \qquad (5.3)$$

and for this reason the result described here will be referred to

as the double-angle formula (although it is equivalent to the
squared-tangent formula (5.2)).   It should be emphasized that
in (5.3) the averages are taken over those two-dimensional
vectors $\vec{k}$ for which $\vec{h}-\vec{k}$ is also two-dimensional so that the
required values for cos $2\phi_{\vec{k}}$ and cos $2\phi_{\vec{h}-\vec{k}}$ (± 1) are known.   Also,
$\vec{h}$ is a fixed three-dimensional vector such that $|E_{\vec{h}}| > 1$ and $\vec{k}$
is so restricted that $|E_{\vec{k}}|$ and $|E_{\vec{h}-\vec{k}}|$ are large, e.g. greater than
unity.

CHAPTER III

THE JOINT PROBABILITY DISTRIBUTION OF TWO
STRUCTURE FACTORS AND RELATED CONDITIONAL
DISTRIBUTIONS AND EXPECTATION VALUES

1. Introduction

Probability methods have been of great importance in the
development of the direct methods of crystal structure analysis.
Historically, the first probability distributions to be found
were derived on the basis that the atomic coordinates were the
primitive random variables and it was assumed that these were
uniformly and independently distributed, i.e. that all positions
of the atoms were equally likely. The reciprocal vectors were
assumed to be fixed. Thus, in the space group $P\bar{1}$, for example,
the joint probability distribution of the pair of normalized
structure factors $E_{\vec{h}}$, $E_{2\vec{h}}$ which are random variables since they
are functions of atomic coordinates led to the answer to the
question: For what fraction of all possible crystal structures
is it the case that, for fixed $\vec{h}$, the sign of $E_{2\vec{h}}$ is the same as
the sign of $|E_{\vec{h}}|^2 - 1$? It is not surprising that the answer to
this question depends on both $|E_{\vec{h}}|$ and $|E_{2\vec{h}}|$.

With the further development of probabilistic methods, it
became possible to fix the crystal structure and to treat the
reciprocal vector as the primitive random variable in deriving
the various probability distributions. From this point of view
the joint probability distribution of the pair of random variables

$E_{\vec{h}}$, $E_{2\vec{h}}$ in the space group $P\bar{1}$ now permitted one to answer the
question: For a fixed crystal structure, for what fraction of
reciprocal vectors $\vec{h}$ is the sign of $E_{2\vec{h}}$ the same as the sign of
$|E_{\vec{h}}|^2$-1? Since the primitive random variable was now the
reciprocal vector $\vec{h}$, there could be no question of a dependence
on $|E_{\vec{h}}|$ and $|E_{2\vec{h}}|$ which are themselves random variables. However,
it proved possible to recover the dependence on $|E_{\vec{h}}|$ and $|E_{2\vec{h}}|$
by asking instead for the conditional probability that the sign
of $E_{2\vec{h}}$ be the sign of $|E_{\vec{h}}|^2$-1, given the magnitudes $|E_{\vec{h}}|$, $|E_{2\vec{h}}|$;
in other words, given a fixed but unknown crystal structure, for
what fraction of those reciprocal vectors $\vec{h}$ such that $|E_{\vec{h}}|$ and
$|E_{2\vec{h}}|$ have fixed, specified values, is it the case that $E_{2\vec{h}}$ and
$|E_{\vec{h}}|^2$-1 have the same sign? It is surely to be expected that the
answer to this question will depend on the given values of $|E_{\vec{h}}|$
and $|E_{2\vec{h}}|$.

This discussion shows clearly that there are two conceptually
distinct kinds of probability distribution. In the first kind
one or more reciprocal vectors $\vec{h}, \vec{k}, \ldots$ are given and the primitive
random variables are the atomic coordinates which are assumed to
be uniformly and independently distributed. One then asks for
the joint probability distribution of the dependent random
variables $E_{\vec{h}}, E_{\vec{k}}, \ldots$. In the second kind of distribution the
crystal structure is assumed to be fixed and one or more
reciprocal vectors $\vec{h}, \vec{h}', \ldots$ are also specified. The reciprocal
vector $\vec{k}$ is taken to be the primitive random variable and its
distribution is assumed to be uniform. One may then ask for the
joint probability distribution of the several structure factors
$E_{\vec{k}}$, $E_{\vec{h}+\vec{k}}$, $E_{\vec{h}'+\vec{k}}, \ldots$ which, as functions of the random variable $\vec{k}$,
are themselves random variables.

In application to crystal structure determination, one is
presented with an unknown but fixed crystal structure, and a
number of normalized structure factor magnitudes $|E|$ are assumed
to be known. Hence it is the second kind of probability

distribution, in which atomic coordinates are fixed and the reciprocal vector $\vec{k}$ is assumed to be the primitive random variable, which is more appropriate for crystal structure analysis. For this reason only this kind of probability distribution will be studied in the sequel. It will be seen that all formulas obtainable by the algebraic method may also be derived by probabilistic methods although the latter machinery tends to be more cumbersome and requires greater mathematical sophistication. However, the probability technique is capable of yielding more general results, provides additional information in the form of precise probability distributions and their parameters, and, by the introduction of conditional probability distributions, yields formulas having no algebraic counterpart. In this chapter the case of two structure factors will be exhaustively investigated. A great variety of probability distributions and related conditional distributions and conditional expectation values will be systematically derived. Many of the latter results have been obtained only recently and even some of the older formulas are here modified so as to secure the greater accuracy which is needed for the present analysis.

It will be assumed, both in this chapter and the next, that the crystal structure is fixed, consists of N identical atoms in the unit cell, and that the space group is P1. The phase $\phi_{\vec{h}}$ of the normalized structure factor $E_{\vec{h}}$ is defined by

$$E_{\vec{h}} = |E_{\vec{h}}|\exp(i\phi_{\vec{h}}) = \frac{1}{N^{1/2}} \sum_{\mu=1}^{N} \exp(2\pi i\vec{h}\cdot\vec{r}_{\mu}) \qquad (1.1)$$

where $\vec{r}_{\mu}$ is the position vector of the atom labeled μ. The symbols $\vec{h}$ and $\vec{h}'$ will denote vectors fixed in reciprocal space while the vector $\vec{k}$, the primitive random variable, will be assumed to range uniformly over the reciprocal vectors subject to conditions to be

precisely specified. It is emphasized again that the class of
probability distributions derived on the basis that the atomic
coordinates are the primitive random variables, uniformly and
independently distributed, are conceptually quite distinct from
the ones to be derived here, although superficial resemblances
have been noted. Only by postulating that atomic coordinates
are fixed and that the primitive random variable is the reciprocal
vector $\vec{k}$, as is done here, is it justified to identify the several
mathematical expectation values to be derived with certain averages
over well defined subsets of reciprocal space.

2. Preliminary Formulas

For convenient reference a number of formulas finding
frequent application are listed here. First, from elementary
trigonometry,

$$\sum_{i=1}^{n} A_i \cos(\phi + \alpha_i) = X \cos(\phi + \xi), \tag{2.1}$$

where

$$X = \left(\sum_{\substack{i,j \\ 1}}^{n} A_i A_j \cos(\alpha_i - \alpha_j) \right)^{1/2}, \tag{2.2}$$

$$X \cos \xi = \sum_{i=1}^{n} A_i \cos \alpha_i, \tag{2.3}$$

$$X \sin \xi = \sum_{i=1}^{n} A_i \sin \alpha_i, \qquad (2.4)$$

so that X and ξ are independent of ϕ.

Next, referring to Watson, Theory of Bessel Functions, (pages 20, 21, 79) the following integral formulas are obtained

$$\frac{1}{\pi} \int_0^{\pi} e^{z \cos \phi} \cos m\phi \, d\phi = I_m(z), \qquad (2.5)$$

$$\frac{i^m}{\pi} \int_0^{\pi} e^{-iz \cos \phi} \cos m\phi \, d\phi = J_m(z), \qquad (2.6)$$

$$\int_0^{\pi} e^{-iz \cos \phi} \sin m\phi \, d\phi = 0, \qquad (2.7)$$

$$\frac{1}{\pi} \int_0^{\pi} \sin(z \sin \phi) \sin m\phi \, d\phi = J_m(z), \quad m \text{ odd}, \qquad (2.8)$$

$$\frac{1}{\pi} \int_0^{\pi} \cos(z \sin \phi)\cos m\phi \, d\phi = J_m(z), \quad m \text{ even}, \qquad (2.9)$$

where $J_m(z)$ is the Bessel function of the first kind and $I_m(z)$ is the Bessel function of imaginary argument.

Next [Watson, page 394, equation (4)], the integral formula,

$$\int_0^{\infty} \exp(-pt^2)J_m(at)t^{m+1}dt = \frac{a^m}{(2p)^{m+1}} \exp\left(-\frac{a^2}{4p}\right), \qquad (2.10)$$

finds important application in the sequel. Successive differentiation of both sides of (2.10) with respect to p leads easily to expressions for

$$\int_0^{\infty} e^{-pt^2} J_m(at)t^{m+2n+1}dt, \ n = 1,2,\ldots, \qquad (2.11)$$

which are also used but are not listed here.

Again (Watson, pages 77 and 394 (eq. 4))

$$\int_0^{\infty} I_n(at)e^{-pt^2}t^{n+1}dt = \frac{a^n}{(2p)^{n+1}} e^{\frac{a^2}{4p}}. \qquad (2.12)$$

In addition to (2.12) there are formulas obtained by differentiating successively both sides of (2.12) with respect to p which yield expressions for

$$\int_0^\infty I_n(at)e^{-pt^2}t^{2m+n+1}dt, \quad m = 1,2,\ldots .$$

(2.13)

Next (Watson, pages 77 and 395 (eq. 1))

$$\int_0^\infty I_m(at)I_m(bt)e^{-pt^2}tdt = \frac{1}{2p}I_m\left(\frac{ab}{2p}\right)e^{\frac{a^2+b^2}{4p}}$$

(2.14)

and successive differentiations of both sides of (2.14) yield formulas for

$$\int_0^\infty I_m(at)I_m(bt)e^{-pt^2}t^{2n+1}dt, \quad n = 1,2,\ldots .$$

(2.15)

Again (Watson, pages 358, 361, equation (7) with $\nu = 0$) one readily obtains the addition formula for Bessel functions in the form

$$I_0\sqrt{Z^2 + z^2 + 2Zz\cos\phi} = \sum_{m=-\infty}^{\infty} I_m(Z)I_m(z)e^{im\phi}.$$

(2.16)

Finally, it is known that (Watson, page 16)

$$J_0(z) = 1 - \frac{z^2}{4} + \frac{z^4}{64} - \frac{z^6}{2304} + \frac{z^8}{147456} - \cdots . \qquad (2.17)$$

Hence

$$\log J_0(z) = - \frac{z^2}{4} - \frac{z^4}{64} - \frac{z^6}{576} - \frac{11z^8}{49152} - \cdots$$

and

$$\prod_{j=1}^{N} J_0(z_j) = \exp \left[\sum_{j=1}^{N} \log J_0(z_j) \right]$$

$$= \exp \left(- \frac{1}{4} \sum_{j=1}^{N} z_j^2 \right) \left[1 - \frac{1}{64} \sum_{j=1}^{N} z_j^4 - \frac{1}{576} \sum_{j=1}^{N} z_j^6 - \right.$$

$$\left. \frac{11}{49152} \sum_{j=1}^{N} z_j^8 + \frac{1}{8192} \left(\sum_{j=1}^{N} z_j^4 \right)^2 - \cdots \right] . \qquad (2.18)$$

Exercises III. 1.

1. Derive equations (2.1) – (2.4)

2. Derive the formulas for (2.11) for n = 1,2,3.

3. Derive the formulas (2.13) for m = 1,2,3.

4. Derive the formulas (2.15) for n = 1,2,3.

3. For Fixed $\vec{h}$ and $\vec{h}'$, the Probability Distribution of the Pair $E_{\vec{h}+\vec{k}}$, $E_{\vec{h}'+\vec{k}}$

Suppose that the reciprocal vectors $\vec{h}$ and $\vec{h}'$ are fixed and that the index $\vec{k}$, the primitive random variable, is uniformly distributed in reciprocal space. Denote by $P(R_0, R_1; \bar{\Phi}_0, \bar{\Phi}_1)$ the joint probability distribution of the quadruple $|E_{\vec{h}+\vec{k}}|$, $|E_{\vec{h}'+\vec{k}}|$, $\phi_{\vec{h}+\vec{k}}$, $\phi_{\vec{h}'+\vec{k}}$. Then the probability that $|E_{\vec{h}+\vec{k}}|$ lie between R_0 and $R_0 + dR_0$, that $|E_{\vec{h}'+\vec{k}}|$ lie between R_1 and $R_1 + dR_1$, that $\phi_{\vec{h}+\vec{k}}$ lie between $\bar{\Phi}_0$ and $\bar{\Phi}_0 + d\bar{\Phi}_0$, and that $\phi_{\vec{h}'+\vec{k}}$ lie between $\bar{\Phi}_1$ and $\bar{\Phi}_1 + d\bar{\Phi}_1$ is given by

$$P(R_0, R_1; \bar{\Phi}_0, \bar{\Phi}_1)dR_0 \, dR_1 \, d\bar{\Phi}_0 \, d\bar{\Phi}_1. \tag{3.1}$$

Hence (Karle and Hauptman, 1958), if $R_0 \geq 0$, $R_1 \geq 0$,

$$P(R_0, R_1; \bar{\Phi}_0, \bar{\Phi}_1) =$$

$$\frac{R_0 R_1}{(2\pi)^4} \int_{\rho_0=0}^{\infty} \int_{\rho_1=0}^{\infty} \int_{\theta_0=0}^{2\pi} \int_{\theta_1=0}^{2\pi} \rho_0 \rho_1 \exp\{-i[R_0\rho_0 \cos(\theta_0-\bar{\Phi}_0) +$$

$$R_1\rho_1 \cos(\theta_1-\bar{\Phi}_1)]\} \times$$

$$\prod_{j=1}^{N} J_0\left\{\frac{1}{N^{1/2}} [\rho_0^2 + \rho_1^2 + 2\rho_0\rho_1 \cos(2\pi(\vec{h}-\vec{h}')\cdot r_j + \theta_0-\theta_1)]^{1/2}\right\} \times$$

$$d\rho_0 \, d\rho_1 \, d\theta_0 \, d\theta_1, \tag{3.2}$$

which, in view of (2.18) and (1.1), reduces, after some
simplification, to

$$P(R_0, R_1; \bar{\Phi}_0, \bar{\Phi}_1) \approx$$

$$\frac{R_0 R_1}{(2\pi)^4} \int\limits_{\rho_0=0}^{\infty} \int\limits_{\rho_1=0}^{\infty} \int\limits_{\theta_0=0}^{2\pi} \int\limits_{\theta_1=0}^{2\pi} \rho_0 \rho_1 \exp\left(-\frac{1}{4}(\rho_0^2 + \rho_1^2)\right) \times$$

$$\exp(-i[R_0\rho_0 \cos(\theta_0 - \bar{\Phi}_0) + R_1\rho_1 \cos(\theta_1 - \bar{\Phi}_1)]) \times$$

$$\exp\left(-\frac{1}{2N^{1/2}} \rho_0 \rho_1 |E_{\vec{h}',-\vec{h}}| \cos(\phi_{\vec{h}'-\vec{h}} + \theta_0 - \theta_1)\right) \times$$

$$\left[1 - \frac{1}{64N}(\rho_0^4 + \rho_1^4 + 4\rho_0^2\rho_1^2)\right] d\rho_0 \, d\rho_1 \, d\theta_0 \, d\theta_1, \qquad (3.3)$$

provided that, if μ, ν, ρ, σ are all different, then

$$r_\mu - r_\nu \neq r_\rho - r_\sigma. \qquad (3.4)$$

The terms in the exponent containing θ_0 may be written, in
view of (2.1) to (2.4),

$$- i\rho_0 \left[R_0 \cos(\theta_0 - \bar{\Phi}_0) - \frac{i\rho_1}{2N^{1/2}} |E_{\vec{h}',-\vec{h}}| \cos(\phi_{\vec{h}'-\vec{h}} + \theta_0 - \theta_1)\right] =$$

$$- i\rho_0 X_0 \cos(\theta_0 + \xi_0), \qquad (3.5)$$

where

$$X_0 = \left[R_0^2 - \frac{iR_0\rho_1}{N^{1/2}} |E_{\vec{h}',-\vec{h}}| \cos(\theta_1 - \Phi_0 - \phi_{\vec{h}',-\vec{h}}) - \frac{\rho_1^2}{4N} |E_{\vec{h}',-\vec{h}}|^2 \right]^{1/2},$$

$$(3.6)$$

$$\cos \xi_0 = \frac{1}{X_0} \left[R_0 \cos \Phi_0 - \frac{i\rho_1}{2N^{1/2}} |E_{\vec{h}',-\vec{h}}| \cos(\theta_1 - \phi_{\vec{h}',-\vec{h}}) \right],$$

$$(3.7)$$

$$\sin \xi_0 = -\frac{1}{X_0} \left[R_0 \sin \Phi_0 - \frac{i\rho_1}{2N^{1/2}} |E_{\vec{h}',-\vec{h}}| \sin(\theta_1 - \phi_{\vec{h}',-\vec{h}}) \right],$$

$$(3.8)$$

so that X_0 and ξ_0 are independent of ρ_0 and θ_0. Hence, employing (2.6), the integration of (3.3) with respect to θ_0 may be carried out:

$$P(R_0, R_1; \Phi_0, \Phi_1) \approx \frac{R_0 R_1}{(2\pi)^3} \int_{\rho_0=0}^{\infty} \int_{\rho_1=0}^{\infty} \int_{\theta_1=0}^{2\pi} \rho_0 \rho_1 \exp\left[-\frac{1}{4}(\rho_0^2 + \rho_1^2)\right] \times$$

$$\exp\left[-iR_1\rho_1 \cos(\theta_1 - \Phi_1)\right] J_0(\rho_0 X_0) \times$$

$$\left[1 - \frac{1}{64N}(\rho_0^4 + \rho_1^4 + 4\rho_0^2\rho_1^2)\right] d\rho_0 \, d\rho_1 \, d\theta_1. \qquad (3.9)$$

Since X_0 is independent of ρ_0, the integration with respect to ρ_0 may be performed, using (2.10) and (2.11):

$$P(R_0, R_1; \bar{\Phi}_0, \bar{\Phi}_1) \approx$$

$$\frac{2R_0 R_1}{(2\pi)^3} \int\limits_{\rho_1=0}^{\infty} \int\limits_{\theta_1=0}^{2\pi} \rho_1 \exp\left[-\frac{1}{4}\rho_1^2 - R_0^2 + \frac{\rho_1^2}{4N} |E_{\vec{h}', -\vec{h}}|^2 + \right.$$

$$\frac{iR_0\rho_1}{N^{1/2}} |E_{\vec{h}', -\vec{h}}| \cos(\theta_1 - \bar{\Phi}_0 - \bar{\Phi}_{\vec{h}', -\vec{h}}) - iR_1\rho_1 \cos(\theta_1 - \bar{\Phi}_1) \right] \times$$

$$\left[1 - \frac{1}{64N}\rho_1^4 - \frac{1}{4N}\rho_1^2(1-R_0^2) - \frac{1}{4N}(2-4R_0^2+R_0^4) \right] d\rho_1 \, d\theta_1.$$

$$(3.10)$$

Collecting the terms of the exponent containing θ_1, one obtains, in view of (2.1)-(2.4).

$$-i\rho_1 \left[R_1 \cos(\theta_1 - \bar{\Phi}_1) - \frac{R_0}{N^{1/2}} |E_{\vec{h}', -\vec{h}}| \cos(\theta_1 - \bar{\Phi}_0 - \phi_{\vec{h}', -\vec{h}}) \right]$$

$$- i\rho_1 X_1 \cos(\theta_1 + \xi_1),$$

$$(3.11)$$

where

$$X_1 = \left[R_1^2 - \frac{2R_0 R_1 |E_{\vec{h}', -\vec{h}}|}{N^{1/2}} \cos(\bar{\Phi}_0 - \bar{\Phi}_1 + \phi_{\vec{h}', -\vec{h}}) + \frac{R_0^2}{N} |E_{\vec{h}', -\vec{h}}|^2 \right]^{1/2},$$

$$(3.12)$$

$$\cos \xi_1 = \frac{1}{X_1} \left[R_1 \cos \bar{\Phi}_1 - \frac{R_0}{N^{1/2}} |E_{\vec{h}', -\vec{h}}| \cos(\bar{\Phi}_0 + \phi_{\vec{h}', -\vec{h}}) \right], \quad (3.13)$$

$$\sin \xi_1 = - \frac{1}{X_1} \left[R_1 \sin \bar{\Phi}_1 - \frac{R_0}{N^{1/2}} |E_{\vec{h}'-\vec{h}}| \sin(\bar{\Phi}_0 + \phi_{\vec{h}'-\vec{h}}) \right], \quad (3.14)$$

so that X_1 and ξ_1 are independent of ρ_1 and θ_1. Hence, using (2.6), the integration with respect to θ_1 may be carried out, leading to

$$P(R_0, R_1; \bar{\Phi}_0, \bar{\Phi}_1) \approx \frac{2R_0 R_1}{(2\pi)^2} \exp(-R_0^2) \; \times$$

$$\int_{\rho_1=0}^{\infty} \rho_1 \exp\left[-\frac{1}{4} \rho_1^2 \left(1 - \frac{|E_{\vec{h}'-\vec{h}}|^2}{N} \right) \right] J_0(\rho_1 X_1) \; \times$$

$$\left[1 - \frac{1}{64N}(2 - 4R_0^2 + R_0^4) - \frac{1}{4N} \rho_1^2 (1 - R_0^2) - \frac{1}{64N} \rho_1^4 \right] d\rho_1. \quad (3.15)$$

Finally, since X_1 is independent of ρ_1, one may employ (2.10) and (2.11) to perform the final integration, with respect to ρ_1, and obtains, after some simplification, the desired distribution

$$P(R_0, R_1; \bar{\Phi}_0, \bar{\Phi}_1) \approx \frac{R_0 R_1}{\pi^2 \left(1 - \frac{|E_{\vec{h}'-\vec{h}}|^2}{N} \right)} \; \times$$

$$\exp\left[- \frac{R_0^2 + R_1^2}{1 - \frac{|E_{\vec{h}'-\vec{h}}|^2}{N}} + \frac{2R_0 R_1 |E_{\vec{h}'-\vec{h}}| \cos(\bar{\Phi}_{\vec{h}'-\vec{h}} + \bar{\Phi}_0 - \bar{\Phi}_1)}{N^{1/2} \left(1 - \frac{|E_{\vec{h}'-\vec{h}}|^2}{N} \right)} \right] \; \times$$

$$\left[1 - \frac{1}{4N}(R_0{}^4 + 4R_0{}^2 R_1{}^2 + R_1{}^4 - 8R_0{}^2 - 8R_1{}^2 + 8) \right] . \tag{3.16}$$

The special case that $\vec{h} = 0$ is of particular importance in the sequel, and in this case (3.16) reduces to

$$P(R_0, R_1; \bar{\Phi}_0, \bar{\Phi}_1) \approx \frac{R_0 R_1}{\pi^2 \left(1 - \frac{|E_{\vec{h}'}|^2}{N} \right)} \times$$

$$\exp \left[- \frac{R_0{}^2 + R_1{}^2}{1 - \frac{|E_{\vec{h}'}|^2}{N}} + \frac{2 R_0 R_1 |E_{\vec{h}'}|^2 \cos(\phi_{\vec{h}'} + \bar{\Phi}_0 - \bar{\Phi}_1)}{N^{1/2} \left(1 - \frac{|E_{\vec{h}'}|^2}{N} \right)} \right] \times$$

$$\left[1 - \frac{1}{4N}(R_0{}^4 + 4R_0{}^2 R_1{}^2 + R_1{}^4 - 8R_0{}^2 - 8R_1{}^2 + 8 \right] . \tag{3.17}$$

4. For Fixed $\vec{h}$ and $\vec{h}'$, the Probability Distribution
        of the Pair $|E_{\vec{h}+\vec{k}}|$, $|E_{\vec{h}'+\vec{k}}|$

As in §3, suppose that the vectors $\vec{h}$ and $\vec{h}'$ are fixed and that the index $\vec{k}$ ranges uniformly over all vectors in reciprocal space. Denote by $P(R_0, R_1)$ the joint probability distribution of the pair of magnitudes $|E_{\vec{h}+\vec{k}}$, $|E_{\vec{h}'+\vec{k}}|$. Thus $P(R_0, R_1) dR_0 dR_1$ is the probability that $|E_{\vec{h}+\vec{k}}|$ lie between R_0 and $R_0 + dR_0$ and that $|E_{\vec{h}'+\vec{k}}|$ lie between R_1 and $R_1 + dR_1$. Then from (3.15) and (2.5) it follows immediately that, if $R_0 \geq 0$, $R_1 \geq 0$,

$$P(R_0,R_1) = \int\limits_{\Phi_0=0}^{2\pi} \int\limits_{\Phi_1=0}^{2\pi} P(R_0,R_1;\bar{\Phi}_0,\bar{\Phi}_1)d\bar{\Phi}_0 \; d\bar{\Phi}_1,$$

$$P(R_0,R_1) \approx \frac{4R_0R_1}{1 - \dfrac{|E_{\vec{h}'-\vec{h}}|^2}{N}} \times$$

$$\exp\left(-\frac{R_0^2+R_1^2}{1 - \dfrac{|E_{\vec{h}'-\vec{h}}|^2}{N}}\right) I_0\left(\frac{2R_0R_1|E_{\vec{h}'-\vec{h}}|}{N^{1/2}\left(1 - \dfrac{|E_{\vec{h}'-\vec{h}}|^2}{N}\right)}\right) \times$$

$$\left[1 - \frac{1}{4N}(R_0^4+4R_0^2R_1^2+R_1^4-8R_0^2-8R_1^2+8)\right]. \tag{4.1}$$

Again, the special case that $\vec{h} = 0$ is immediately obtained from (4.1) and has important application later,

$$P(R_0,R_1) \approx \frac{4R_0R_1}{1 - \dfrac{|E_{\vec{h}'}|^2}{N}} \times$$

$$\exp\left(-\frac{R_0^2+R_1^2}{1 - \dfrac{|E_{\vec{h}'}|^2}{N}}\right) I_0\left(\frac{2R_0R_1|E_{\vec{h}'}|}{N^{1/2}\left(1 - \dfrac{|E_{\vec{h}'}|^2}{N}\right)}\right) \times$$

$$\left[1 - \frac{1}{4N}(R_0^4+4R_0^2R_1^2+R_1^4-8R_0^2-8R_1^2+8)\right]. \tag{4.2}$$

5. The Conditional Distribution of $E_{\vec{h}+\vec{k}}$, Given $E_{\vec{k}}$

Suppose that the reciprocal vector $\vec{h}$ is fixed and that $\vec{k}$
ranges uniformly over those vectors in reciprocal space for which
the (complex) normalized structure factor $E_{\vec{k}}$ has an assigned
value, i.e. for which the magnitude, $|E_{\vec{k}}|$, and phase, $\phi_{\vec{k}}$, have
specified, fixed values. Denote by $P(R_1;\bar{\Phi}_1\,|\,|E_{\vec{k}}|,\phi_{\vec{k}}) =$
$P(R_1;\bar{\Phi}_1\,|\,R_0 = |E_{\vec{k}}|, \bar{\Phi}_0 = \phi_{\vec{k}})$ the conditional joint probability
distribution of the magnitude, $|E_{\vec{h}+\vec{k}}|$, and the phase, $\phi_{\vec{h}+\vec{k}}$,
respectively, of $E_{\vec{h}+\vec{k}}$, given $|E_{\vec{k}}|$ and $\phi_{\vec{k}}$. Thus $P(R_1;\bar{\Phi}_1\,|\,|E_{\vec{k}}|,\phi_{\vec{k}})$
$dR_1\,d\bar{\Phi}_1$ is the conditional probability that $|E_{\vec{h}+\vec{k}}|$ lie between
R_1 and $R_1 + dR_1$ and that $\phi_{\vec{h}+\vec{k}}$ lie between $\bar{\Phi}_1$ and $\bar{\Phi}_1 + d\bar{\Phi}_1$, given
$|E_{\vec{k}}|$ and $\phi_{\vec{k}}$. Then $P(R_1;\bar{\Phi}_1\,|\,|E_{\vec{k}}|,\phi_{\vec{k}})$ is obtained from (3.16) by
replacing R_0 by $|E_{\vec{k}}|$, $\bar{\Phi}_0$ by $\phi_{\vec{k}}$, and multiplying by a suitable
normalizing factor:

$$P(R_1;\bar{\Phi}_1\,\big|\,|E_{\vec{k}}|,\phi_{\vec{k}}) \approx \frac{R_1}{K_1} \exp\left\{ - \frac{R_1^2}{1 - \dfrac{|E_{\vec{h}}|^2}{N}} + \right.$$

$$\left. \frac{2R_1|E_{\vec{h}}E_{\vec{k}}|\cos(\phi_{\vec{h}}+\phi_{\vec{k}}-\bar{\Phi}_1)}{N^{1/2}\left(1 - \dfrac{|E_{\vec{h}}|^2}{N}\right)} \right\}\left\{ 1 - \frac{1}{4N}\left(R_1^4 + \right.\right.$$

$$\left.\left. 4R_1^2(|E_{\vec{k}}|^2 - 2) + |E_{\vec{k}}|^4 - 8|E_{\vec{k}}|^2 + 8\right) \right\} , \qquad (5.1)$$

where

$$K_1 = \int_{R_1=0}^{\infty} \int_{\bar\Phi_1=0}^{2\pi} R_1 \, \exp\left\{-\frac{R_1^2}{1-\dfrac{|E_{\vec{h}}|^2}{N}} + \frac{2R_1|E_{\vec{h}}E_{\vec{k}}|\cos(\phi_{\vec{h}}+\phi_{\vec{k}}-\bar\Phi_1)}{N^{1/2}\left(1-\dfrac{|E_{\vec{h}}|^2}{N}\right)}\right\} \times$$

$$\left\{1 - \frac{1}{4N}\left(R_1^4 + 4R_1^2(|E_{\vec{k}}|^2-2) + |E_{\vec{k}}|^4 - 8|E_{\vec{k}}|^2 + 8\right)\right\} dR_1 \, d\bar\Phi_1.$$

$$(5.2)$$

Referring to equation (2.5), the integration with respect to $\bar\Phi_1$ may be carried out:

$$K_1 = 2\pi \int_{R_1=0}^{\infty} R_1 \, \exp\left(-\frac{R_1^2}{1-\dfrac{|E_{\vec{h}}|^2}{N}}\right) I_0\left(\frac{2|E_{\vec{h}}E_{\vec{k}}|R_1}{N^{1/2}\left(1-\dfrac{|E_{\vec{h}}|^2}{N}\right)}\right) \times$$

$$\left\{1 - \frac{1}{4N}\left(R_1^4 + 4R_1^2(|E_{\vec{k}}|^2-2) + |E_{\vec{k}}|^4 - 8|E_{\vec{k}}|^2 + 8\right)\right\} dR_1, \qquad (5.3)$$

which, in view of (2.12) and (2.13), finally reduces to

$$K_1 \approx \pi\left(1 - \frac{|E_{\vec{h}}|^2}{N}\right)\exp\left\{\frac{|E_{\vec{h}}E_{\vec{k}}|^2}{N\left(1-\dfrac{|E_{\vec{h}}|^2}{N}\right)}\right\}\left\{1 - \frac{1}{4N}(|E_{\vec{k}}|^4 - 4|E_{\vec{k}}|^2 + 2)\right\}.$$

$$(5.4)$$

Combining (5.4) and (5.1) yields the desired distribution

$$P(R_1;\bar{\Phi}_1\big|\,|E_{\vec{k}}|,\phi_{\vec{k}}) \approx \frac{R_1}{\pi\left(1 - \dfrac{|E_{\vec{h}}|^2}{N}\right)} \exp\left\{\frac{1}{1 - \dfrac{|E_{\vec{h}}|^2}{N}}\left[-R_1^2 + \right.\right.$$

$$\frac{2R_1|E_{\vec{h}}E_{\vec{k}}|}{N^{1/2}}\cos(\phi_{\vec{h}} + \phi_{\vec{k}} - \bar{\Phi}_1) - \frac{|E_{\vec{h}}E_{\vec{k}}|^2}{N}\bigg]\bigg\} \times$$

$$\left\{1 - \frac{1}{4N}\Big(R_1^4 + 4R_1^2(|E_{\vec{k}}|^2 - 2) - 4|E_{\vec{k}}|^2 + 6\Big)\right\}. \tag{5.5}$$

6. The Conditional Distribution of $|E_{\vec{h}+\vec{k}}|$, Given $|E_{\vec{k}}|$

Again, let the reciprocal vector $\vec{h}$ be fixed and assume that $\vec{k}$ ranges uniformly over those vectors in reciprocal space for which $|E_{\vec{k}}|$ has a specified, fixed value. Denote by $P(R_1\big|\,|E_{\vec{k}}|) = P(R_1\big|R_0 = |E_{\vec{k}}|)$ the conditional probability distribution of the magnitude, $|E_{\vec{h}+\vec{k}}|$, given $|E_{\vec{k}}|$. Thus $P(R_1\big|\,|E_{\vec{k}}|)dR_1$ is the conditional probability that $|E_{\vec{h}+\vec{k}}|$ lie between R_1 and R_1+dR_1, given $|E_{\vec{k}}|$. Then $P(R_1\big|\,|E_{\vec{k}}|)$ is obtained from (5.2) by replacing R_0 by $|E_{\vec{k}}|$ and multiplying by a suitable normalizing factor.

$$P(R_1\big|\,|E_{\vec{k}}|) \approx \frac{R_1}{K_2} \exp\left\{-\frac{R_1^2}{1 - \dfrac{|E_{\vec{h}}|^2}{N}}\right\} I_0\left\{\frac{2|E_{\vec{h}}E_{\vec{k}}|R_1}{N^{1/2}\left(1 - \dfrac{|E_{\vec{h}}|^2}{N}\right)}\right\} \times$$

$$\left\{1 - \frac{1}{4N}\Big(R_1^4 + 4R_1^2(|E_{\vec{k}}|^2 - 2) + |E_{\vec{k}}|^4 - 8|E_{\vec{k}}|^2 + 8\Big)\right\}, \tag{6.1}$$

where

$$
K_2 = \int\limits_{R_1=0}^{\infty} R_1 \exp\left\{-\frac{R_1^{\,2}}{1-\dfrac{|E_{\vec{h}}|^2}{N}}\right\} I_0\left\{\frac{2|E_{\vec{h}}E_{\vec{k}}|R_1}{N^{1/2}\left(1-\dfrac{|E_{\vec{h}}|^2}{N}\right)}\right\} \times
$$

$$
\left\{1-\frac{1}{4N}\Big(R_1^{\,4}+4R_1^{\,2}(|E_{\vec{k}}|^2-2)+|E_{\vec{k}}|^4-8|E_{\vec{k}}|^2+8\Big)\right\} . \tag{6.2}
$$

Comparison of (6.2) with (5.3) shows that $K_2 = 1/2\pi\, K_1$ so that, in view of (5.4), the desired distribution is

$$
P(R_1\big|\,|E_{\vec{k}}|) \approx \frac{2R_1}{1-\dfrac{|E_{\vec{h}}|^2}{N}} \exp\left\{\frac{1}{1-\dfrac{|E_{\vec{h}}|^2}{N}}\left(-R_1^{\,2}-\frac{|E_{\vec{h}}E_{\vec{k}}|^2}{N}\right)\right\} \times
$$

$$
I_0\left\{\frac{2|E_{\vec{h}}E_{\vec{k}}|R_1}{N^{1/2}\left(1-\dfrac{|E_{\vec{h}}|^2}{N}\right)}\right\}\left\{1-\frac{1}{4N}\Big(R_1^{\,4}+4R_1^{\,2}(|E_{\vec{k}}|^2-2)-4|E_{\vec{k}}|^2+6\Big)\right\}. \tag{6.3}
$$

Comparison of (5.5) with (6.3) reveals that the latter could have been derived from the former by integrating (5.5) with respect to $\bar{\Phi}_1$. The resulting expression for $P(R_1\big|E_{\vec{k}})$ turns out to be independent of $\phi_{\vec{k}}$ and thus must be identical with $P(R_1\big|\,|E_{\vec{k}}|)$ and the calculation does show, in fact, that it coincides with (6.3).

7. The Conditional Distribution of the Pair, $\phi_{\vec{k}}$, $\phi_{\vec{h}+\vec{k}}$,
Given $|E_{\vec{k}}|$ and $|E_{\vec{h}+\vec{k}}|$

As usual, assume that the reciprocal vector $\vec{h}$ is fixed, but suppose now that $\vec{k}$ ranges uniformly over those vectors in reciprocal space for which $|E_{\vec{k}}|$ and $|E_{\vec{h}+\vec{k}}|$ have fixed, specified values. Denote by $P(\bar{\Phi}_0,\bar{\Phi}_1\,\big|\,|E_{\vec{k}}|,|E_{\vec{h}+\vec{k}}|) = P(\bar{\Phi}_0,\bar{\Phi}_1\,\big|\,R_0 = |E_{\vec{k}}|, R_1 = |E_{\vec{h}+\vec{k}}|)$ the conditional joint probability distribution of the pair of phases, $\phi_{\vec{k}}, \phi_{\vec{h}+\vec{k}}$, given $|E_{\vec{k}}|$ and $|E_{\vec{h}+\vec{k}}|$. Thus $P(\bar{\Phi}_0,\bar{\Phi}_1\,\big|\,|E_{\vec{k}}|,|E_{\vec{h}+\vec{k}}|)d\bar{\Phi}_0 d\bar{\Phi}_1$ is the conditional probability that $\phi_{\vec{k}}$ lie between $\bar{\Phi}_0$ and $\bar{\Phi}_0 + d\bar{\Phi}_0$ and that $\phi_{\vec{h}+\vec{k}}$ lie between $\bar{\Phi}_1$ and $\bar{\Phi}_1 + d\bar{\Phi}_1$, given $|E_{\vec{k}}|$ and $|E_{\vec{h}+\vec{k}}|$. Then $P(\bar{\Phi}_0,\bar{\Phi}_1\,\big|\,|E_{\vec{k}}|,|E_{\vec{h}+\vec{k}}|)$ is obtained from (3.16) by replacing R_0 by $|E_{\vec{k}}|$, R_1 by $|E_{\vec{h}+\vec{k}}|$, and multiplying by a suitable constant:

$$P(\bar{\Phi}_0,\bar{\Phi}_1\,\big|\,|E_{\vec{k}}|,|E_{\vec{h}+\vec{k}}|) \approx \frac{1}{K_3}\,\exp\left\{\frac{2\,|E_{\vec{h}}E_{\vec{k}}E_{\vec{h}+\vec{k}}|}{N^{1/2}\left(1-\dfrac{|E_{\vec{h}}|^2}{N}\right)}\cos(\phi_{\vec{h}}+\bar{\Phi}_0-\bar{\Phi}_1)\right\}$$

$$(7.1)$$

where

$$K_3 = \int\limits_{\bar{\Phi}_0=0}^{2\pi}\int\limits_{\bar{\Phi}_1=0}^{2\pi}\exp\left\{\frac{2\,|E_{\vec{h}}E_{\vec{k}}E_{\vec{h}+\vec{k}}|}{N^{1/2}\left(1-\dfrac{|E_{\vec{h}}|^2}{N}\right)}\cos(\phi_{\vec{h}}+\bar{\Phi}_0-\bar{\Phi}_1)\right\}d\bar{\Phi}_0 d\bar{\Phi}_1,$$

$$(7.2)$$

$$K_3 = 4\pi^2\,I_0\left(\frac{2\,|E_{\vec{h}}E_{\vec{k}}E_{\vec{h}+\vec{k}}|}{N^{1/2}\left(1-\dfrac{|E_{\vec{h}}|^2}{N}\right)}\right),$$

$$(7.3)$$

in view of (2.5). Introducing the abbreviation

$$A = \frac{2 \left| E_{\vec{h}} E_{\vec{k}} E_{\vec{h}+\vec{k}} \right|}{N^{1/2} \left(1 - \frac{\left| E_{\vec{h}} \right|^2}{N} \right)} \approx \frac{2 \left| E_{\vec{h}} E_{\vec{k}} E_{\vec{h}+\vec{k}} \right|}{N^{1/2}} \qquad (7.4)$$

(7.1) becomes

$$P(\bar{\Phi}_0, \bar{\Phi}_1 \mid \left| E_{\vec{k}} \right|, \left| E_{\vec{h}+\vec{k}} \right|) \approx \frac{1}{4\pi^2 I_0(A)} \exp \left\{ A \cos(\phi_{\vec{h}} + \bar{\Phi}_0 - \bar{\Phi}_1) \right\}, \quad (7.5)$$

the desired distribution.

8. The Conditional Distribution of $\phi_{\vec{h}} + \phi_{\vec{k}} + \phi_{-\vec{h}-\vec{k}}$, Given A

Next, suppose that the vector $\vec{h}$ is fixed and that $\vec{k}$ ranges uniformly over those vectors in reciprocal space such that A (Eq. (7.4)) has a fixed value. Denote by $P(\Omega \mid A)$ the conditional distribution of $\omega = \phi_{\vec{h}} + \phi_{\vec{k}} + \phi_{-\vec{h}-\vec{k}}$, given A. Thus $P(\Omega \mid A)d\Omega$ is the conditional probability that $\phi_{\vec{h}} + \phi_{\vec{k}} + \phi_{-\vec{h}-\vec{k}}$ lie between Ω and $\Omega + d\Omega$, given A. Then $P(\Omega \mid A)$ is readily found from (7.5) by making the transformation

$$\Omega = \phi_{\vec{h}} + \bar{\Phi}_0 - \bar{\Phi}_1. \qquad (8.1)$$

Clearly, for each fixed value $\bar{\Phi}_1$ of $\phi_{\vec{h}+\vec{k}}$ in the interval $(0,2\pi)$, the probability that ω lie between Ω and $\Omega + d\Omega$ is equal to the probability that $\phi_{\vec{k}}$ lie between $\Omega - \phi_{\vec{h}} + \bar{\Phi}_1$ and $\Omega - \phi_{\vec{h}} + \bar{\Phi}_1 + d\Omega$, or, in view of (7.5), is equal to

$$\frac{1}{4\pi^2 I_0(A)} \exp (A \cos \Omega) \, d\Omega. \qquad (8.2)$$

Integrating (8.2) with respect to $\bar{\Phi}_1$ **from** 0 to 2π, the desired distribution is found to be

$$P(\Omega|A) \approx \frac{1}{2\pi I_0(A)} \exp (A \cos \Omega). \qquad (8.3)$$

9.   The Conditional Distribution of $\mathrm{Cos}(\phi_{\vec{h}} + \phi_{\vec{k}} + \phi_{-\vec{h}-\vec{k}})$,
Given A

If $\vec{h}$ is fixed and $\vec{k}$ ranges uniformly over the vectors in reciprocal space for which A has a specified value, then the conditional distribution, $P_c(x|A)$, of $\cos(\phi_{\vec{h}} + \phi_{\vec{k}} + \phi_{-\vec{h}-\vec{k}})$, given A, is found from (8.3) by applying the transformation

$$x = \cos \Omega, \quad dx = - \sin \Omega d\Omega, \quad d\Omega = \pm \frac{dx}{\sqrt{1-x^2}}. \qquad (9.1)$$

Then $\cos(\phi_{\vec{h}} + \phi_{\vec{k}} + \phi_{-\vec{h}-\vec{k}}) = \cos \omega$ lies in the interval $(x, x + dx)$ if and only if ω lies in the interval $(\Omega, \Omega - dx/\sqrt{1-x^2})$ or in the interval $(2\pi - \Omega, 2\pi - \Omega + dx/\sqrt{1-x^2})$ the probability of which, in view of (8.3) and (9.1) is

$$\frac{1}{\pi I_0(A)} \exp (Ax) \frac{dx}{\sqrt{1-x^2}}. \qquad (9.2)$$

It follows that $P_c(x|A)$, the conditional probability distribution of $\cos(\phi_{\vec{h}} + \phi_{\vec{k}} + \phi_{-\vec{h}-\vec{k}})$, given A, is

$$P_c(x|A) \approx \frac{\exp (Ax)}{\pi I_0(A) \sqrt{1-x^2}}. \qquad (9.3)$$

Thus the conditional probability that $\cos(\phi_{\vec{h}} + \phi_{\vec{k}} + \phi_{-\vec{h}-\vec{k}})$ lie between x and x + dx, given A, is $P_c(x|A)dx$ where $P_c(x|A)$ is given by (9.3). The distribution (9.3) has found important application in recently secured techniques of phase determination via the least squares analysis of the structure invariants. The cumulative distribution function,

$$\int_{x}^{1} \frac{e^{Ax}}{\pi I_0(A)\sqrt{1-x^2}} \, dx, \tag{9.4}$$

is displayed in Table 1 and Table 2 exhibits several of its parameters (§§14 and 15). For a more complete tabulation, see Fisher, Hancock and Hauptman, 1970.

10. The Conditional Expectation of $E_{\vec{h}+\vec{k}}$, Given $E_{\vec{k}}$

Fix the vector $\vec{h}$ and suppose that $\vec{k}$ ranges uniformly over those reciprocal vectors for which $E_{\vec{k}}$ has a specified (complex) value. Denote by $\boldsymbol{\mathcal{E}}(E_{\vec{h}+\vec{k}}|E_{\vec{k}})$ the conditional expected value of $E_{\vec{h}+\vec{k}}$, given $E_{\vec{k}}$. Then, in view of (5.5),

$$\boldsymbol{\mathcal{E}}(E_{\vec{h}+\vec{k}}|E_{\vec{k}}) = \int_{R_1=0}^{\infty} \int_{\Phi_1=0}^{2\pi} R_1 \exp(i\Phi_1) \, P(R_1, \Phi_1 | |E_{\vec{k}}|, \phi_{\vec{k}}) dR_1 d\Phi_1$$

$$\tag{10.1}$$

Table 1

The Conditional Probability That $\cos(\phi_{\vec{h}} + \phi_{\vec{k}} + \phi_{-\vec{h}-\vec{k}}) > x$,

Given $A = \dfrac{2}{N^{1/2}} |E_{\vec{h}}\, E_{\vec{k}}\, E_{\vec{h}+\vec{k}}|$

X	A										
	0.0	0.5	1.0	1.5	2.0	2.5	3.0	3.5	4.0	4.5	5.0
0.998	0.0199	0.0311	0.0431	0.0547	0.0652	0.0744	0.0827	0.0902	0.0970	0.1034	0.1093
0.980	0.0635	0.0984	0.1359	0.1718	0.2040	0.2323	0.2573	0.2797	0.3000	0.3188	0.3362
0.950	0.1009	0.1553	0.2134	0.2683	0.3170	0.3592	0.3961	0.4285	0.4575	0.4839	0.5080
0.910	0.1358	0.2077	0.2835	0.3542	0.4158	0.4683	0.5132	0.5519	0.5859	0.6161	0.6432
0.840	0.1823	0.2754	0.3717	0.4594	0.5336	0.5949	0.6455	0.6877	0.7234	0.7540	0.7805
0.760	0.2250	0.3353	0.4468	0.5456	0.6266	0.6911	0.7424	0.7835	0.8168	0.8442	0.8670
0.680	0.2617	0.3849	0.5065	0.6114	0.6949	0.7590	0.8081	0.8458	0.8752	0.8985	0.9170
0.600	0.2949	0.4279	0.5563	0.6642	0.7474	0.8091	0.8546	0.8883	0.9135	0.9325	0.9472
0.520	0.3257	0.4662	0.5989	0.7075	0.7888	0.8471	0.8886	0.9180	0.9392	0.9546	0.9659
0.440	0.3547	0.5009	0.6360	0.7438	0.8221	0.8765	0.9138	0.9392	0.9568	0.9691	0.9777
0.360	0.3825	0.5328	0.6687	0.7746	0.8493	0.8996	0.9327	0.9546	0.9691	0.9788	0.9853
0.280	0.4094	0.5625	0.6980	0.8010	0.8717	0.9178	0.9472	0.9658	0.9777	0.9853	0.9903
0.200	0.4357	0.5903	0.7244	0.8238	0.8903	0.9323	0.9583	0.9740	0.9837	0.9897	0.9935
0.120	0.4615	0.6166	0.7483	0.8438	0.9060	0.9441	0.9668	0.9802	0.9881	0.9928	0.9956
0.040	0.4870	0.6416	0.7702	0.8613	0.9191	0.9536	0.9735	0.9848	0.9912	0.9949	0.9970
0.000	0.4998	0.6537	0.7804	0.8692	0.9249	0.9576	0.9763	0.9866	0.9924	0.9957	0.9975
-0.040	0.5125	0.6656	0.7903	0.8767	0.9303	0.9613	0.9787	0.9883	0.9935	0.9964	0.9980
-0.120	0.5381	0.6887	0.8089	0.8905	0.9399	0.9677	0.9829	0.9909	0.9951	0.9974	0.9986
-0.200	0.5639	0.7111	0.8263	0.9028	0.9481	0.9730	0.9861	0.9929	0.9963	0.9981	0.9990
-0.280	0.5901	0.7329	0.8426	0.9140	0.9552	0.9773	0.9888	0.9944	0.9972	0.9986	0.9993
-0.360	0.6170	0.7545	0.8580	0.9241	0.9614	0.9810	0.9909	0.9956	0.9979	0.9990	0.9995
-0.440	0.6448	0.7759	0.8727	0.9333	0.9669	0.9841	0.9926	0.9965	0.9984	0.9993	0.9997
-0.520	0.6738	0.7974	0.8869	0.9419	0.9718	0.9868	0.9940	0.9973	0.9988	0.9995	0.9998
-0.600	0.7046	0.8192	0.9008	0.9500	0.9762	0.9891	0.9952	0.9979	0.9991	0.9996	0.9998
-0.680	0.7378	0.8419	0.9146	0.9577	0.9803	0.9911	0.9962	0.9984	0.9993	0.9997	0.9999
-0.760	0.7746	0.8660	0.9288	0.9653	0.9841	0.9930	0.9970	0.9988	0.9995	0.9998	0.9999
-0.840	0.8172	0.8929	0.9439	0.9731	0.9879	0.9947	0.9978	0.9991	0.9996	0.9999	0.9999
-0.920	0.8716	0.9258	0.9617	0.9819	0.9919	0.9966	0.9986	0.9994	0.9998	0.9999	1.0000
-0.980	0.9360	0.9634	0.9813	0.9912	0.9961	0.9984	0.9994	0.9998	0.9999	1.0000	1.0000

Table 1 (Continued)

X	A										
	5.5	6.0	6.5	7.0	7.5	8.0	8.5	9.0	9.5	10.0	10.5
0.998	0.1150	0.1203	0.1255	0.1304	0.1351	0.1396	0.1440	0.1483	0.1524	0.1565	0.1604
0.980	0.3525	0.3678	0.3824	0.3962	0.4094	0.4219	0.4340	0.4456	0.4567	0.4674	0.4778
0.950	0.5302	0.5508	0.5700	0.5880	0.6049	0.6208	0.6358	0.6500	0.6634	0.6762	0.6883
0.910	0.6676	0.6899	0.7102	0.7288	0.7460	0.7618	0.7765	0.7900	0.8026	0.8144	0.8253
0.840	0.8036	0.8238	0.8417	0.8575	0.8716	0.8841	0.8953	0.9053	0.9143	0.9223	0.9296
0.760	0.8861	0.9022	0.9158	0.9274	0.9373	0.9457	0.9530	0.9592	0.9646	0.9692	0.9732
0.680	0.9319	0.9439	0.9537	0.9617	0.9683	0.9737	0.9781	0.9818	0.9848	0.9873	0.9894
0.600	0.9584	0.9672	0.9740	0.9794	0.9836	0.9869	0.9896	0.9917	0.9933	0.9947	0.9957
0.520	0.9743	0.9805	0.9852	0.9887	0.9914	0.9934	0.9950	0.9961	0.9970	0.9977	0.9982
0.440	0.9839	0.9883	0.9915	0.9938	0.9954	0.9967	0.9975	0.9982	0.9987	0.9990	0.9993
0.360	0.9898	0.9929	0.9950	0.9965	0.9976	0.9983	0.9988	0.9991	0.9994	0.9996	0.9997
0.280	0.9935	0.9957	0.9971	0.9980	0.9987	0.9991	0.9994	0.9996	0.9997	0.9998	0.9999
0.200	0.9958	0.9973	0.9983	0.9989	0.9993	0.9995	0.9997	0.9998	0.9999	0.9999	0.9999
0.120	0.9973	0.9983	0.9990	0.9994	0.9996	0.9998	0.9998	0.9999	0.9999	1.0000	1.0000
0.040	0.9983	0.9990	0.9994	0.9996	0.9998	0.9999	0.9999	1.0000	1.0000	1.0000	1.0000
0.000	0.9986	0.9992	0.9995	0.9997	0.9998	0.9999	0.9999	1.0000	1.0000	1.0000	1.0000
-0.040	0.9989	0.9994	0.9996	0.9998	0.9999	0.9999	1.0000	1.0000	1.0000	1.0000	1.0000
-0.120	0.9992	0.9996	0.9998	0.9999	0.9999	1.0000	1.0000	1.0000	1.0000	1.0000	1.0000
-0.200	0.9995	0.9997	0.9999	0.9999	1.0000	1.0000	1.0000	1.0000	1.0000	1.0000	1.0000
-0.280	0.9997	0.9998	0.9999	1.0000	1.0000	1.0000	1.0000	1.0000	1.0000	1.0000	1.0000
-0.360	0.9998	0.9999	0.9999	1.0000	1.0000	1.0000	1.0000	1.0000	1.0000	1.0000	1.0000
-0.440	0.9998	0.9999	1.0000	1.0000	1.0000	1.0000	1.0000	1.0000	1.0000	1.0000	1.0000
-0.520	0.9999	1.0000	1.0000	1.0000	1.0000	1.0000	1.0000	1.0000	1.0000	1.0000	1.0000
-0.600	0.9999	1.0000	1.0000	1.0000	1.0000	1.0000	1.0000	1.0000	1.0000	1.0000	1.0000
-0.680	0.9999	1.0000	1.0000	1.0000	1.0000	1.0000	1.0000	1.0000	1.0000	1.0000	1.0000
-0.760	1.0000	1.0000	1.0000	1.0000	1.0000	1.0000	1.0000	1.0000	1.0000	1.0000	1.0000
-0.840	1.0000	1.0000	1.0000	1.0000	1.0000	1.0000	1.0000	1.0000	1.0000	1.0000	1.0000
-0.920	1.0000	1.0000	1.0000	1.0000	1.0000	1.0000	1.0000	1.0000	1.0000	1.0000	1.0000
-0.980	1.0000	1.0000	1.0000	1.0000	1.0000	1.0000	1.0000	1.0000	1.0000	1.0000	1.0000

Table 2

The Conditional Expected Values, Given A, of
$$Cos(\phi_{\vec{h}} + \phi_{\vec{k}} + \phi_{-\vec{h}-\vec{k}}), \; Cos^2(\phi_{\vec{h}} + \phi_{\vec{k}} + \phi_{-\vec{h}-\vec{k}})$$
$$Sin^2(\phi_{\vec{h}} + \phi_{\vec{k}} + \phi_{-\vec{h}-\vec{k}}),$$
and Variances of
$$Cos(\phi_{\vec{h}} + \phi_{\vec{k}} + \phi_{-\vec{h}-\vec{k}}), \text{ and } Sin(\phi_{\vec{h}} + \phi_{\vec{k}} + \phi_{-\vec{h}-\vec{k}})$$

A	$\varepsilon(\cos Y)$	$\varepsilon(\cos^2 Y)$	$\varepsilon(\sin^2 Y)$	Var(cos Y)	Var(sin Y)
0.00	0.000000	0.500000	0.500000	0.500000	0.500000
0.50	0.242500	0.515001	0.484999	0.456195	0.484999
1.00	0.446390	0.553610	0.446390	0.354346	0.446390
1.50	0.596133	0.602578	0.397422	0.247203	0.397422
2.00	0.697775	0.651113	0.348887	0.164223	0.348887
2.50	0.764997	0.694001	0.305999	0.108781	0.305999
3.00	0.809985	0.730005	0.269995	0.073929	0.269995
3.50	0.841104	0.759685	0.240315	0.052229	0.240315
4.00	0.863523	0.784119	0.215881	0.038448	0.215881
4.50	0.880331	0.804371	0.195629	0.029388	0.195629
5.00	0.893383	0.821323	0.178677	0.023190	0.178677
5.50	0.903817	0.835670	0.164330	0.018784	0.164330
6.00	0.912359	0.847940	0.152060	0.015541	0.152060
6.50	0.919488	0.858540	0.141460	0.013082	0.141460
7.00	0.925532	0.867781	0.132219	0.011171	0.132219
7.50	0.930725	0.875903	0.124097	0.009655	0.124097
8.00	0.935235	0.883096	0.116904	0.008430	0.116904
8.50	0.939192	0.889507	0.110493	0.007426	0.110493
9.00	0.942690	0.895257	0.104743	0.006592	0.104743
9.50	0.945806	0.900441	0.099559	0.005892	0.099559
10.00	0.948600	0.905140	0.094860	0.005298	0.094860
10.50	0.951119	0.909417	0.090583	0.004790	0.090583
11.00	0.953402	0.913327	0.086673	0.004352	0.086673
11.50	0.955481	0.916915	0.083085	0.003972	0.083085
12.00	0.957381	0.920218	0.079782	0.003639	0.079782
12.50	0.959126	0.923270	0.076730	0.003347	0.076730
13.00	0.960734	0.926097	0.073903	0.003088	0.073903
13.50	0.962219	0.928724	0.071276	0.002859	0.071276
14.00	0.963596	0.931172	0.068828	0.002654	0.068828
14.50	0.964877	0.933457	0.066543	0.002470	0.066543
15.00	0.966070	0.935595	0.064405	0.002305	0.064405

$$\mathcal{E}(E_{\vec{h}+\vec{k}}|E_{\vec{k}}) \approx \int\limits_{R_1=0}^{\infty} \int\limits_{\bar{\Phi}_1=0}^{2\pi} \frac{R_1{}^2 \exp\left[i(\phi_{\vec{h}}+\phi_{\vec{k}})\right]}{\pi\left(1-\dfrac{|E_{\vec{h}}|^2}{N}\right)} \quad \times$$

$$\exp\left\{\frac{1}{1-\dfrac{|E_{\vec{h}}|^2}{N}}\left(-R_1{}^2 - \frac{|E_{\vec{h}}E_{\vec{k}}|^2}{N}\right)\right\} \quad \times$$

$$\exp\left\{\frac{2R_1|E_{\vec{h}}E_{\vec{k}}|}{N^{1/2}\left(1-\dfrac{|E_{\vec{h}}|^2}{N}\right)}\cos(\bar{\Phi}_1 - \phi_{\vec{h}} - \phi_{\vec{k}}) + i(\bar{\Phi}_1 - \phi_{\vec{h}} - \phi_{\vec{k}})\right\} \times$$

$$\left\{1 - \frac{1}{4N}\left(R_1{}^4 + 4R_1{}^2(|E_{\vec{k}}|^2 - 2) - 4|E_{\vec{k}}|^2 + 6\right)\right\} dR_1 d\bar{\Phi}_1 \ . \tag{10.2}$$

Referring to (2.5) and (2.7), the integration with respect to $\bar{\Phi}_1$ may be carried out:

$$\mathcal{E}(E_{\vec{h}+\vec{k}}|E_{\vec{k}}) \approx \int\limits_{R_1=0}^{\infty} \frac{2R_1{}^2 \exp\left[i(\phi_{\vec{h}}+\phi_{\vec{k}})\right]}{1-\dfrac{|E_{\vec{h}}|^2}{N}} \quad \times$$

$$\exp\left\{\frac{1}{1-\dfrac{|E_{\vec{h}}|^2}{N}}\left(-R_1{}^2 - \frac{|E_{\vec{h}}E_{\vec{k}}|^2}{N}\right)\right\} \quad \times$$

$$I_1\left\{\frac{2R_1|E_{\vec{h}}E_{\vec{k}}|}{N^{1/2}\left(1-\dfrac{|E_{\vec{h}}|^2}{N}\right)}\right\} \quad \times$$

$$\left\{ 1 - \frac{1}{4N} \left[R_1^{\;4} + 4R_1^{\;2} (|E_{\vec{k}}|^2 - 2) - 4|E_{\vec{k}}|^2 + 6 \right] \right\} dR_1 \qquad (10.3)$$

which, in view of (2.12) and (2.13), finally reduces to

$$\mathcal{E}(E_{\vec{h}+\vec{k}} | E_{\vec{k}}) \approx \frac{1}{N^{1/2}} \; |E_{\vec{h}} E_{\vec{k}}| \; \exp\left[i(\phi_{\vec{h}} + \phi_{\vec{k}}) \right] \left\{ 1 - \frac{1}{N} \; (|E_{\vec{k}}|^2 - 1) \right\} \quad (10.4)$$

or to

$$\mathcal{E}(E_{\vec{h}+\vec{k}} | E_{\vec{k}}) \approx \frac{1}{N^{1/2}} \; E_{\vec{h}} E_{\vec{k}} \left\{ 1 - \frac{1}{N} \; (|E_{\vec{k}}|^2 - 1) \right\} , \qquad (10.5)$$

or, for large N, simply to

$$\mathcal{E}(E_{\vec{h}+\vec{k}} | E_{\vec{k}}) \approx \frac{1}{N^{1/2}} \; E_{\vec{h}} E_{\vec{k}}. \qquad (10.6)$$

Eq. (10.6), then, gives the conditional average value of $E_{\vec{h}+\vec{k}}$ when, for fixed $\vec{h}$, $\vec{k}$ ranges uniformly over those vectors in reciprocal space for which $E_{\vec{k}}$ has a specified (complex) value.

11.   The Conditional Expectation of $|E_{\vec{h}+\vec{k}}|^2 - 1$, Given $|E_{\vec{k}}|$

Suppose that the vector $\vec{h}$ is fixed and that $\vec{k}$ ranges uniformly over those vectors in reciprocal space for which $|E_{\vec{k}}|$ has a preassigned value. Denote by $\mathcal{E}\left[(|E_{\vec{h}+\vec{k}}|^2 - 1) \Big| |E_{\vec{k}}| \right]$ the conditional expected value of $|E_{\vec{h}+\vec{k}}|^2 - 1$, given $|E_{\vec{k}}|$. Then, from (6.3),

$$\mathcal{E}\left((|E_{\vec{h}+\vec{k}}|^2 - 1)\Big|\,|E_{\vec{k}}|\right) = \int_{R_1=0}^{\infty} (R_1^2 - 1)P(R_1\Big|\,|E_{\vec{k}}|)dR_1 \approx$$

$$-1 + \int_{R_1=0}^{\infty} \frac{2R_1^3}{1 - \dfrac{|E_{\vec{h}}|^2}{N}} \exp\left\{\frac{1}{1 - \dfrac{|E_{\vec{h}}|^2}{N}}\left(-R_1^2 - \frac{|E_{\vec{h}}E_{\vec{k}}|^2}{N}\right)\right\} \times$$

$$I_0\left\{\frac{2|E_{\vec{h}}E_{\vec{k}}|R_1}{N^{1/2}\left(1 - \dfrac{|E_{\vec{h}}|^2}{N}\right)}\right\}\left\{1 - \frac{1}{4N}\Big(R_1^4 + 4R_1^2(|E_{\vec{k}}|^2 - 2) - \right.$$

$$\left. 4|E_{\vec{k}}|^2 + 6\Big)\right\} dR_1. \tag{11.1}$$

In view of (2.12) and (2.13), this integration is readily carried out and leads eventually to

$$\mathcal{E}\left((|E_{\vec{h}+\vec{k}}|^2 - 1)\Big|\,|E_{\vec{k}}|\right) \approx \frac{1}{N}(|E_{\vec{h}}|^2 - 1)(|E_{\vec{k}}|^2 - 1), \tag{11.2}$$

which should be compared with (10.6). Thus (11.2) is the conditional average value of $|E_{\vec{h}+\vec{k}}|^2 - 1$ when, for fixed $\vec{h}$, $\vec{k}$ ranges uniformly over those reciprocal vectors for which $|E_{\vec{k}}|$ has a specified value.

Next, let the vectors $\vec{h}$ and $\vec{h}'$ be fixed and assume that $\vec{k}$ ranges uniformly over those vectors in reciprocal space for which $|E_{\vec{h}'+\vec{k}}|$ has a preassigned value. Employing the corresponding generalization of (6.3), the conditional distribution of

$|E_{\vec{h}+\vec{k}}|$ given $|E_{\vec{h}'+\vec{k}}|$, one obtains the following generalization of (11.2),

$$\mathcal{E}\left((|E_{\vec{h}+\vec{k}}|^2 - 1)\Big| |E_{\vec{h}'+\vec{k}}| \right) \approx \frac{1}{N}(|E_{\vec{h}-\vec{h}'}|^2 - 1)(|E_{\vec{h}'+\vec{k}}|^2 - 1), \quad (11.3)$$

which has important application in the sequel. Eq. (11.3) gives the conditional average value of $|E_{\vec{h}+\vec{k}}|^2 - 1$ when, for fixed $\vec{h}$ and $\vec{h}'$, $\vec{k}$ ranges uniformly over those reciprocal vectors for which $|E_{\vec{h}'+\vec{k}}|$ has a specified value. Thus (11.2) is the special case $\vec{h}' = 0$ of (11.3).

12. The Conditional Expectation of $\phi_{\vec{h}} + \phi_{\vec{k}} + \phi_{-\vec{h}-\vec{k}}$, Given A

Let $\vec{h}$ be fixed and $\vec{k}$ range uniformly over the region in reciprocal space for which A (Eq. (7.4)) has a fixed value. Restrict $\phi_{\vec{h}} + \phi_{\vec{k}} + \phi_{-\vec{h}-\vec{k}}$ to lie in the interval $(-\pi, +\pi]$, which may always be done since the phases are multiple-valued (modulo 2π). Subject to this restriction, denote by $\mathcal{E}\left((\phi_{\vec{h}} + \phi_{\vec{k}} + \phi_{-\vec{h}-\vec{k}}) \Big| A \right)$ the conditional expectation of $\phi_{\vec{h}} + \phi_{\vec{k}} + \phi_{-\vec{h}-\vec{k}}$, given A. Then, from (8.3),

$$\mathcal{E}\left((\phi_{\vec{h}} + \phi_{\vec{k}} + \phi_{-\vec{h}-\vec{k}}) \Big| A \right) = \frac{1}{2\pi I_0(A)} \int_{\Omega=-\pi}^{\pi} \Omega \exp(A \cos \Omega)d\Omega \quad (12.1)$$

$$\mathcal{E}\left((\phi_{\vec{h}} + \phi_{\vec{k}} + \phi_{-\vec{h}-\vec{k}}) \Big| A \right) = 0 \quad (12.2)$$

since the integrand is an odd function of Ω.

13. The Conditional Expectation of
Sin $m(\phi_{\vec{h}} + \phi_{\vec{k}} + \phi_{-\vec{h}-\vec{k}})$, Given A

Assume that $\vec{h}$ is fixed and $\vec{k}$ ranges uniformly over the vectors in reciprocal space for which A (Eq. (7.4)) has a preassigned value. Denote by $\mathcal{E}\left(\sin m(\phi_{\vec{h}} + \phi_{\vec{k}} + \phi_{-\vec{h}-\vec{k}})\Big|A\right)$ the conditional expectation of $\sin m(\phi_{\vec{h}} + \phi_{\vec{k}} + \phi_{-\vec{h}-\vec{k}})$, given A, where m is an arbitrary parameter. Then, in view of (8.3),

$$\mathcal{E}\left(\sin m(\phi_{\vec{h}} + \phi_{\vec{k}} + \phi_{-\vec{h}-\vec{k}})\Big|A\right) \approx \frac{1}{2\pi I_0(A)} \int_{\Omega=0}^{2\pi} \sin m\Omega \, \exp\,(A\cos\,\Omega)d\Omega,$$

$$(13.1)$$

or, referring to (2.7),

$$\mathcal{E}\left(\sin m(\phi_{\vec{h}} + \phi_{\vec{k}} + \phi_{-\vec{h}-\vec{k}})\Big|A\right) = 0. \qquad (13.2)$$

14. The Conditional Expectation of
Cos $m(\phi_{\vec{h}} + \phi_{\vec{k}} + \phi_{-\vec{h}-\vec{k}})$, Given A

Again let $\vec{h}$ be fixed and suppose that $\vec{k}$ ranges uniformly over those reciprocal vectors corresponding to a fixed value of A. Denote by $\mathcal{E}\left(\cos m(\phi_{\vec{h}} + \phi_{\vec{k}} + \phi_{-\vec{h}-\vec{k}})\Big|A\right)$ the conditional expectation of $\cos m(\phi_{\vec{h}} + \phi_{\vec{k}} + \phi_{-\vec{h}-\vec{k}})$, given A, where m is an arbitrary parameter. Referring to (8.3) and (2.5), it follows that

$$\mathcal{E}\left(\cos\ m(\phi_{\vec{h}} + \phi_{\vec{k}} + \phi_{-\vec{h}-\vec{k}})\ \middle|\ A\right) \approx$$

$$\frac{1}{2\pi I_0(A)} \int\limits_{\Omega=0}^{2\pi} \cos m\Omega\ \exp(A\cos\Omega)d\Omega \qquad (14.1)$$

$$\mathcal{E}\left(\cos\ m(\phi_{\vec{h}} + \phi_{\vec{k}} + \phi_{-\vec{h}-\vec{k}})\ \middle|\ A\right) \approx \frac{I_m(A)}{I_0(A)}\ . \qquad (14.2)$$

The case that $m = 1$,

$$\mathcal{E}\left(\cos\ (\phi_{\vec{h}} + \phi_{\vec{k}} + \phi_{-\vec{h}-\vec{k}})\ \middle|\ A\right) \approx \frac{I_1(A)}{I_0(A)}\ , \qquad (14.3)$$

is of particular importance since it is needed in the development of recently derived procedures of phase determination and the case that $m = 2$ leads to a useful generalization of the squared tangent formula.

15.   The Conditional Variances

Following the derivations of §§13 and 14, one readily obtains also the related conditional variances. Only the final results are given here.

$$\mathcal{E}\left(\cos^2\ (\phi_{\vec{h}} + \phi_{\vec{k}} + \phi_{-\vec{h}-\vec{k}})\ \middle|\ A\right) \approx 1 - \frac{I_1(A)}{A I_0(A)}\ . \qquad (15.1)$$

$$\text{Var}\left(\sin\ (\phi_{\vec{h}} + \phi_{\vec{k}} + \phi_{-\vec{h}-\vec{k}})\ \Big|\ A\right) =$$

$$\epsilon\left(\sin^2\ (\phi_{\vec{h}} + \phi_{\vec{k}} + \phi_{-\vec{h}-\vec{k}})\ \Big|\ A\right) \approx \frac{I_1(A)}{AI_0(A)}\ . \qquad (15.2)$$

$$\text{Var}\left(\cos\ (\phi_{\vec{h}} + \phi_{\vec{k}} + \phi_{-\vec{h}-\vec{k}})\ \Big|\ A\right) \approx 1 - \frac{I_1(A)}{AI_0(A)} - \frac{I_1^{\ 2}(A)}{I_0^{\ 2}(A)}\ . \qquad (15.3)$$

It turns out that

$$\text{Var}\left(\cos\ (\phi_{\vec{h}} + \phi_{\vec{k}} + \phi_{-\vec{h}-\vec{k}})\Big|A\right) < \text{Var}\left(\sin\ (\phi_{\vec{h}} + \phi_{\vec{k}} + \phi_{-\vec{h}-\vec{k}})\Big|A\right)$$

$$(15.4)$$

for all positive values of A and that both variances decrease monotonically with increasing A (Table 2).

CHAPTER IV

PROBABILITY DISTRIBUTION OF THREE STRUCTURE
FACTORS AND APPLICATIONS

1. Introduction

The program begun in the previous chapter is extended here
to the case of three structure factors. Owing to the great
variety of probability distributions which now exist, it is
necessary to restrict severely the number which can be studied
here. Thus only the three most important of these distributions
are considered and only the barest outline of the analysis is
presented.

The notation and assumptions of Chapter III are retained.
The symbols $\vec{h}_1$, $\vec{h}_2$, $\vec{h}_3$, subject to

$$\vec{h}_1 + \vec{h}_2 + \vec{h}_3 = 0, \tag{1.1}$$

will denote vectors fixed in reciprocal space while the reciprocal
vector $\vec{k}$ will again denote the primitive random variable.

2. The Intuitive Background

Suppose that the space group is $P\bar{1}$. Let $\vec{h}_1$, $\vec{h}_2$, $\vec{h}_3$ be three
fixed vectors in reciprocal space satisfying (1.1) and let $\vec{k}$ be
an arbitrary reciprocal vector. Denote by ϕ the phase of the

normalized structure factor E. It is known that the structure invariant

$$\phi_{\vec{h}_1} + \phi_{\vec{h}_2} + \phi_{\vec{h}_3} \qquad\qquad (2.1)$$

is always more likely to be 0 than π and that the larger $\left| E_{\vec{h}_1} E_{\vec{h}_2} E_{\vec{h}_3} \right|$ is the more likely it is that (1.2) is equal to 0. (Compare with Eq. (9.3) of Chapter III when the space group is P1). Consider, then, the quadruple of structure invariants

$$\phi_{\vec{h}_1} + \phi_{\vec{h}_2} + \phi_{\vec{h}_3} \qquad\qquad (2.2)$$

$$\phi_{-\vec{k}} + \phi_{-\vec{h}_1} + \phi_{\vec{h}_1+\vec{k}} \qquad\qquad (2.3)$$

$$\phi_{\vec{k}} + \phi_{-\vec{h}_3} + \phi_{\vec{h}_3-\vec{k}} \qquad\qquad (2.4)$$

$$\phi_{-\vec{h}_2} + \phi_{-\vec{h}_1-\vec{k}} + \phi_{-\vec{h}_3+\vec{k}} \qquad\qquad (2.5)$$

the sum of all of which is evidently identically equal to 0. Suppose that the fixed vectors $\vec{h}_1$, $\vec{h}_2$, $\vec{h}_3$ and the variable vector $\vec{k}$ are so chosen that each of $\left| E_{\vec{h}_1} \right|$, $\left| E_{\vec{h}_2} \right|$, $\left| E_{\vec{h}_3} \right|$, $\left| E_{\vec{k}} \right|$, and $\left| E_{\vec{h}_1+\vec{k}} \right|$ is large. Assume too that, under these circumstances, $\left| E_{-\vec{h}_3+\vec{k}} \right|$ also turns out to be large relatively frequently, i.e. for a large fraction of the vectors $\vec{k}$. For each such vector $\vec{k}$, then, it is unlikely that any of the four invariants (2.2)-(2.5) is equal to π. Since the sum of the four structure invariants (2.2)-(2.5) is identically equal to 0, either all four of them are equal to 0, or two of them are equal to 0 and two are equal

to π, or all four of them are equal to π. Hence there are four
ways in which the invariant (2.2) is equal to 0 and four ways in
which (2.2) is equal to π. In each of the latter four cases,
however, precisely one or all three of the remaining invariants
(2.3), (2.4), (2.5) would also have to be equal to π, an extremely
unlikely event under the circumstances described. It is plausible
to conclude then that the invariant (2.2) is equal to 0 if, when
$\vec{k}$ is such that $|E_{\vec{k}}|$ and $|E_{\vec{h}_1 + \vec{k}}|$ are both large, it turns out
that $|E_{-\vec{h}_3 + \vec{k}}|$ is also large for a substantial fraction of these
vectors $\vec{k}$. Conversely, since the structure invariant (2.2) may,
on occasion, be equal to π, it is plausible to assume that this
event occurs precisely in the circumstance that, for almost all
vectors $\vec{k}$ such that $|E_{\vec{k}}|$ and $|E_{\vec{h}_1 + \vec{k}}|$ are both large, it turns
out that $|E_{-\vec{h}_3 + \vec{k}}|$ is small; for in this case it is not too
unlikely that one of (2.4), (2.5) is in fact equal to π.

Let t be an arbitrary, fixed number greater than unity, e.g.
t = 1.5. One may then summarize the heuristic argument just
given by concluding that the structure invariant (2.2) is equal
to 0 or π according as the conditional average

$$\left\langle (|E_{-\vec{h}_3 + \vec{k}}|^2 - 1) \,\Big|\, |E_{\vec{k}}| > t, \ |E_{\vec{h}_1 + \vec{k}}| > t \right\rangle_{\vec{k}} , \qquad (2.6)$$

of $|E_{-\vec{h}_3 + \vec{k}}|^2 - 1$ taken over all vectors $\vec{k}$ such that $|E_{\vec{k}}| > t$ and
$|E_{\vec{h}_1 + \vec{k}}| > t$, is relatively large or small respectively.

Although the heuristic argument presented here is most
transparent in the case that the space group is $P\bar{1}$, a similar kind
of argument also holds if the space group is P1. However, one
now employs the conditional probability distribution of
$\cos(\phi_{\vec{h}} + \phi_{\vec{k}} + \phi_{-\vec{h}-\vec{k}})$, given A, (Eq. (9.3) of Chapter III) and
concludes that the larger the conditional average (2.6) is, the

larger will be the value of $\left|E_{\vec{h}_1\vec{h}_2\vec{h}_3}\right|\cos(\phi_{h_1}+\phi_{h_2}+\phi_{h_3})$. Only the space group P1 is considered in this chapter but a like result holds for all the noncentrosymmetric space groups.

Not only does the heuristic argument strongly suggest the existence of a relationship between the structure invariant (2.2) and suitable conditional averages of $\left|E_{-\vec{h}_3+\vec{k}}\right|^2 - 1$, but (2.6) indicates also the path that a rigorous analysis should take. First, in order to calculate the required conditional expectation, it is necessary to derive a certain conditional probability distribution which depends on the joint probability distribution of the three structure factor magnitudes $\left|E_{-\vec{h}_3+\vec{k}}\right|$, $\left|E_{\vec{k}}\right|$, $\left|E_{\vec{h}_1+\vec{k}}\right|$. The latter, in turn, requires the joint probability distribution of the magnitudes and phases of $E_{-\vec{h}_3+\vec{k}}$, $E_{\vec{k}}$, $E_{\vec{h}_1+\vec{k}}$. In addition, the corresponding distributions of two structure factors $E_{\vec{k}}$ and $E_{\vec{h}+\vec{k}}$ turn out to be needed. In this way one is not only led to the desired relationship involving (2.6) but, with the machinery thus created, it is then possible to derive also an important generalization of Eq. (4.28) of Chapter II.

3. The Probability Distribution of $E_{-\vec{h}_3+\vec{k}}$, $E_{\vec{k}}$, $E_{\vec{h}_1+\vec{k}}$

Introduce the abbreviation

$$E_i = E_{\vec{h}_i}, \quad |E_i| = \left|E_{\vec{h}_i}\right|, \quad \phi_i = \phi_{\vec{h}_i}, \quad i = 1.2,3 \qquad (3.1)$$

in which $\vec{h}_1$, $\vec{h}_2$, $\vec{h}_3$ are fixed vectors satisfying (1.1). Let the vector $\vec{k}$ range uniformly through reciprocal space. Denote by

$P(R_1,R_2,R_3;\bar{\Phi}_1,\bar{\Phi}_2,\bar{\Phi}_3)$ the joint probability distribution of the magnitudes R_1,R_2,R_3 and the phases $\bar{\Phi}_1,\bar{\Phi}_2,\bar{\Phi}_3$ of the respective structure factors $E_{-\vec{h}_3+\vec{k}}, E_{\vec{k}}, E_{\vec{h}_1+\vec{k}}$. Thus $P(R_1,R_2,R_3;\bar{\Phi}_1,\bar{\Phi}_2,\bar{\Phi}_3)$ $dR_1 dR_2 dR_3 d\bar{\Phi}_1 d\bar{\Phi}_2 d\bar{\Phi}_3$ is the probability that $|E_{-\vec{h}_3+\vec{k}}|$ lie between R_1 and R_1+dR_1, $|E_{\vec{k}}|$ lie between R_2 and R_2+dR_2, $|E_{\vec{h}_1+\vec{k}}|$ lie between R_3 and R_3+dR_3, $\phi_{-\vec{h}_3+\vec{k}}$ lie between $\bar{\Phi}_1$ and $\bar{\Phi}_1+d\bar{\Phi}_1$, $\phi_{\vec{k}}$ lie between $\bar{\Phi}_2$ and $\bar{\Phi}_2+d\bar{\Phi}_2$, and $\phi_{\vec{h}_1+\vec{k}}$ lie between $\bar{\Phi}_3$ and $\bar{\Phi}_3+d\bar{\Phi}_3$. Then (Karle and Hauptman 1958) if $R_1 \geq 0, R_2 \geq 0, R_2 \geq 0$,

$$P(R_1,R_2,R_3;\bar{\Phi}_1,\bar{\Phi}_2,\bar{\Phi}_3) =$$

$$\frac{R_1 R_2 R_3}{(2\pi)^6} \int_{\rho_1=0}^{\infty} \int_{\rho_2=0}^{\infty} \int_{\rho_3=0}^{\infty} \int_{\theta_1=0}^{2\pi} \int_{\theta_2=0}^{2\pi} \int_{\theta_3=0}^{2\pi} \rho_1 \rho_2 \rho_3 \times$$

$$\exp\left\{-i[R_1\rho_1 \cos(\theta_1-\bar{\Phi}_1)+R_2\rho_2 \cos(\theta_2-\bar{\Phi}_2)+R_3\rho_3 \cos(\theta_3-\bar{\Phi}_3)]\right\} \times$$

$$\prod_{j=1}^{N} J_0\left\{\frac{1}{N^{1/2}} [\rho_1^2+\rho_2^2+\rho_3^2+2\rho_1\rho_2 \cos(2\pi\vec{h}_3\cdot\vec{r}_j+\theta_1-\theta_2) +\right.$$

$$\left. 2\rho_2\rho_3 \cos(2\pi\vec{h}_1\cdot\vec{r}_j+\theta_2-\theta_3)+2\rho_3\rho_1 \cos(2\pi\vec{h}_2\cdot\vec{r}_j+\theta_3-\theta_1)]^{1/2}\right\} \times$$

$$d\rho_1 d\rho_2 d\rho_3 d\theta_1 d\theta_2 d\theta_3, \tag{3.2}$$

where N is the number of atoms, assumed identical, in the unit cell. As in the derivation of (3.3) (Chapter III) one finds after some calculation, employing (2.18) and (1.1) of Chapter III, that

$$P(R_1,R_2,R_3;\Phi_1,\Phi_2,\Phi_3) \approx$$

$$\frac{R_1R_2R_3}{(2\pi)^6} \int\limits_{\rho_1=0}^{\infty} \int\limits_{\rho_2=0}^{\infty} \int\limits_{\rho_3=0}^{\infty} \int\limits_{\theta_1=0}^{2\pi} \int\limits_{\theta_2=0}^{2\pi} \int\limits_{\theta_3=0}^{2\pi} \rho_1\rho_2\rho_3 \; X$$

$$\left\{ 1 - \frac{1}{64N} [\rho_1^4+\rho_2^4+\rho_3^4+4\rho_1^2\rho_2^2+4\rho_2^2\rho_3^2+4\rho_3^2\rho_1^2] - \right.$$

$$\frac{1}{32N^{3/2}} [2\rho_1\rho_2(\rho_1^2+\rho_2^2+2\rho_3^2)|E_3|\cos(\theta_1-\theta_2+\phi_3) +$$

$$2\rho_2\rho_3(2\rho_1^2+\rho_2^2+\rho_3^2)|E_1|\cos(\theta_2-\theta_3+\phi_1) +$$

$$2\rho_3\rho_1(\rho_1^2+2\rho_2^2+\rho_3^2)|E_2|\cos(\theta_3-\theta_1+\phi_2)+\rho_1^2\rho_2^2|E_{2\vec{h}_3}| \; X$$

$$\cos(2\theta_1-2\theta_2+\phi_{2\vec{h}_3})+\rho_2^2\rho_3^2|E_{2\vec{h}_1}|\cos(2\theta_2-2\theta_3+\phi_{2\vec{h}_1}) +$$

$$\rho_3^2\rho_1^2|E_{2\vec{h}_2}|\cos(2\theta_3-2\theta_1+\phi_{2\vec{h}_2})+2\rho_1^2\rho_2\rho_3|E_{\vec{h}_2-\vec{h}_3}| \; X$$

$$\cos(2\theta_1-\theta_2-\theta_3-\phi_{\vec{h}_2-\vec{h}_3})+2\rho_1\rho_2^2\rho_3|E_{\vec{h}_3-\vec{h}_1}| \; X$$

$$\cos(2\theta_2-\theta_3-\theta_1-\phi_{\vec{h}_3-\vec{h}_1})+2\rho_1\rho_2\rho_3^2|E_{\vec{h}_1-\vec{h}_2}| \; X$$

$$\left. \cos(2\theta_3-\theta_1-\theta_2-\phi_{\vec{h}_1-\vec{h}_2})] \right\} \; X$$

$$\exp\left\{-\frac{1}{4}(\rho_1{}^2+\rho_2{}^2+\rho_3{}^2)-i[R_1\rho_1\cos(\theta_1-\bar{\Phi}_1)+\right.$$

$$\left. R_2\rho_2\cos(\theta_2-\bar{\Phi}_2)+R_3\rho_3\cos(\theta_3-\bar{\Phi}_3)]\right\}\times$$

$$\exp\left\{-\frac{1}{2N^{1/2}}[\rho_1\rho_2|E_3|\cos(\theta_1-\theta_2+\phi_3)+\rho_2\rho_3|E_1|\times\right.$$

$$\left. \cos(\theta_2-\theta_3+\phi_1)+\rho_3\rho_1|E_2|\cos(\theta_3-\theta_1+\phi_2)]\right\}\times$$

$$d\rho_1 d\rho_2 d\rho_3 d\theta_1 d\theta_2 d\theta_3. \tag{3.3}$$

Next, proceed as in Chapter III. Employ (2.1)-(2.4) to collect all terms in the exponent which contain θ_1 and use (2.6) and (2.7) to carry out the integration with respect to θ_1. The integration with respect to ρ_1 is then performed, employing (2.10) and (2.11). The remaining integrations with respect to $\theta_2,\rho_2,\theta_3,\rho_3$, in that order, are done in the same way and finally, after a very long analysis, one obtains the desired distribution,

$$P(R_1,R_2,R_3;\bar{\Phi}_1,\bar{\Phi}_2,\bar{\Phi}_3)\approx\frac{R_1R_2R_3}{\pi^3\Delta}\times$$

$$\exp\left\{-\frac{1}{\Delta}\left[R_1{}^2\left(1-\frac{|E_1|^2}{N}\right)+R_2{}^2\left(1-\frac{|E_2|^2}{N}\right)+\right.\right.$$

$$\left.\left. R_3{}^2\left(1-\frac{|E_3|^2}{N}\right)\right]\right\}\times$$

$$\exp\left\{\frac{2}{N^{1/2}\Delta}\left[R_1R_2|E_3|\cos(\overline{\Phi}_1-\overline{\Phi}_2+\phi_3)+R_2R_3|E_1|\times\right.\right.$$

$$\left.\cos(\overline{\Phi}_2-\overline{\Phi}_3+\phi_1)+R_3R_1|E_2|\cos(\overline{\Phi}_3-\overline{\Phi}_1+\phi_2)]\right\}\times$$

$$\exp\left\{-\frac{2}{N\Delta}\left[R_1R_2|E_1E_2|\cos(\overline{\Phi}_1-\overline{\Phi}_2-\phi_1-\phi_2)+R_2R_3|E_2E_3|\times\right.\right.$$

$$\left.\cos(\overline{\Phi}_2-\overline{\Phi}_3-\phi_2-\phi_3)+R_3R_1|E_3E_1|\cos(\overline{\Phi}_3-\overline{\Phi}_1-\phi_3-\phi_1)]\right\}\times$$

$$\left\{1-\frac{1}{4N}\left(R_1^4+R_2^4+R_3^4+4R_1^2R_2^2+4R_2^2R_3^2+4R_3^2R_1^2-\right.\right.$$

$$\left.12R_1^2-12R_2^2-12R_3^2+18\right)+\frac{2}{N^{3/2}}\left(R_1^2+R_2^2+R_3^2-4\right)\times$$

$$[R_1R_2|E_3|\cos(\overline{\Phi}_1-\overline{\Phi}_2+\phi_3)+R_2R_3|E_1|\cos(\overline{\Phi}_2-\overline{\Phi}_3+\phi_1)+$$

$$R_3R_1|E_2|\cos(\overline{\Phi}_3-\overline{\Phi}_1+\phi_2)]-\frac{1}{2N^{3/2}}\times$$

$$[R_1^2R_2^2|E_{2\vec{h}_3}|\cos(2\overline{\Phi}_1-2\overline{\Phi}_2+\phi_{2\vec{h}_3})+R_2^2R_3^2|E_{2\vec{h}_1}|\times$$

$$\cos(2\overline{\Phi}_2-2\overline{\Phi}_3+\phi_{2\vec{h}_1})+R_3^2R_1^2|E_{2\vec{h}_2}|\cos(2\overline{\Phi}_3-2\overline{\Phi}_1+\phi_{2\vec{h}_2})]-$$

$$\frac{R_1R_2R_3}{N^{3/2}}[R_1|E_{\vec{h}_2-\vec{h}_3}|\cos(2\overline{\Phi}_1-\overline{\Phi}_2-\overline{\Phi}_3-\phi_{\vec{h}_2-\vec{h}_3})+$$

$$R_2|E_{\vec{h}_3-\vec{h}_1}|\cos(2\overline{\Phi}_2-\overline{\Phi}_3-\overline{\Phi}_1-\phi_{\vec{h}_3-\vec{h}_1})+$$

$$\left.R_3|E_{\vec{h}_1-\vec{h}_2}|\cos(2\overline{\Phi}_3-\overline{\Phi}_1-\overline{\Phi}_2-\phi_{\vec{h}_1-\vec{h}_2})]\right\},\qquad(3.4)$$

where

$$\Delta = 1 - \frac{1}{N} (|E_1|^2 + |E_2|^2 + |E_3|^2) + \frac{2|E_1E_2E_3|}{N^{3/2}} \cos \phi \geq 0 \qquad (3.5)$$

and

$$\phi = \phi_1 + \phi_2 + \phi_3, \qquad (3.6)$$

and the inequality (3.5) is a consequence of some early work of Karle and Hauptman (1950).

4. The Probability Distribution of $|E_{-\vec{h}_3+\vec{k}}|$, $|E_{\vec{k}}|$, $|E_{\vec{h}_1+\vec{k}}|$

As in §3, suppose that the vectors $\vec{h}_1, \vec{h}_2, \vec{h}_3$ are fixed and satisfy (1.1). Let the vector $\vec{k}$ range uniformly through reciprocal space and denote by $P(R_1, R_2, R_3)$ the joint probability distribution of the magnitudes R_1, R_2, R_3 of the respective structure factors $E_{-\vec{h}_3+\vec{k}}$, $E_{\vec{k}}$, $E_{\vec{h}_1+\vec{k}}$. Thus $P(R_1, R_2, R_3)dR_1 dR_2 dR_3$ is the probability that $|E_{-\vec{h}_3+\vec{k}}|$ lie between R_1 and R_1+dR_1, $|E_{\vec{k}}|$ lie between R_2 and R_2+dR_2, and $|E_{\vec{h}_1+\vec{k}}|$ lie between R_3 and R_3+dR_3. Then if $R_1 \geq 0, R_2 \geq 0, R_3 \geq 0$,

$$P(R_1, R_2, R_3) = \int\limits_{\Phi_1=0}^{2\pi} \int\limits_{\Phi_2=0}^{2\pi} \int\limits_{\Phi_3=0}^{2\pi} P(R_1, R_2, R_3; \Phi_1, \Phi_2, \Phi_3) d\Phi_1 d\Phi_2 d\Phi_3,$$

$$(4.1)$$

where $P(R_1,R_2,R_3;\bar{\Phi}_1,\bar{\Phi}_2,\bar{\Phi}_3)$ is given by (3.4). In order to carry out the integration with respect to $\bar{\Phi}_1$, all terms in the exponent of (3.4) which involve $\bar{\Phi}_1$ are combined by means of (2.1)-(2.4) of Chapter III to yield

$$\frac{2R_1}{\Delta N^{1/2}}\left[[R_2|E_3|\cos(\bar{\Phi}_1-\bar{\Phi}_2+\phi_3)+R_3|E_2|\cos(\bar{\Phi}_1-\bar{\Phi}_3-\phi_2) - \right.$$

$$\frac{R_2|E_1E_2|}{N^{1/2}}\cos(\bar{\Phi}_1-\bar{\Phi}_2-\phi_1-\phi_2) - \frac{R_3|E_1E_3|}{N^{1/2}}\times$$

$$\left. \cos(\bar{\Phi}_1-\bar{\Phi}_3+\phi_1+\phi_3)\right] = \frac{2R_1}{\Delta N^{1/2}}\ Y_1\ \cos(\bar{\Phi}_1+\eta_1), \qquad (4.2)$$

where

$$Y_1 = \left[R_2^{\ 2}|E_3|^2+R_3^{\ 2}|E_2|^2+2R_2R_3|E_2E_3|\cos(\bar{\Phi}_2-\bar{\Phi}_3-\phi_2-\phi_3) - \right.$$

$$\frac{2|E_1E_2E_3|}{N^{1/2}}(R_2^{\ 2}+R_3^{\ 2})\cos\ \phi - \frac{2R_2R_3|E_1|}{N^{1/2}}\ (|E_2|^2+|E_3|^2)\times$$

$$\cos(\bar{\Phi}_2-\bar{\Phi}_3+\phi_1) + \frac{R_2^{\ 2}|E_1E_2|^2}{N} + \frac{R_3^{\ 2}|E_1E_3|^2}{N} +$$

$$\left. \frac{2R_2R_3|E_1^{\ 2}E_2E_3|}{N}\ \cos(\bar{\Phi}_2-\bar{\Phi}_3+2\phi_1+\phi_2+\phi_3)\right]^{1/2} , \qquad (4.3)$$

$$\cos \eta_1 = \frac{1}{Y_1}\left[R_2|E_3|\cos(\bar{\Phi}_2-\phi_3)+R_3|E_2|\cos(\bar{\Phi}_3+\phi_2) - \right.$$

$$\left. \frac{R_2|E_1E_2|}{N^{1/2}}\cos(\bar{\Phi}_2+\phi_1+\phi_2) - \frac{R_3|E_1E_3|}{N^{1/2}}\cos(\bar{\Phi}_3-\phi_1-\phi_3)\right], \quad (4.4)$$

$$\sin \eta_1 = -\frac{1}{Y_1}\left[R_2|E_3|\sin(\bar{\Phi}_2-\phi_3)+R_3|E_2|\sin(\bar{\Phi}_3+\phi_2) - \right.$$

$$\left. \frac{R_2|E_1E_2|}{N^{1/2}}\sin(\bar{\Phi}_2+\phi_1+\phi_2) - \frac{R_3|E_1E_3|}{N^{1/2}}\sin(\bar{\Phi}_3-\phi_1-\phi_3)\right], \quad (4.5)$$

so that Y_1 and η_1 are independent of $\bar{\Phi}_1$. Hence the integration with respect to $\bar{\Phi}_1$ in (4.1), using (2.5) (Chapter III), may be carried out and leads to

$$P(R_1,R_2,R_3) \approx \frac{2R_1R_2R_3}{\pi^2\Delta} \times$$

$$\exp\left\{-\frac{1}{\Delta}\left[R_1^2\left(1-\frac{|E_1|^2}{N}\right) + R_2^2\left(1-\frac{|E_2|^2}{N}\right) + \right.\right.$$

$$\left.\left. R_3^2\left(1-\frac{|E_3|^2}{N}\right)\right]\right\} \int_{\bar{\Phi}_2=0}^{2\pi}\int_{\bar{\Phi}_3=0}^{2\pi}\exp\left\{\frac{1}{\Delta}\left[\frac{2R_2R_3|E_1|}{N^{1/2}}\times\right.\right.$$

$$\left.\left. \cos(\bar{\Phi}_2-\bar{\Phi}_3+\phi_1) - \frac{2R_2R_3|E_2E_3|}{N}\cos(\bar{\Phi}_2-\bar{\Phi}_3-\phi_2-\phi_3)\right]\right\} \times$$

$$\left\{ I_0 \left(\frac{2R_1 Y_1}{\Delta N^{1/2}} \right) \left[1 - \frac{1}{4N} (R_1^4 + R_2^4 + R_3^4 + 4R_1^2 R_2^2 + 4R_2^2 R_3^2 + \right. \right.$$

$$\left. 4R_3^2 R_1^2 - 12R_1^2 - 12R_2^2 - 12R_3^2 + 18 \right] + \frac{2}{N^{3/2}} (R_1^2 + R_2^2 + R_3^2 - 4) \times$$

$$\left[I_1 \left(\frac{2R_1 Y_1}{\Delta N^{1/2}} \right) R_1 R_2 |E_3| \cos(\eta_1 + \bar{\Phi}_2 - \phi_3) + I_0 \left(\frac{2R_1 Y_1}{\Delta N^{1/2}} \right) \times \right.$$

$$\left. R_2 R_3 |E_1| \cos(\bar{\Phi}_2 - \bar{\Phi}_3 + \phi_1) + I_1 \left(\frac{2R_1 Y_1}{\Delta N^{1/2}} \right) R_3 R_1 |E_2| \cos(\eta_1 + \bar{\Phi}_3 + \phi_2) \right] -$$

$$\frac{1}{2N^{3/2}} \left[I_2 \left(\frac{2R_1 Y_1}{\Delta N^{1/2}} \right) R_1^2 R_2^2 |E_{2\vec{h}_3}| \cos(2\eta_1 + 2\bar{\Phi}_2 + \phi_{2\vec{h}_3}) + \right.$$

$$I_0 \left(\frac{2R_1 Y_1}{\Delta N^{1/2}} \right) R_2^2 R_3^2 |E_{2\vec{h}_1}| \cos(2\bar{\Phi}_2 - 2\bar{\Phi}_3 + \phi_{2\vec{h}_1}) +$$

$$\left. I_2 \left(\frac{2R_1 Y_1}{\Delta N^{1/2}} \right) R_3^2 R_1^2 |E_{2\vec{h}_2}| \cos(2\eta_1 + 2\bar{\Phi}_3 + \phi_{2\vec{h}_2}) \right] -$$

$$\frac{R_1 R_2 R_3}{N^{3/2}} \left[I_2 \left(\frac{2R_1 Y_1}{\Delta N^{1/2}} \right) R_1 |E_{\vec{h}_2 - \vec{h}_3}| \cos(2\eta_1 - \bar{\Phi}_2 - \bar{\Phi}_3 - \phi_{\vec{h}_2 - \vec{h}_3}) + \right.$$

$$I_1 \left(\frac{2R_1 Y_1}{\Delta N^{1/2}} \right) R_2 |E_{\vec{h}_3 - \vec{h}_1}| \cos(\eta_1 + 2\bar{\Phi}_2 - \bar{\Phi}_3 - \phi_{\vec{h}_3 - \vec{h}_1}) +$$

$$\left. \left. I_1 \left(\frac{2R_1 Y_1}{\Delta N^{1/2}} \right) R_3 |E_{\vec{h}_1 - \vec{h}_2}| \cos(\eta_1 - \bar{\Phi}_2 + 2\bar{\Phi}_3 - \phi_{\vec{h}_1 - \vec{h}_2}) \right] \right\} d\bar{\Phi}_2 d\bar{\Phi}_3 .$$

$$(4.6)$$

Next, it is clear that the terms of order $1/N^{3/2}$ inside the braces of (4.6) contribute only to terms of order $1/N^2$ or higher in the final result. This is obvious for those terms containing I_1 or I_2 as a factor since these Bessel functions are themselves of order $1/N^{1/2}$ or higher. Owing to the cyclic symmetry, the same conclusion holds also for the two terms involving I_0. Since only terms of order $1/N^{3/2}$ or lower are included in the final result, it remains therefore to consider only those terms inside the braces of (4.6) which are of order 1 or $1/N$.

In order to perform the integration with respect to $\bar{\Phi}_2$ in (4.6) it is convenient, since Y_1 depends on $\bar{\Phi}_2$, to analyze Y_1 by first collecting the terms of Y_1^2 which depend on $\bar{\Phi}_2$ and utilizing, as usual, (2.1)-(2.4) of Chapter III. In view of (4.3), then,

$$
2R_2R_3 \left[|E_2E_3|\cos(\bar{\Phi}_2-\bar{\Phi}_3-\phi_2-\phi_3) - \frac{|E_1|}{N^{1/2}} (|E_2|^2+|E_3|^2) \times \right.
$$

$$
\left. \cos(\bar{\Phi}_2-\bar{\Phi}_3+\phi_1) + \frac{|E_1^2E_2E_3|}{N} \cos(\bar{\Phi}_2-\bar{\Phi}_3+2\phi_1+\phi_2+\phi_3) \right] =
$$

$$
2R_2R_3Y' \cos(\bar{\Phi}_2 + \eta'), \tag{4.7}
$$

where, using (3.6),

$$
Y' = \left[|E_2E_3|^2 - 2|E_1E_2E_3|(|E_2|^2+|E_3|^2)\left(\frac{1}{N^{1/2}} + \frac{|E_1|^2}{N^{3/2}}\right) \cos \phi + \right.
$$

$$
\left. \frac{|E_1|^2}{N}(|E_2|^2+|E_3|^2)^2 + \frac{2|E_1E_2E_3|^2}{N} \cos 2\phi + \frac{|E_1^2E_2E_3|^2}{N^2} \right]^{1/2},
$$

$$
\tag{4.8}
$$

$$\cos \eta' = \frac{1}{Y'} \left[|E_2 E_3| \cos(\bar{\Phi}_3 + \phi_2 + \phi_3) - \frac{|E_1|}{N^{1/2}} (|E_2|^2 + |E_3|^2) \times \right.$$

$$\left. \cos(\bar{\Phi}_3 - \phi_1) + \frac{|E_1{}^2 E_2 E_3|}{N} \cos(\bar{\Phi}_3 - 2\phi_1 - \phi_2 - \phi_3) \right] , \qquad (4.9)$$

$$\sin \eta' = \frac{1}{Y'} \left[|E_2 E_3| \sin(\bar{\Phi}_3 + \phi_2 + \phi_3) - \frac{|E_1|}{N^2} (|E_2|^2 + |E_3|^2) \times \right.$$

$$\left. \sin(\bar{\Phi}_3 - \phi_1) + \frac{|E_1{}^2 E_2 E_3|}{N} \sin(\bar{\Phi}_3 - 2\phi_1 - \phi_2 - \phi_3) \right] , \qquad (4.10)$$

so that η' is independent of $\bar{\Phi}_2$. Then, from (4.3),

$$Y_1 = \left[R_2{}^2 \left(|E_3|^2 - \frac{2|E_1 E_2 E_3|}{N^{1/2}} \cos \phi + \frac{|E_1 E_2|^2}{N} \right) + 2R_2 R_3 Y' \times \right.$$

$$\left. \cos(\bar{\Phi}_2 + \eta') + R_3{}^2 \left(|E_2|^2 - \frac{2|E_1 E_2 E_3|}{N^{1/2}} \cos \phi + \frac{|E_3 E_1|^2}{N} \right) \right]^{1/2} ,$$

$$(4.11)$$

in which, as is easily verified from (4.8),

$$Y' = Z_2 Z_3 \qquad (4.12)$$

where

$$Z_2 = \left(|E_2|^2 - \frac{2|E_1 E_2 E_3|}{N^{1/2}} \cos \phi + \frac{|E_3 E_1|^2}{N} \right)^{1/2} \qquad (4.13)$$

and

$$Z_3 = \left(|E_3|^2 - \frac{2|E_1 E_2 E_3|}{N^{1/2}} \cos \phi + \frac{|E_1 E_2|^2}{N} \right)^{1/2}, \qquad (4.14)$$

so that Z_2 and Z_3 are constants (independent of $\bar{\Phi}_2$). Hence, from (4.11)

$$Y_1 = [R_2^2 Z_3^2 + 2R_2 R_3 Z_2 Z_3 \cos(\bar{\Phi}_2 + \eta') + R_3^2 Z_2^2]^{1/2}, \qquad (4.15)$$

so that, in view of (2.16) (Chapter III),

$$I_0 \left(\frac{2R_1 Y_1}{\Delta N^{1/2}} \right) = \sum_{m=-\infty}^{\infty} I_m \left(\frac{2R_1 R_2 Z_3}{\Delta N^{1/2}} \right) I_m \left(\frac{2R_1 R_3 Z_2}{\Delta N^{1/2}} \right) \exp i \, m(\bar{\Phi}_2 + \eta')$$

$$(4.16)$$

which expresses the dependence of $I_0 \left(\dfrac{2R_1 Y_1}{\Delta N^{1/2}} \right)$ on $\bar{\Phi}_2$ in terms of the exponential function.

Next, that portion of the exponent of (4.6) which involves $\bar{\Phi}_2$ may be written, in view of (2.1)-(2.4) of Chapter III,

$$\frac{2R_2 R_3}{\Delta N^{1/2}} \left[|E_1| \cos(\bar{\Phi}_2 - \bar{\Phi}_3 + \phi_1) - \frac{|E_2 E_3|}{\Delta N^{1/2}} \cos(\bar{\Phi}_2 - \bar{\Phi}_3 - \phi_2 - \phi_3) \right] =$$

$$\frac{2R_2 R_3}{\Delta N^{1/2}} \, Y_2 \, \cos(\bar{\Phi}_2 + \eta_2), \qquad (4.17)$$

where

$$Y_2 = \left(|E_1|^2 - \frac{2|E_1 E_2 E_3|}{N^{1/2}} \cos \phi + \frac{|E_2 E_3|^2}{N} \right)^{1/2} = Z_1$$

(definition of Z_1),

(4.18)

$$\cos \eta_2 = \frac{1}{Z_1} \left[|E_1| \cos(\bar\Phi_3 - \phi_1) - \frac{|E_2 E_3|}{N^{1/2}} \cos(\bar\Phi_3 + \phi_2 + \phi_3) \right], \qquad (4.19)$$

$$\sin \eta_2 = -\frac{1}{Z_1} \left[|E_1| \sin(\bar\Phi_3 - \phi_1) - \frac{|E_2 E_3|}{N^{1/2}} \sin(\bar\Phi_3 + \phi_2 + \phi_3) \right], \qquad (4.20)$$

and η_2 is independent of $\bar\Phi_2$.

In view of (4.16) and (4.17) and of (2.5) and (2.7) of Chapter III, the integration of (4.6) with respect to $\bar\Phi_2$ is readily carried out:

$$P(R_1, R_2, R_3) \approx \frac{4R_1 R_2 R_3}{\pi \Delta} \times$$

$$\exp \left\{ -\frac{1}{\Delta} \left[R_1^2 \left(1 - \frac{|E_1|^2}{N} \right) + R_2^2 \left(1 - \frac{|E_2|^2}{N} \right) + \right. \right.$$

$$\left. \left. R_3^2 \left(1 - \frac{|E_3|^2}{N} \right) \right] \right\} \left[1 - \frac{1}{4N} (R_1^4 + R_2^4 + R_3^4 + 4R_1^2 R_2^2 + 4R_2^2 R_3^2 + \right.$$

$$\left. 4R_3^2 R_1^2 - 12R_1^2 - 12R_2^2 - 12R_3^2 + 18) \right] \int\limits_{\bar\Phi_3 = 0}^{2\pi} \sum_{m=-\infty}^{\infty} I_m \left(\frac{2R_1 R_2 Z_3}{\Delta N^{1/2}} \right) \times$$

$$I_m \left(\frac{2R_2 R_3 Z_1}{\Delta N^{1/2}} \right) I_m \left(\frac{2R_3 R_1 Z_2}{\Delta N^{1/2}} \right) \exp[im(\eta' - \eta_2)] d\bar{\Phi}_3. \tag{4.21}$$

Write

$$\eta = \eta' - \eta_2. \tag{4.22}$$

Then, in view of (4.9), (4.10), (4.19), and (4.20), the following expressions for $\cos \eta$ and $\sin \eta$ are readily derived:

$$\cos \eta = \frac{1}{Z_1 Z_2 Z_3} \left[|E_1 E_2 E_3| \cos \phi - \frac{1}{N^{1/2}} (|E_1 E_2|^2 + |E_2 E_3|^2 + |E_3 E_1|^2) + \right.$$

$$\left. \frac{1}{N} |E_1 E_2 E_3| (|E_1|^2 + |E_2|^2 + |E_3|^2) \cos \phi - \frac{|E_1 E_2 E_3|^2}{N^{3/2}} \cos 2\phi \right], \tag{4.23}$$

$$\sin \eta = - \frac{1}{Z_1 Z_2 Z_3} \left[|E_1 E_2 E_3| \sin \phi - \frac{1}{N} |E_1 E_2 E_3| (|E_1|^2 + |E_2|^2 + |E_3|^2) \times \right.$$

$$\left. \sin \phi + \frac{|E_1 E_2 E_3|^2}{N^{3/2}} \sin 2\phi \right], \tag{4.24}$$

and are found to be independent of $\bar{\Phi}_3$. The final integration with respect to $\bar{\Phi}_3$ in (4.21) is thus trival and the formula for $P(R_1, R_2, R_3)$, (correct to and including terms of order $1/N^{3/2}$), is found to be

$$P(R_1, R_2, R_3) \approx \frac{8R_1 R_2 R_3}{\Delta} \exp \left\{ -\frac{1}{\Delta} \left[R_1^2 \left(1 - \frac{|E_1|^2}{N} \right) + \right. \right.$$

$$R_2{}^2 \left(1 - \frac{|E_2|^2}{N} \right) + R_3{}^2 \left(1 - \frac{|E_3|^2}{N} \right) \Big] \Big] \Big\} \times$$

$$\left[1 - \frac{1}{4N} \ (R_1{}^4 + R_2{}^4 + R_3{}^4 + 4R_1{}^2 R_2{}^2 + 4R_2{}^2 R_3{}^2 + 4R_3{}^2 R_1{}^2 - \right.$$

$$12R_1{}^2 - 12R_2{}^2 - 12R_3{}^2 + 18) \Big] \sum_{m=-\infty}^{\infty} I_m \left(\frac{2R_1 R_2 Z_3}{N^{1/2} \Delta} \right) \times$$

$$I_m \left(\frac{2R_2 R_3 Z_1}{N^{1/2} \Delta} \right) I_m \left(\frac{2R_3 R_1 Z_2}{N^{1/2} \Delta} \right) \ \exp[im\eta], \tag{4.25}$$

where Δ, ϕ, Z_1, Z_2, Z_3, cos η and sin η are given by (3.5), (3.6), (4.18), (4.13), (4.14), (4.23), (4.24) respectively.

It should be emphasized that (4.25), as well as (4.1) of Chapter III, is not only correct to and including terms of order $1/N^{3/2}$, but the forms of these distributions are such that, for large N, they remain essentially positive for all permitted values of the parameters $|E_{\vec{h}', -\vec{h}}|$, $|E_1|$, $|E_2|$, $|E_3|$, ϕ_1, ϕ_2, ϕ_3 and for all values of the variables R_0, R_1, R_2, R_3, no matter how large. In addition, they tend to zero with increasing R_0, R_1, R_2, or R_3 and are suitably normalized, as any well behaved probability distribution should be. Hence they are accurate over the whole range of values of parameters and variables and (4.25), in particular, permits the derivation of the conditional distribution, with the necessary accuracy, which is secured in the next section.

5. The Conditional Distribution of $|E_{-\vec{h}_3+\vec{k}}|$,
Given $|E_{\vec{k}}|$ and $|E_{\vec{h}_1+\vec{k}}|$

Suppose now that $\vec{h}_1$, $\vec{h}_2$, $\vec{h}_3$ are fixed reciprocal vectors satisfying (1.1) and that $\vec{k}$ ranges uniformly over those vectors in reciprocal space for which $|E_{\vec{k}}|$ and $|E_{\vec{h}_1+\vec{k}}|$ have fixed, specified values (not necessarily equal). Denote by $P(R_1 \mid |E_{\vec{k}}|$, $|E_{\vec{h}_1+\vec{k}}|) = P(R_1 \mid R_2=|E_{\vec{k}}|, R_3=|E_{\vec{h}_1+\vec{k}}|)$ the conditional distribution of $|E_{-\vec{h}_3+\vec{k}}|$, given $|E_{\vec{k}}|$ and $|E_{\vec{h}_1+\vec{k}}|$. Thus $P(R_1 \mid |E_{\vec{k}}|, |E_{\vec{h}_1+\vec{k}}|)dR_1$ is the conditional probability that $|E_{-\vec{h}_3+\vec{k}}|$ lie between R_1 and R_1+dR_1, given $|E_{\vec{k}}|$ and $|E_{\vec{h}_1+\vec{k}}|$. Again, $P(R_1 \mid |E_{\vec{k}}|, |E_{\vec{h}_1+\vec{k}}|)$ is obtained from (4.25) by replacing R_2 by $|E_{\vec{k}}|$, R_3 by $|E_{\vec{h}_1+\vec{k}}|$, and multiplying by a suitable constant:

$$P(R_1 \mid |E_{\vec{k}}|, |E_{\vec{h}_1+\vec{k}}|) \approx \frac{R_1}{K_4} \exp\left\{ -\frac{1}{\Delta}\left(1 - \frac{|E_1|^2}{N}\right) R_1{}^2 \right\} \times$$

$$\left\{ \sum_{m=-\infty}^{\infty} I_m\left(\frac{2R_1|E_{\vec{k}}|Z_3}{N^{1/2}\Delta}\right) I_m\left(\frac{2R_1|E_{\vec{h}_1+\vec{k}}|Z_2}{N^{1/2}\Delta}\right) I_m\left(\frac{2|E_{\vec{k}}E_{\vec{h}_1+\vec{k}}|Z_1}{N^{1/2}\Delta}\right) \times \right.$$

$$\left. \exp[im\eta] \right\} \left\{ 1 - \frac{1}{4N}\left(R_1{}^4 + 4R_1{}^2(|E_{\vec{k}}|^2 + |E_{\vec{h}_1+\vec{k}}|^2 - 3) + \right. \right.$$

$$\left. \left. |E_{\vec{k}}|^4 + |E_{\vec{h}_1+\vec{k}}|^4 + 4|E_{\vec{k}}E_{\vec{h}_1+\vec{k}}|^2 - 12|E_{\vec{k}}|^2 - 12|E_{\vec{h}_1+\vec{k}}|^2 + 18\right) \right\} \quad (5.1)$$

where

$$K_4 = \int_{R_1=0}^{\infty} R_1 \exp\left\{-\frac{1}{\Delta}\left(1-\frac{|E_1|^2}{N}\right)R_1{}^2\right\} \times$$

$$\left\{\sum_{m=-\infty}^{\infty} I_m\left(\frac{2R_1|E_{\vec{k}}|z_3}{N^{1/2}\Delta}\right) I_m\left(\frac{2R_1|E_{\vec{h}_1+\vec{k}}|z_2}{N^{1/2}\Delta}\right) I_m\left(\frac{2|E_{\vec{k}}E_{\vec{h}_1+\vec{k}}|z_1}{N^{1/2}\Delta}\right) \times\right.$$

$$\exp[im\eta]\left.\right\}\left\{1-\frac{1}{4N}\left(R_1{}^4+4R_1{}^2(|E_{\vec{k}}|^2+|E_{\vec{h}_1+\vec{k}}|^2-3)+\right.\right.$$

$$\left.\left.|E_{\vec{k}}|^4+|E_{\vec{h}_1+\vec{k}}|^4+4|E_{\vec{k}}E_{\vec{h}_1+\vec{k}}|^2-12|E_{\vec{k}}|^2-12|E_{\vec{h}_1+\vec{k}}|^2+18\right)\right\}dR_1.$$

$$(5.2)$$

In view of Chapter III, one finds, after some calculation, that

$$K_4 \approx \frac{\Delta}{2\left(1-\frac{|E_1|^2}{N}\right)} \exp\left\{\frac{1}{N\Delta\left(1-\frac{|E_1|^2}{N}\right)}\left(|E_{\vec{k}}|^2 z_3{}^2+\right.\right.$$

$$\left.\left.|E_{\vec{h}_1+\vec{k}}|^2 z_2{}^2\right)\right\} I_0\left\{\frac{2|E_{\vec{k}}E_{\vec{h}_1+\vec{k}}|}{N^{1/2}\Delta}\left(z_1{}^2+\frac{2z_1 z_2 z_3}{N^{1/2}\left(1-\frac{|E_1|^2}{N}\right)}\times\right.\right.$$

$$\left.\left.\cos\eta+\frac{z_2{}^2 z_3{}^2}{N\left(1-\frac{|E_1|^2}{N}\right)^2}\right)^{1/2}\right\}\left\{1-\frac{1}{4N}\left(|E_{\vec{k}}|^4+|E_{\vec{h}_1+\vec{k}}|^4+\right.\right.$$

$$4|E_{\vec{k}}E_{\vec{h}_1+\vec{k}}|^2-8|E_{\vec{k}}|^2-8|E_{\vec{h}_1+\vec{k}}|^2+8\Bigg)\Bigg\}\;.\tag{5.3}$$

Combining (5.3) with (5.1) leads finally to the desired conditional distribution of $|E_{-\vec{h}_3+\vec{k}}|$, given $|E_{\vec{k}}|$ and $|E_{\vec{h}_1+\vec{k}}|$,

$$P(R_1\Big|\,|E_{\vec{k}}|,|E_{\vec{h}_1+\vec{k}}|)\approx$$

$$\frac{2R_1\left(1-\dfrac{|E_1|^2}{N}\right)}{\Delta I_0\left\{\dfrac{2|E_{\vec{k}}E_{\vec{h}_1+\vec{k}}|}{N^{1/2}\Delta}\left(z_1{}^2+\dfrac{2z_1z_2z_3}{N^{1/2}\left(1-\dfrac{|E_1|^2}{N}\right)}\cos\eta+\dfrac{z_2{}^2z_3{}^2}{N\left(1-\dfrac{|E_1|^2}{N}\right)^2}\right)^{1/2}\right\}}\times$$

$$\exp\left\{-\frac{1}{\Delta}\left(1-\frac{|E_1|^2}{N}\right)R_1{}^2-\frac{1}{N\Delta\left(1-\dfrac{|E_1|^2}{N}\right)}\times\right.$$

$$\left.\left(|E_{\vec{k}}|^2z_3{}^2+|E_{\vec{h}_1+\vec{k}}|^2z_2{}^2\right)\right\}\left\{\sum_{m=-\infty}^{\infty}I_m\left(\frac{2R_1|E_{\vec{k}}|z_3}{N^{1/2}\Delta}\right)\times\right.$$

$$\left.I_m\left(\frac{2R_1|E_{\vec{h}_1+\vec{k}}|z_2}{N^{1/2}\Delta}\right)I_m\left(\frac{2|E_{\vec{k}}E_{\vec{h}_1+\vec{k}}|z_1}{N^{1/2}\Delta}\right)\exp[imn]\right\}\times$$

$$\left\{1-\frac{1}{4N}\left(R_1{}^4+4R_1{}^2(|E_{\vec{k}}|^2+|E_{\vec{h}_1+\vec{k}}|^2-3)-4|E_{\vec{k}}|^2-4|E_{\vec{h}_1+\vec{k}}|^2+10\right)\right\},$$

$$\tag{5.4}$$

where Δ, ϕ, Z_1, Z_2, Z_3, $\cos \eta$, and $\sin \eta$ are given by (3.5), (3.6), (4.18), (4.13), (4.14), (4.23), and (4.24) respectively. It is instructive to compare (the doubly conditioned) (5.4) with the singly conditioned distribution of Chapter III and with the known, (unconditional) probability distribution of a single structure factor magnitude

$$P(R) \approx 2Re^{-R^2} \left\{ 1 - \frac{1}{4N} \ (R^4 - 4R^2 + 2) + \right.$$

$$\left. \frac{1}{288N^2} \ (9R^8 - 112R^6 + 360R^4 - 288R^2 + 24) \right\} \ . \qquad (5.5)$$

6. The Conditional Expectation of $\left|E_{-\vec{h}_3 + \vec{k}}\right|^2 - 1$,

Given $\left|E_{\vec{k}}\right|$ and $\left|E_{\vec{h}_1 + \vec{k}}\right|$

Suppose finally that the three reciprocal vectors $\vec{h}_1$, $\vec{h}_2$, $\vec{h}_3$ satisfying (1.1) are given and that $\vec{k}$ ranges uniformly over reciprocal space in such a way that $\left|E_{\vec{k}}\right|$ and $\left|E_{\vec{h}_1 + \vec{k}}\right|$ have specified, fixed values (not necessarily equal). Denote by $\varepsilon \left((\left|E_{-\vec{h}_3 + \vec{k}}\right|^2 - 1) \left| \left|E_{\vec{k}}\right|, \left|E_{\vec{h}_1 + \vec{k}}\right| \right. \right)$ the conditional expectation of $\left|E_{-\vec{h}_3 + \vec{k}}\right|^2 - 1$, given $\left|E_{\vec{k}}\right|$ and $\left|E_{\vec{h}_1 + \vec{k}}\right|$. Then

$$\varepsilon \left((\left|E_{-\vec{h}_3 + \vec{k}}\right|^2 - 1) \left| \left|E_{\vec{k}}\right|, \left|E_{\vec{h}_1 + \vec{k}}\right| \right. \right) = \int_{R_1 = 0}^{\infty} (R_1^2 - 1) P(R_1 \left| \left|E_{\vec{k}}\right|, \left|E_{\vec{h}_1 + \vec{k}}\right| \right.) dR_1$$

$$(6.1)$$

in which $P(R_1 \left| \left|E_{\vec{k}}\right|, \left|E_{\vec{h}_1 + \vec{k}}\right| \right.)$ is given by (5.4). The integration is carried out in the same way as (5.2) utilizing (2.12)-(2.16)

(Chapter III) and, after a lengthy calculation, one finally obtains

$$\mathcal{E}\left(\left(|E_{-\vec{h}_3+\vec{k}}|^2-1\right)\Big|\,|E_{\vec{k}}|,|E_{\vec{h}_1+\vec{k}}|\right) \approx \frac{1}{N}\left\{\left(|E_{\vec{k}}|^2-1\right)\left(|E_3|^2-1\right)+\right.$$

$$\left.\left(|E_{\vec{h}_1+\vec{k}}|^2-1\right)\left(|E_2|^2-1\right)\right\}+\frac{2}{N^{3/2}}\,|E_1E_2E_3|\left(|E_{\vec{k}}|^2-1\right)\times$$

$$\left(|E_{\vec{h}_1+\vec{k}}|^2-1\right)\cos(\phi_1+\phi_2+\phi_3),\qquad\qquad(6.2)$$

in which the notation of (1.6) is used. Referring to (11.2) and (11.3) of Chapter III, Eq. (6.2) may be written in the following form,

$$\mathcal{E}\left(\left(|E_{-\vec{h}_3+\vec{k}}|^2-1\right)\Big|\,|E_{\vec{k}}|,|E_{\vec{h}_1+\vec{k}}|\right)\approx\mathcal{E}\left(\left(|E_{-\vec{h}_3+\vec{k}}|^2-1\right)\Big|\,|E_{\vec{k}}|\right)+$$

$$\mathcal{E}\left(\left(|E_{-\vec{h}_3+\vec{k}}|^2-1\right)\Big|\,|E_{\vec{h}_1+\vec{k}}|\right)+\frac{2}{N^{3/2}}\,|E_1E_2E_3|\left(|E_{\vec{k}}|^2-1\right)\times$$

$$\left(|E_{\vec{h}_1+\vec{k}}|^2-1\right)\cos(\Phi_1+\phi_2+\phi_3),\qquad\qquad(6.3)$$

the chief result of the present chapter. It should perhaps be emphasized that the left side of (6.3) is the conditional average of $|E_{-\vec{h}_3+\vec{k}}|^2-1$ as $\vec{k}$ ranges over those vectors in reciprocal space for which $|E_{\vec{k}}|$ and $|E_{\vec{h}_1+\vec{k}}|$ have fixed, specified values and is thus doubly conditioned. The first term on the right side is the conditional average of $|E_{-\vec{h}_3+\vec{k}}|^2-1$ as $\vec{k}$ ranges over those vectors in reciprocal space such that $|E_{\vec{k}}|$ has a specified value while the second term on the right is the conditional average of

$|E_{-\vec{h}_3+\vec{k}}|^2-1$ as $\vec{k}$ ranges over the reciprocal vectors such that $|E_{\vec{h}_1+\vec{k}}|$ has a specified value. Hence the set of vectors contributing to the average on the left is the set-theoretic intersection (or the set of common vectors) of the set consisting of the contributors to the first average on the right and that consisting of the contributors to the second average on the right.

7. A Generalization of (4.28) of Chapter II

Employing (4.25) and the appropriate integral formulas it is straightforward, but tedious, to evaluate the expectation value of

$$\left(|E_{\vec{k}}|^p - \overline{|E|^p} \right) \left(|E_{\vec{h}_1+\vec{k}}|^q - \overline{|E|^q} \right) \left(|E_{-\vec{h}_3+\vec{k}}|^r - \overline{|E|^r} \right) ,$$

where

$$\overline{|E|^p} = \left\langle |E_{\vec{k}}|^p \right\rangle_{\vec{k}} \tag{7.1}$$

is the expectation value of $|E_{\vec{k}}|^p$, etc. In this way one readily derives the following generalization of Eq. (4.28) of Chapter II:

$$|E_1 E_2 E_3| \cos(\phi_1+\phi_2+\phi_3) \approx \frac{4N^{3/2}}{pqr\Gamma\left(\frac{p+2}{2}\right)\Gamma\left(\frac{q+2}{2}\right)\Gamma\left(\frac{r+2}{2}\right)} \times$$

$$\left\langle \left(|E_{\vec{k}}|^p - \overline{|E|^p} \right) \left(|E_{\vec{h}_1+\vec{k}}|^q - \overline{|E|^q} \right) \left(|E_{-\vec{h}_3+\vec{k}}|^r - \overline{|E|^r} \right) \right\rangle_{\vec{k}} +R$$

$$\tag{7.2}$$

where

$$R = -\frac{1}{4N^{1/2}}\left[(q-2)|E_1E_2|^2+(r-2)|E_2E_3|^2+(p-2)|E_3E_1|^2 - \right.$$

$$\left.(p+q)|E_1|^2-(q+r)|E_2|^2-(r+p)|E_3|^2+(p+q+r+2)\right], \qquad (7.3)$$

and Γ signifies the Gamma Function. Clearly the case $p=q=r=2$ of (7.2) and (7.3) reduces to (4.28) of Chapter II while the case $p=q=r=\frac{1}{2}$ leads to (4.31) and (4.32) of Chapter II.

PART B

TECHNIQUES OF IMPLEMENTATION

Even with a good understanding of the theoretical basis of
the direct methods of phase determination, the practising
crystallographer will find that the step from the theory to the
actual applications is neither small nor trivial. In fact the
path leading from theory to practice is, in general, not even
unique. For these reasons, it will be helpful to describe those
extensions of the theory and the various means of implementing
them which have been found by experience to be useful. The reader
may very well think of others. Part B, then, may be regarded as a
bridge crossing the gap between the mathematical basis and the
applications of the direct methods.

In Chapter V are described those techniques which have been
devised for calculating initial values for the cosine seminvariants and for improving these initially calculated values. It must
be emphasized at the outset that the weak link in the methods described in this book is the inability to calculate, *ab initio*,
these cosines with great accuracy. It is only the enormous overdeterminacy of the problem which permits one to derive accurate
values for individual phases from imperfectly calculated cosines.
Thus, an important direction for future research in these methods
is to find better methods for calculating accurately the values of
the cosine seminvariants.

Chapter VI is devoted to a description of methods found to
be useful in calculating the values of a basic set of phases
assuming as known the previously calculated values of the cosine
seminvariants. Here the chief problem is how to use large

185

numbers of cosine seminvariants, calculated with less than perfect accuracy, in order to evaluate a small number, typically several dozen, of individual phases.

Once the values of a sufficiently broad base of phases have been accurately determined, it is usually a straightforward matter to expand this set into the several hundred or so needed to calculate an initial E-map. How this is done by means of the tangent techniques is the subject matter of Chapter VII. This chapter also contains a rigorous derivation of these procedures.

CHAPTER V

THE CALCULATION OF THE

COSINE SEMINVARIANTS

1. The Space Group Dependent

Cosine Seminvariants

Generally speaking, the implementation of the several formulas for the cosine seminvariants $\cos \phi$, $\cos(\phi_1 + \phi_2)$, and, in the space group $P2_12_12_1$, $\cos 2\phi$, as given in Chapter II, and their generalizations and analogues in the space groups of higher symmetry, presents few if any serious obstacles. However a few remarks may prove helpful.

1.1. Cos ϕ. As a general rule the Σ_1 formulas and their generalizations (e.g. the formula for $\cos \phi_{2h,2k,2\ell}$ in $P2_12_12_1$, not given explicitly in Chapter II) have practical utility primarily in the case that the magnitudes of calculated cosines are so large as to imply that $\cos \phi = \pm 1$. In this case probability formulas are available (using the methods of Chapter III) which are often useful. For example (cf. Eq. (2.23) of Chapter II), in the space group $P2_12_12_1$ the probability that $\phi_{0\ 2k\ 2\ell}$ be zero is given by

$$P_+ \approx \frac{1}{2} + \frac{1}{2} \tanh \left\{ \frac{\sigma_3}{2\sigma_2^{3/2}} \left| E_{0\ 2k\ 2\ell} \right| \sum_h (-1)^{h+k} (\left| E_{hk\ell} \right|^2 - 1) \right\}$$

(1.1)

187

in which, as usual,

$$\sigma_n = \sum_{\mu=1}^{N} Z_\mu^{\ n}, \qquad\qquad (1.2)$$

N is the number of atoms in the unit cell, and Z_μ is the atomic number of the μ th atom. If all atoms are identical, then, clearly,

$$\frac{\sigma_3}{\sigma_2^{\ 3/2}} = \frac{1}{N^{1/2}} . \qquad\qquad (1.3)$$

1.1.1. Renormalization. It may happen occasionally that the components of many interatomic vectors (or their projections) are rational numbers with small denominators (at least approximately). In such cases, as inspection of the derivation of the Σ_1 formulas shows, these formulas are not even approximately valid. Such a situation is easy to detect, when it exists, because the average values of $|E|^2$ taken over certain well defined subsets of reciprocal vectors will differ significantly from unity. Before applying the Σ_1 formulas then, all that is necessary to do is to renormalize (i.e. re-scale) the $|E|^2$ values in the affected subsets in such a way as to restore their average values to unity. For further details and applications the reader is referred to the literature (e.g. Hauptman & Karle, 1959 and Hauptman, Karle & Karle, 1960).

1.2. $\text{Cos}(\phi_1+\phi_2)$. As with the Σ_1 formulas, the formulas for the cosine seminvariants $\cos(\phi_1+\phi_2)$ are useful mostly in the case that the magnitudes of the calculated values for these cosines are so large as to imply that $\cos(\phi_1+\phi_2) = \pm1$. An important exception occurs when two or more contributors to this formula are large but tend to cancel out, so that the formula itself gives no strong indication of the value of $\cos(\phi_1+\phi_2)$. Further study

of the invariants associated with these contributors and the
calculated values of their cosines may resolve the ambiguity.
Thus, if the space group is $P2_1$, inspection of Eq. (3.57) of
Chapter II suggests a further study of the pair of cosine
invariants:

$$\cos\left(\phi_{h_1\ 0\ \ell_1} + \phi_{\frac{1}{2}(-h_1-h_2),k,\frac{1}{2}(-\ell_1-\ell_2)} + \phi_{\frac{1}{2}(-h_1+h_2),\bar{k},\frac{1}{2}(-\ell_1+\ell_2)}\right),$$

$$(1.4)$$

$$\cos\left(\phi_{h_2\ 0\ \ell_2} + \phi_{\frac{1}{2}(-h_1-h_2),k,\frac{1}{2}(-\ell_1-\ell_2)} + \phi_{\frac{1}{2}(h_1-h_2),\bar{k},\frac{1}{2}(\ell_1-\ell_2)}\right),$$

$$(1.5)$$

where $\left|E_{\frac{1}{2}(h_1+h_2),k,\frac{1}{2}(\ell_1+\ell_2)}\right|$ and $\left|E_{\frac{1}{2}(h_1-h_2),k,\frac{1}{2}(\ell_1-\ell_2)}\right|$ are
assumed to be large so that the corresponding contributor to
$\cos(\phi_1+\phi_2)$, Eq. (3.57) of Chapter II, is large and positive if k
is even, large and negative if k is odd. In any case, the space
group dependent relations among the phases require that (1.4) and
(1.5) be either equal to each other or be negatives of each other.
If it is assumed that one of the major contributors to $\cos(\phi_1+\phi_2)$
has k even and another k odd, so that these two terms tend to
cancel out, then one pair of cosines, (1.4), (1.5), will be equal
to each other and another pair will be negatives of each other.
In short, either just one or, much less likely, precisely three of
these four cosines will be negative. Often the negative cosine(s)
can be identified by the cosine calculation to be described in the
next section and the ambiguity resolved. Even if only one
contributor to $\cos(\phi_1+\phi_2)$ dominates the sum, it is still useful to
consider the calculated values of the associated cosine invariants

(1.4) and (1.5) in order to confirm (or possibly question) the calculated value for $\cos(\phi_1+\phi_2)$. The method described here is related to the coincidence method of Grant, Howells, and Rogers (1957) but differs from the latter in its dependence on the calculated values of the cosine invariants associated with the dominant contributors in the formula for $\cos(\phi_1+\phi_2)$.

2. The Universal Cosine Invariants, $\cos(\phi_1+\phi_2+\phi_3)$

In Chapter I, §6, it was shown that accurate values of the cosine invariants, $\cos(\phi_1+\phi_2+\phi_3)$, lead unambiguously to the values of the individual phases, ϕ.

2.1. Modified triple product procedure. In order to calculate the values of the required cosine invariants, $\cos(\phi_1+\phi_2+\phi_3)$, one is led to study, in view of §4 of Chapter II and §7 of Chapter IV, the formula

$$\cos(\phi_1+\phi_2+\phi_3) \approx \frac{K\psi}{|E_1 E_2 E_3|} + \frac{R_3}{|E_1 E_2 E_3|} \tag{2.1}$$

where

$$\psi = \left\langle \left(|E_{\vec{\ell}}|^P - \overline{|E|^P} \right) \left(|E_{\vec{h}_1+\vec{\ell}}|^P - \overline{|E|^P} \right) \left(|E_{-\vec{h}_3+\vec{\ell}}|^P - \overline{|E|^P} \right) \right\rangle_{\vec{\ell}} , \tag{2.2}$$

$$\overline{|E|^P} = \left\langle |E_\ell|^P \right\rangle_\ell , \tag{2.3}$$

R_3 is a term which depends in a known way on the normalized structure factor magnitudes $|E_1|$, $|E_2|$, $|E_3|$, and, as usual,

$$\phi_i = \phi_{\vec{h}_i}, \quad |E_i| = |E_{\vec{h}_i}|, \quad i = 1,2,3, \tag{2.4}$$

$$\vec{h}_1 + \vec{h}_2 + \vec{h}_3 = 0. \tag{2.5}$$

In (2.3) the average is taken over all vectors $\vec{\ell}$ in reciprocal space while in (2.2) the average is taken over all reciprocal vectors $\vec{\ell}$ such that $|E_{\vec{\ell}}| \geq t$ where t may be zero but, in order to achieve significant savings in computation time without sacrificing accuracy, t is chosen to have some fixed specified value usually in the range 1 to 2. The value $p = \frac{1}{2}$ has been found in practice to be most useful and in this case R_3 is given by

$$R_3 = \frac{\sigma_3}{4\sigma_2^{3/2}} \left[\frac{3}{2} \left(|E_1 E_2|^2 + |E_2 E_3|^2 + |E_3 E_1|^2 \right) + \right.$$

$$\left. |E_1|^2 + |E_2|^2 + |E_3|^2 - \frac{7}{2} \right], \tag{2.6}$$

where, as usual,

$$\sigma_n = \sum_{j=1}^{N} z_j^{\,n} \, , \quad n = 2,3, \tag{2.7}$$

Z_j is the atomic number, and N is the number of atoms in the unit cell. The parameter K in (2.1) is a sliding scale factor which is expected to be a function of A,

$$A = \frac{2\sigma_3}{\sigma_2^{3/2}} |E_1 E_2 E_3| . \tag{2.8}$$

2.1.1. Determination of K. The distribution of the cosine invariants $\cos(\phi_1 + \phi_2 + \phi_3)$ (§9, Chapter III) is a function of A only and the scale factors K are chosen in such a way as to make the distribution of the calculated invariants agree as closely as possible with the theoretical distribution as given in §9 of

Chapter III. The theoretical distributions of the cosines for several values of A are illustrated in Fig. 1. Their most striking feature is that, for large A, most cosines are positive and the proportion of cosines whose value is approximately unity is large.

Although several procedures for calculating the sliding scale factor K have been devised, the one in most common use consists in arranging the computed averages ψ (2.2) in increasing order of A and grouping them into sets having 200 to 300 members in each set. Thus, in each set the values of A are essentially constant. The value of K in each such set is determined in such a way as to force the frequency distribution of calculated cosine values and the theoretical distrubution of the $\cos(\phi_1+\phi_2+\phi_3)$ to coincide. Once K is known as a function of A, the values of the cosine invariants are found from (2.1). Details of the actual procedure used in determining K for the two androstane derivatives, $C_{19}H_{30}O_2$ and $C_{19}H_{32}O_2$, space group $P2_1$, are described in Chapter VIII.

2.2. MDKS procedure. Let $t \geq 0$ be specified. If one averages both sides of Eq. (6.3) of Chapter IV over those reciprocal vectors $\vec{k}$ such that $|E_{\vec{k}}| \geq t$ and $|E_{\vec{h}_1+\vec{k}}| \geq t$, one finds

$$\frac{2}{N^{3/2}} |E_1 E_2 E_3| \cos(\phi_1+\phi_2+\phi_3) = \frac{D-S}{B} \qquad (2.9)$$

where

$$D = \left\langle \left(|E_{-\vec{h}_3+\vec{k}}|^2 - 1 \right) \Big| |E_{\vec{k}}| > t, \ |E_{\vec{h}_1+\vec{k}}| > t \right\rangle_{\vec{k}}, \qquad (2.10)$$

$$S = S_1 + S_2, \qquad (2.11)$$

$$S_1 = \left\langle \left(|E_{-\vec{h}_3+\vec{k}}|^2 - 1 \right) \Big| |E_{\vec{k}}| > t \right\rangle_{\vec{k}}, \qquad (2.12)$$

Fig. 1.

Conditional Probability Distribution
of $\cos(\phi_1 + \phi_2 + \phi_3)$, Given A

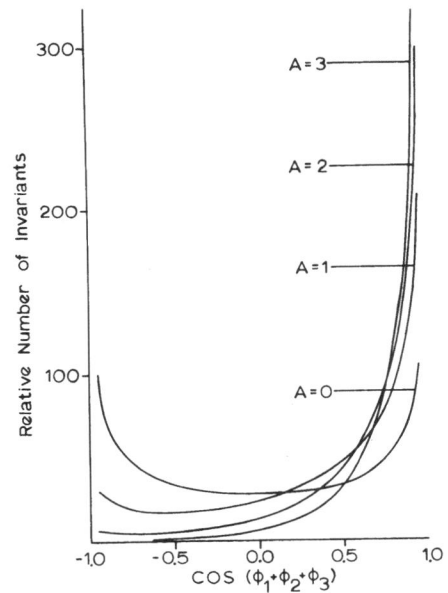

$$S_2 = \left\langle \left(\left| E_{-\vec{h}_3+\vec{k}} \right|^2 - 1 \right) \right| \left| E_{\vec{h}_1+\vec{k}} \right| > t \right\rangle_{\vec{k}}, \tag{2.13}$$

$$B = \left\langle \left(\left| E_{\vec{k}} \right|^2 - 1 \right) \left(\left| E_{\vec{h}_1+\vec{k}} \right|^2 - 1 \right) \right| \left| E_{\vec{k}} \right| > t, \left| E_{\vec{h}_1+\vec{k}} \right| > t \right\rangle_{\vec{k}}. \tag{2.14}$$

Just as it was necessary to introduce the scaling parameter K in (2.1) in order to take into account the approximate overlap in the Patterson function which is usually present, so now, for the same reason, one must introduce positive scaling parameters M and K into (2.9). Furthermore, owing to the symmetry of the left side of (2.9), one may permute the subscripts 1,2,3 of (2.10) to (2.14). After carrying out these operations it is finally found that

$$\cos(\phi_1 + \phi_2 + \phi_3) \approx M(D - KS) \tag{2.15}$$

where M and K are (positive) scaling parameters which remain to be determined. Now D is given by

$$D = \frac{\displaystyle\sum_{i=1}^{3} m_i D_i}{\displaystyle\sum_{i=1}^{3} m_i} \tag{2.16}$$

where

$$D_i = \left\langle \left(\left| E_{-\vec{h}_i+\vec{k}} \right|^2 - 1 \right) \right| \left| E_{\vec{k}} \right| \geq t, \left| E_{\vec{h}_j+\vec{k}} \right| \geq t \right\rangle_{\vec{k}}, \tag{2.17}$$

$i = 1,2,3$ when $j = 2,3,1$ respectively, m_i is the number of contributors to the average (2.17), and t is a fixed real number, usually greater than unity, e.g. $t = 1.5$. Each of the three averages (2.17) is extended over all vectors $\vec{k}$ in reciprocal space such that $|E_{\vec{k}}| \geq t$ and $|E_{h_j+k}| \geq t$.

Again S in (2.15) is given by

$$ S = \frac{\displaystyle\sum_{i=1}^{3} n_i S_i}{\displaystyle\sum_{i=1}^{3} n_i} \tag{2.18} $$

where

$$ S_i = 2 \left\langle \left(\left| E_{-\vec{h}_i+\vec{k}} \right|^2 - 1 \right) \middle| \; |E_{\vec{k}}| \geq t \right\rangle_{\vec{k}}, \tag{2.19} $$

$i = 1,2,3$, n_i is the number of contributors to the average (2.19), and t has already been defined to be an arbitrary fixed number exceeding unity. Each of the three averages (2.19) is extended over all vectors in reciprocal space such that $|E_{\vec{k}}| \geq t$.

The parameters K and M of (2.15) are determined as functions of A, (2.8), in such a way that the resulting empirical distribution of computed cosines (2.16) agrees with the theoretical conditional distribution of the cosine invariants, given A (§9, Chapter III). The method for doing this is described in the following sections.

2.2.1. Determination of K. For fixed A, the probability that

$$ \cos(\phi_1 + \phi_2 + \phi_3) < 0 \tag{2.20} $$

may be read from Table 1, Chapter III. (For phases obeying
centrosymmetric statistics, e.g. $\phi_{h\,0\,\ell}$ in $P2_1$, the probability,
P_-, that (2.20) hold is given by

$$P_- = 1/(1+e^A). \qquad (2.21)$$

It so happens that this value for P_- agrees rather well, for most
values of A, with P_- defined in the following sentence). In short,
for each fixed value of A, the fraction, P_-, of cosine invariants,
$\cos(\phi_1+\phi_2+\phi_3)$, which is negative is known. The computed values
of D and S from (2.16) and (2.18) are arranged in increasing order
of A and grouped into sets of 100 to 300 members. In each such
set the value of A is essentially constant. In the set defined
by A, K(>0) is determined in such a way that that fraction of the
quantities

$$D - KS \qquad (2.22)$$

which is negative agrees with the known theoretical fraction, P_-.
It is to be noted that in the three or four groups corresponding
to the very largest values of A, there may be only 20 to 50
members since in this range the values of A change rapidly.

2.2.2. Determination of M. Once K is determined as a
function of A, the value of M(>0) is readily determined for each
fixed A by observing that the value μ,

$$\mu = \left\langle \cos^2(\phi_1+\phi_2+\phi_3)\,\bigg|\,A \right\rangle = M^2 \left\langle (D-KS)^2 \right\rangle , \qquad (2.23)$$

where the average is taken over all members of the set corresponding to the fixed value of A, may be read from Table 2 of Chapter
III and is therefore known as a function of A. Since D and S are
known from (2.16) and (2.18), $\left\langle (D-KS)^2 \right\rangle$ is also known, for each

fixed A, and M may be found from

$$M = \left[\frac{\mu}{\left\langle (D-KS)^2 \right\rangle} \right]^{1/2} \tag{2.24}$$

as a function of A. Once K and M have been found as functions of A, the values of the cosines are calculated from (2.15), where D and S are given by (2.16) and (2.18). Details of the first applications of the MDKS formula are described in Chapter IX.

The first attempt to solve crystal structures via the systematic analysis of a large number of calculated cosine invariants was made in 1969 when the structure of estriol was determined. Since then, some twenty noncentrosymmetric structures have been solved by one variant or another of this basic theme. In the initial applications, the further analysis of the calculated cosines was carried out by the principle of least-squares described in §1 of Chapter VI. Later, methods for implementing calculated cosine invariants were found which appear to be at least as effective, and these are described in §§2-4 of Chapter VI.

3. Relationships among the Structure Invariants and Their Applications

The existence of relationships among the phases implies that certain identities among the structure invariants must also exist. These identities, when used in conjunction with the formulas for calculating cosine invariants given in §2, often permit increased accuracy in cosine calculation and, particularly when cosine values are ±1, often enable one to calculate cosines not definitely established by the methods of §2.

3.1. The space group independent relation; quadruples. In view of $\phi_{-\vec{h}} = - \phi_{\vec{h}}$, it follows that, if $\bar{\Phi}_1$, $\bar{\Phi}_2$, $\bar{\Phi}_3$, $\bar{\Phi}_4$ are defined by

$$\Phi_1 = \phi_{\vec{h}_1} + \phi_{\vec{h}_2} + \phi_{\vec{h}_3}, \qquad (3.1)$$

$$\Phi_2 = \phi_{-\vec{h}_1} + \phi_{-\vec{k}} + \phi_{\vec{h}_1+\vec{k}}, \qquad (3.2)$$

$$\Phi_3 = \phi_{-\vec{h}_3} + \phi_{\vec{k}} + \phi_{\vec{h}_3-\vec{k}}, \qquad (3.3)$$

$$\Phi_4 = \phi_{-\vec{h}_1-\vec{k}} + \phi_{-\vec{h}_2} + \phi_{-\vec{h}_3+\vec{k}}, \qquad (3.4)$$

and if $\vec{h}_1$, $\vec{h}_2$, $\vec{h}_3$ satisfy

$$\vec{h}_1 + \vec{h}_2 + \vec{h}_3 = 0, \qquad (3.5)$$

then Φ_1, Φ_2, Φ_3, and Φ_4 are structure invariants which satisfy

$$\Phi_1 + \Phi_2 + \Phi_3 + \Phi_4 \equiv 0 \qquad (3.6)$$

identically. Since $\cos \Phi_1, \ldots, \cos \Phi_4$ may be independently
calculated, at least approximately, by the methods of §2,
(3.6) suggests an obvious way to modify the initially calculated
values of the cosine invariants so as to satisfy (3.6) exactly.
It turns out that, in practice, the only cases in which it has
been possible to implement (3.6) are those in which three of
$\cos \Phi_1, \ldots, \cos \Phi_4$ are calculated to have values ±1 or the cosine
invariants, because of the space group symmetries, have the values
±1. Then (3.6) has been found to be useful in confirming initially
calculated values of the cosines and in evaluating certain cosines
not unambiguously calculated by the methods of §2. Several
examples taken from the valinomycin structure ($C_{54}N_6O_{18}H_{90}$), space

group $P2_1$, (Chapter XII) illustrate the use of the quadruple relationship:

Ex. 1. $\phi_{6\ 0\ 4} = 0$. The Σ_1 formula did not give a definitive indication for this phase and the calculation of $\cos(\phi_{3\ 0\ 2} + \phi_{3\ 0\ 2} + \phi_{\bar{6}\ 0\ \bar{4}})$ was likewise ambiguous. However the invariant $\phi_{3\ 0\ 2} + \phi_{3\ 0\ 2} + \phi_{\bar{6}\ 0\ \bar{4}}$ was a member of four quadruples each of which strongly indicated that the value of this invariant was zero. The results are shown in Table 1 in which each quadruple of reciprocal vectors is to be identified with (3.1)-(3.4). All cosines were calculated using the MDKS formula with the indicated value of t. Cosines calculated to be significantly greater than +1 were naturally assumed to have the value unity.

Ex. 2. $\phi_{3\ 0\ 2} + \phi_{\bar{4}\ 0\ \bar{4}} + \phi_{1\ 0\ 2} = 0$. This invariant was a member of the three quadruples shown in Table 2. Although the calculated cosine value of +1.55 indicates quite definitely that the true value of this cosine is +1, it is nevertheless desirable to have the confirmation given in Table 2. It should be noted that the second invariant of the first quadruple is already known to be zero from Ex. 1 so that the small value of its calculated cosine (+0.46) is irrelevant. Particularly noteworthy are the two negative cosines in the third quadruple.

Ex. 3. $\phi_{2\ 0\ 0} + \phi_{4\ 0\ 4} + \phi_{\bar{6}\ 0\ \bar{4}} = 0$. In view of Tables 1 and 2, Table 3, confirming the calculated value of this cosine invariant, is self-explanatory. The confirmation supplied by the quadruples relationship is particularly important here because of the relatively small A value of 0.843. Again the two negative cosines in the third quadruple are noteworthy.

Ex. 4. $\phi_{5\ 0\ 2} + \phi_{4\ 0\ 4} + \phi_{\bar{9}\ 0\ \bar{6}} = \pi$. Although this cosine invariant is definitely calculated to be -1, the confirmation

Table 1. The use of quadruples to determine
cos $(\phi_{3\,0\,2} + \phi_{3\,0\,2} + \phi_{\bar{6}\,0\,\bar{4}})$ = +1 for valinomycin

$\vec{h}_1$	$\vec{h}_2$	$\vec{h}_3$	A	Calculated Cosine	t
3 0 2	3 0 2	$\bar{6}$ 0 $\bar{4}$	3.738	+0.23	1.5
5 0 2	$\bar{3}$ 0 $\bar{2}$	$\bar{2}$ 0 0	3.888	+1.86	1.5
$\bar{5}$ 0 $\bar{2}$	6 0 4	$\bar{1}$ 0 $\bar{2}$	1.253	+1.77	1.5
$\bar{3}$ 0 $\bar{2}$	2 0 0	1 0 2	1.421	+1.97	1.5
3 0 2	3 0 2	$\bar{6}$ 0 $\bar{4}$	3.738	+0.46	1.3
$\overline{14}$ 0 15	$\bar{3}$ 0 $\bar{2}$	17 0 $\overline{13}$	1.270	+1.33	1.3
14 0 $\overline{15}$	6 0 4	$\overline{20}$ 0 11	0.532	+2.46	1.3
$\bar{3}$ 0 $\bar{2}$	$\overline{17}$ 0 13	20 0 $\overline{11}$	0.711	+3.15	1.3
3 0 2	3 0 2	$\bar{6}$ 0 $\bar{4}$	3.738	+0.46	1.3
2 0 0	$\bar{3}$ 0 $\bar{2}$	1 0 2	1.421	+2.28	1.3
$\bar{2}$ 0 0	6 0 4	$\bar{4}$ 0 $\bar{4}$	0.843	+1.20	1.3
$\bar{3}$ 0 $\bar{2}$	$\bar{1}$ 0 $\bar{2}$	4 0 4	1.322	+1.55	1.3
3 0 2	3 0 2	$\bar{6}$ 0 $\bar{4}$	3.738	+0.45	1.7
$\overline{19}$ 0 14	$\bar{3}$ 0 $\bar{2}$	22 0 $\overline{12}$	1.329	+1.83	1.7
19 0 $\overline{14}$	6 0 4	$\overline{25}$ 0 10	0.677	+1.95	1.7
$\bar{3}$ 0 $\bar{2}$	$\overline{22}$ 0 12	25 0 $\overline{10}$	1.116	+3.10	1.7

Table 2. The use of quadruples to confirm
$\cos (\phi_{3\,0\,2} + \phi_{\bar{4}\,0\,\bar{4}} + \phi_{1\,0\,2}) = +1$ for valinomycin

$\vec{h}_1$	$\vec{h}_2$	$\vec{h}_3$	A	Calculated Cosine	t
3 0 2	$\bar{4}$ 0 $\bar{4}$	1 0 2	1.322	+1.55	1.3
3 0 $\bar{2}$	$\bar{3}$ 0 $\bar{2}$	6 0 4	3.738	+0.46*	1.3
3 0 2	$\bar{1}$ 0 $\bar{2}$	$\bar{2}$ 0 0	1.421	+2.28	1.3
4 0 4	$\bar{6}$ 0 $\bar{4}$	2 0 0	0.843	+1.20	1.3
3 0 2	$\bar{4}$ 0 $\bar{4}$	1 0 2	1.322	+1.55	1.3
5 0 2	$\bar{3}$ 0 $\bar{2}$	$\bar{2}$ 0 0	3.889	+1.92	1.3
$\bar{5}$ 0 $\bar{2}$	$\bar{1}$ 0 $\bar{2}$	6 0 4	1.253	+1.01	1.3
4 0 4	2 0 0	$\bar{6}$ 0 $\bar{4}$	0.843	+1.20	1.3
3 0 2	$\bar{4}$ 0 $\bar{4}$	1 0 2	1.322	+1.55	1.3
5 0 2	4 0 4	$\bar{9}$ 0 $\bar{6}$	1.014	-1.82	1.3
$\bar{5}$ 0 $\bar{2}$	$\bar{1}$ 0 $\bar{2}$	6 0 4	1.253	+1.01	1.3
$\bar{3}$ 0 $\bar{2}$	9 0 6	$\bar{6}$ 0 $\bar{4}$	1.048	-0.90	1.3

* Ex. 1

Table 3. The use of quadruples to confirm

$$\cos\left(\phi_{2\,0\,0} + \phi_{4\,0\,4} + \phi_{\bar{6}\,0\,\bar{4}}\right) = +1 \text{ for valinomycin}$$

$\vec{h}_1$	$\vec{h}_2$	$\vec{h}_3$	A	Calculated Cosine	t
2 0 0	4 0 4	$\bar{6}$ 0 $\bar{4}$	0.843	+1.05	1.5
3 0 2	$\bar{2}$ 0 0	$\bar{1}$ 0 $\bar{2}$	1.421	+1.97	1.5
$\bar{3}$ 0 $\bar{2}$	6 0 4	$\bar{3}$ 0 $\bar{2}$	3.738	+0.23*	1.5
$\bar{4}$ 0 $\bar{4}$	1 0 2	3 0 2	1.322	+1.68	1.5
2 0 0	4 0 4	$\bar{6}$ 0 $\bar{4}$	0.843	+1.05	1.5
$\bar{3}$ 0 $\bar{2}$	$\bar{2}$ 0 0	5 0 2	3.888	+1.86	1.5
3 0 2	$\bar{4}$ 0 $\bar{4}$	1 0 2	1.322	+1.68	1.5
6 0 4	$\bar{5}$ 0 $\bar{2}$	$\bar{1}$ 0 $\bar{2}$	1.253	+1.77	1.5
2 0 0	4 0 4	$\bar{6}$ 0 $\bar{4}$	0.843	+1.20	1.3
$\bar{3}$ 0 $\bar{2}$	$\bar{2}$ 0 0	5 0 2	3.888	+1.92	1.3
3 0 2	6 0 4	$\bar{9}$ 0 $\bar{6}$	1.048	-0.90	1.3
$\bar{4}$ 0 $\bar{4}$	$\bar{5}$ 0 $\bar{2}$	9 0 6	1.014	-1.82	1.3
2 0 0	4 0 4	$\bar{6}$ 0 $\bar{4}$	0.843	+1.05	1.5
5 0 2	$\bar{4}$ 0 $\bar{4}$	$\bar{1}$ 0 2	1.262	+1.31	1.5
$\bar{5}$ 0 $\bar{2}$	6 0 4	$\bar{1}$ 0 $\bar{2}$	1.253	+1.77	1.5
$\bar{2}$ 0 0	1 0 $\bar{2}$	1 0 2	0.496	+3.37	1.5
2 0 0	4 0 4	$\bar{6}$ 0 $\bar{4}$	0.843	+1.05	1.5
24 0 $\bar{8}$	$\bar{4}$ 0 $\bar{4}$	$\overline{20}$ 0 12	0.656	+2.30	1.5
$\overline{24}$ 0 8	6 0 4	18 0 $\overline{12}$	1.311	+1.80	1.5
2 0 0	20 0 $\overline{12}$	$\overline{18}$ 0 12	0.704	+2.03	1.5

* Ex. 1

afforded by Table 4 is desirable because negative cosines are relatively rare and they play a key role in the phase determination process.

Ex. 5. $\phi_{3\ 0\ 2} + \phi_{6\ 0\ 4} + \phi_{\bar{9}\ 0\ \bar{6}} = \pi$. Calculated values for this cosine invariant are not so definitive (cos = -0.90, t = 1.3; cos = -0.65, t = 1.5; cos = -0.48, t = 1.7) but do indicate that the true value is -1. Thus the strong corroboration provided by Table 5 is particularly significant.

The examples given here were selected because they clearly illustrate the role played by the quadruples in confirming initially calculated values of some cosine invariants and in the evaluation of others not unambiguously calculated by the methods of §2. In reality most quadruples are not so well behaved. However, one usually finds some quadruples which permit the calculation of new cosine invariants and these in turn may be used with other quadruples to calculate still more cosines, etc. (cf. Tables 3 and 5). Of particular significance are those cosines which are definitely determined to be negative. While these are few in number, the identification of some of the negative cosines is of critical importance in initiating the phase determination process, particularly for very complex noncentrosymmetric structures. Further details of application will be found in Chapter X.

3.2. <u>Space group dependent relations; pairs.</u> Although the same methods are applicable to all space groups not in the triclinic system, it will be assumed here that the space group is $P2_1$. Suppose that $h_1 + h_2$ and $\ell_1 + \ell_2$ are both even so that, if ϕ_1 and ϕ_2 are defined by

$$\phi_1 = \phi_{h_1\ 0\ \ell_1}, \qquad\qquad (3.7)$$

Table 4. The use of quadruples to corroborate that
cos $(\phi_{5\ 0\ 2} + \phi_{4\ 0\ 4} + \phi_{\bar{9}\ 0\ \bar{6}})$ = -1 for valinomycin

$\vec{h}_1$	$\vec{h}_2$	$\vec{h}_3$	A	Calculated Cosine	t
5 0 2	4 0 4	$\bar{9}$ 0 $\bar{6}$	1.014	-1.82	1.3
3 0 2	$\bar{5}$ 0 $\bar{2}$	2 0 0	3.888	+1.92	1.3
$\bar{3}$ 0 $\bar{2}$	9 0 6	$\bar{6}$ 0 $\bar{4}$	1.048	-0.90	1.3
$\bar{4}$ 0 $\bar{4}$	$\bar{2}$ 0 0	6 0 4	0.843	+1.20	1.3
5 0 2	4 0 4	$\bar{9}$ 0 $\bar{6}$	1.014	-1.82	1.3
3 0 2	$\bar{4}$ 0 $\bar{4}$	1 0 2	1.322	+1.55	1.3
$\bar{3}$ 0 $\bar{2}$	9 0 6	$\bar{6}$ 0 $\bar{4}$	1.048	-0.90	1.3
$\bar{5}$ 0 $\bar{2}$	$\bar{1}$ 0 $\bar{2}$	6 0 4	1.253	+1.01	1.3
5 0 2	4 0 4	$\bar{9}$ 0 $\bar{6}$	1.014	-1.82	1.3
23 0 $\bar{3}$	$\bar{4}$ 0 $\bar{4}$	$\overline{19}$ 0 7	0.845	+1.44	1.3
$\overline{23}$ 0 3	9 0 6	14 0 $\bar{9}$	0.515	+0.80	1.3
$\bar{5}$ 0 $\bar{2}$	19 0 $\bar{7}$	$\overline{14}$ 0 9	1.009	-1.40	1.3

Table 5. The use of quadruples to determine

$$\cos (\phi_{3\ 0\ 2} + \phi_{6\ 0\ 4} + \phi_{\bar{9}\ 0\ \bar{6}}) = -1 \text{ for valinomycin}$$

$\vec{h}_1$	$\vec{h}_2$	$\vec{h}_3$	A	Calculated Cosine	t
3 0 2	6 0 4	$\bar{9}$ 0 $\bar{6}$	1.048	−0.48	1.7
5 0 2	$\bar{3}$ 0 $\bar{2}$	$\bar{2}$ 0 0	3.888	+2.25	1.7
$\bar{5}$ 0 $\bar{2}$	9 0 6	$\bar{4}$ 0 $\bar{4}$	1.014	−0.82*	1.7
$\bar{6}$ 0 $\bar{4}$	2 0 0	4 0 4	0.843	+1.85	1.7
3 0 2	6 0 4	$\bar{9}$ 0 $\bar{6}$	1.048	−0.65	1.5
5 0 2	$\bar{6}$ 0 $\bar{4}$	1 0 2	1.253	+1.77	1.5
$\bar{5}$ 0 $\bar{2}$	9 0 6	$\bar{4}$ 0 $\bar{4}$	1.014	−1.60*	1.5
$\bar{3}$ 0 $\bar{2}$	$\bar{1}$ 0 $\bar{2}$	4 0 4	1.322	+1.68	1.5
3 0 2	6 0 4	$\bar{9}$ 0 $\bar{6}$	1.048	−0.65	1.5
8 0 $\bar{5}$	$\bar{3}$ 0 $\bar{2}$	$\bar{5}$ 0 7	1.410	+1.89	1.5
$\bar{8}$ 0 5	$\bar{6}$ 0 $\bar{4}$	14 0 $\bar{1}$	0.633	+0.90	1.5
9 0 6	5 0 $\bar{7}$	$\overline{14}$ 0 1	0.182	−2.18	1.5
3 0 2	6 0 4	$\bar{9}$ 0 $\bar{6}$	1.048	−0.48	1.7
$\overline{13}$ 0 15	$\bar{6}$ 0 $\bar{4}$	19 0 $\overline{11}$	0.691	+1.37	1.7
13 0 $\overline{15}$	9 0 6	$\overline{22}$ 0 9	0.411	−1.32	1.7
$\bar{3}$ 0 $\bar{2}$	$\overline{19}$ 0 11	22 0 $\bar{9}$	0.883	+2.90	1.7

* Ex. 4

$$\phi_2 = \phi_{h_2\,0\,\ell_2},\tag{3.8}$$

then $\cos(\phi_1+\phi_2) = \pm 1$ is a cosine seminvariant. Construct the structure invariants

$$\bar{\Phi}_1 = \phi_{\frac{1}{2}(-h_1-h_2),\bar{k},\frac{1}{2}(-\ell_1-\ell_2)} + \phi_{\frac{1}{2}(-h_1+h_2),k,\frac{1}{2}(-\ell_1+\ell_2)} + \phi_{h_1\,0\,\ell_1},$$

$$\tag{3.9}$$

$$\bar{\Phi}_2 = \phi_{\frac{1}{2}(-h_1-h_2),\bar{k},\frac{1}{2}(-\ell_1-\ell_2)} + \phi_{\frac{1}{2}(h_1-h_2),k,\frac{1}{2}(\ell_1-\ell_2)} + \phi_{h_2\,0\,\ell_2}.$$

$$\tag{3.10}$$

Then, because of the space group dependent relationships among the phases and the fact that $\phi_{h\,0\,\ell} = 0$ or π, it follows readily that

$$\cos\bar{\Phi}_1 = \pm\cos\bar{\Phi}_2.\tag{3.11}$$

If A values are large then initial values for both cosine invariants $\cos\bar{\Phi}_1$ and $\cos\bar{\Phi}_2$ may be calculated by §2 and the correct relation (3.11) may often be established. In this way one may be led to the value of $\cos(\phi_1+\phi_2)$ and improved values for $\cos\bar{\Phi}_1$ and $\cos\bar{\Phi}_2$. Occasionally, in addition to the pair (3.9), (3.10), there is another **similar** pair, with k replaced by k', thus defining $\bar{\Phi}_1'$ and $\bar{\Phi}_2'$, having large A values and k+k' odd (cf. §1.2). In such a case the space group symmetries require that precisely one (or, rarely, three) of the four cosines $\cos\bar{\Phi}_1$, $\cos\bar{\Phi}_2$, $\cos\bar{\Phi}_1'$, $\cos\bar{\Phi}_2'$ be negative. Calculated values of these four cosines will often reveal the negative one(s) and then the space group dependent relations among the cosines will yield improved cosine values.

Details of the use of the "pair" relation for $\cos(\phi_1+\phi_2)$ are described in Chapter XI.

   3.3. Space group dependent relations; Σ_1 cosines. Just as the analysis of the individual contributors to $\cos(\phi_1+\phi_2)$ and their associated cosine invariants often serves as an aid in evaluating $\cos(\phi_1+\phi_2)$, so a study of the individual contributors to the cosine seminvariant $\cos\phi$ occasionally permits the evaluation of $\cos\phi$, particularly in the case that two or more major contributors tend to cancel so that the Σ_1 formula is not useful. Thus, if the space group is $P2_1$, then, under conditions described in Chapter II (Ex. II .1),

$$E_{2h\ 0\ 2\ell} = N^{1/2} \left\langle (-1)^k \left(|E_{hk\ell}|^2 - 1 \right) \right\rangle_k. \qquad (3.12)$$

Clearly, owing to the space group symmetry, if

$$\overline{\Phi} = \phi_{\overline{h}k\overline{\ell}} + \phi_{\overline{h}\overline{k}\overline{\ell}} + \phi_{2h\ 0\ 2\ell} \qquad (3.13)$$

$$\overline{\Phi}' = \phi_{\overline{h}k'\overline{\ell}} + \phi_{\overline{h}\overline{k}'\overline{\ell}} + \phi_{2h\ 0\ 2\ell} , \qquad (3.14)$$

then

$$\cos \overline{\Phi} = \cos \overline{\Phi}' \text{ if k+k' is even,} \qquad (3.15)$$

$$\cos \overline{\Phi} = - \cos \overline{\Phi}' \text{ if k+k' is odd.} \qquad (3.16)$$

Thus the value of a single cosine, $\cos \overline{\Phi}$, corresponding to a major contributor to (3.12), determines the value of $\cos \phi_{2h\ 0\ 2\ell} = \cos\phi$. If two of the major contributors to (3.12) cancel, then the corresponding cosines, $\cos \overline{\Phi}$ and $\cos \overline{\Phi}'$, satisfy (3.16). In either case, calculated values of the appropriate cosine invariants,

(3.13) and (3.14), often facilitate the determination of $\cos \phi$ and, on occasion, may even cast doubt on the Σ_1 calculation itself.

4.   Space Group $P2_12_12_1$; "Mixed" Triples

The methods given in this chapter for calculating the cosine invariants, $\cos(\phi_1 + \phi_2 + \phi_3)$, are strongly dependent on the conditional distribution of the cosine invariants, given A (§9 of Chapter III).   Therefore the fact that, in $P2_12_12_1$ and other selected space groups, there exist classes of cosine invariants which obey different distribution laws has important implications for cosine calculations.   Thus, in $P2_12_12_1$, if two of the phases ϕ_1, ϕ_2, ϕ_3 are two-dimensional and the third is three-dimensional, the conditional probability distribution of $\cos(\phi_1 + \phi_2 + \phi_3)$, given A, has the same form as Eq. (9.3) of Chapter III except that A of (9.3) is now replaced by 2A.   Again, if each of ϕ_1, ϕ_2, ϕ_3 is two-dimensional of the form

$$\left.\begin{array}{l} \phi_1 = \phi_{0\,k_1\,\ell_1} \\[2ex] \phi_2 = \phi_{h_2\,0\,\ell_2} \\[2ex] \phi_3 = \phi_{h_3\,k_3\,0} \end{array}\right\} \tag{4.1}$$

and if, because of the space group symmetries, $\cos(\phi_1 + \phi_2 + \phi_3) = \pm 1$, then the conditional probability, given A, that $\cos(\phi_1 + \phi_2 + \phi_3) = -1$ is given by

$$P_- = \frac{1}{1 + e^{2A}} , \tag{4.2}$$

rather than the anticipated (2.21) which is appropriate for the usual "centrosymmetric" cosine invariants. Clearly then, in order to calculate these "mixed" cosine invariants by the methods of this chapter, it is necessary to replace A by 2A where appropriate.

CHAPTER VI

USING CALCULATED COSINES TO DETERMINE
VALUES OF INITIAL PHASES

Once the values of a sufficiently large set of phases have
been accurately found, it is possible to determine the values of
additional phases by means of the tangent techniques to be
described in the next chapter. The present chapter is devoted to
a description of methods which have been found to be useful for
finding the values of the initial phases via the calculated values
of the cosine invariants.

1. The Least-Squares Analysis of
the Cosine Invariants

The iterative process whereby the values of a basic set of
phases, which are presumed to be known, lead, via calculated
values of cosine invariants, to the values of unknown phases, $\phi_{\vec{h}}$,
has been described in §6 of Chapter I. If two or more cosine
invariants,

$$\cos(\phi_{\vec{h}} + \phi_{\vec{k}} + \phi_{-\vec{h}-\vec{k}}),\qquad\qquad (1.1)$$

(where the $\phi_{\vec{k}}$ and $\phi_{-\vec{h}-\vec{k}}$ are presumed known) which involve a common
vector $\vec{h}$, have been computed, they will not generally yield the
same value for the unknown phase $\phi_{\vec{h}}$ for two reasons. First, there
are always some experimental errors in the measurement of the

211

normalized structure factor magnitudes. In addition, the methods
of Chapter V for calculating cosines are only approximate because
of errors of finite sampling in (2.2), (2.16), (2.18) (Chapter V)
and because the approximate overlap of Patterson peaks in real
structures introduces additional errors. Consequently, the best
value for $\phi_{\vec{h}}$ may be found, in accordance with the principle of
least squares, by minimizing the function

$$\bar{\Phi} = \frac{\sum\limits_{\vec{k}} w_{\vec{k}}[\cos(\phi_{\vec{h}} + \phi_{\vec{k}} + \phi_{-\vec{h}-\vec{k}}) - c_{\vec{k}}]^2}{\sum\limits_{\vec{k}} w_{\vec{k}}}, \qquad (1.2)$$

where the values $c_{\vec{k}}$ of several structure invariants,

$$\cos(\phi_{\vec{h}} + \phi_{\vec{k}} + \phi_{-\vec{h}-\vec{k}}) = c_{\vec{k}}, \qquad (1.3)$$

have been found from §2, Chapter V, and each invariant has been
assigned a weight

$$w_{\vec{k}} = |E_{\vec{h}} E_{\vec{k}} E_{\vec{h}+\vec{k}}| \sqrt{n} \qquad (1.4)$$

where n is the number of contributors in (2.2) or (2.17) (Chapter
V). The minima in the function $\bar{\Phi}$ may be located most readily by
varying $\phi_{\vec{h}}$ from 0 to 2π in small increments (e.g. 0.01 radians)
and evaluating the function at each point. The first time a cosine
invariant is used which differs significantly from ±1, there will
be two equal minima in the function $\bar{\Phi}$ and either of the corresponding values of $\phi_{\vec{h}}$ may be chosen, thus specifying the enantiomorph.
If the calculated values of the cosine invariants are internally
consistent and if the enantiomorph determining phase is involved
directly or indirectly in all subsequent invariants whose values
are not ±1, the function $\bar{\Phi}$ will have just one minimum for each

additional phase. In practice, however, because the invariants
are not computed with perfect accuracy and are not perfectly
consistent with each other, there will be cases where two relative
minima occur. If these minima are unequal, the smaller one is
chosen. If, on the other hand, they are nearly equal, it is
necessary to carry both phases through the remainder of the
computations unless one of the phases can, at some intermediate
stage, be shown to be less likely to be correct. Thus, even with
the use of the structure invariants, it may not be possible to
eliminate all the phase ambiguities. Criteria for judging the
accuracy of phases calculated by means of (1.2) and for
distinguishing among the several ambiguities (if they exist)
readily suggest themselves. Thus the individual residual for the
vector $\vec{h}$,

$$R_{\vec{h}} = \bar{\Phi}^{1/2}_{\text{min}} \, , \tag{1.5}$$

is simply the square root of the minimum value of $\bar{\Phi}$. The cycle
residual,

$$R_{\text{cycle}} = \left[\frac{\sum\limits_{\vec{h}} R_{\vec{h}}^2 \left(\sum\limits_{\vec{k}} A_{\vec{k}} \right)}{\sum\limits_{\vec{h}} \left(\sum\limits_{\vec{k}} A_{\vec{k}} \right)} \right]^{1/2} , \tag{1.6}$$

is a weighted average of the individual residuals for all vectors
$\vec{h}$ for which phases were calculated during a cycle. Another measure
of the accuracy of phases calculated during a single cycle is the
centrosymmetric residual,

$$R_{\text{centro}} = \left[\frac{\sum\limits_{\vec{h}} d_{\vec{h}}^2 \left(\sum\limits_{\vec{k}} A_{\vec{k}} \right)}{\sum\limits_{\vec{h}} \left(\sum\limits_{\vec{k}} A_{\vec{k}} \right)} \right]^{1/2} , \tag{1.7}$$

where $d_{\vec{h}}$ is the deviation of the calculated value of a phase of a purely real or purely imaginary structure factor (e.g. $E_{h\,0\,\ell}$ in $P2_1$) from the nearer of its two allowed values, or, in the case of phases determined by fixation of the origin or by use of the Σ_1 formula, the deviation from the known value. These residuals will presumably be small when the use of the structure invariants leads to correct phases. Experience has shown that cycle residuals of the order of 0.18 and centrosymmetric residuals of the order of 0.3 radians are acceptable.

It occasionally happens that the true value of a cosine invariant, used for enantiomorph selection because its calculated value is not ±1, is indeed ±1 so that the enantiomorph is not in fact selected. If, during the same or a later cycle, only one phase has substantially different values for the two enantiomorphs, the enantiomorph would then be unambiguously selected, unwittingly, at that time. Precisely this situation occurred in two applications, once with estriol, the initial application of this technique, and again with epiandrosterone (Chapter VIII). However, even if, unsuspected by the investigator, enantiomorph selection were to occur during a cylce in which two or more phases, one of which alone would suffice for enantiomorph selection, are simultaneously determined, no damage would be done. In such a situation, all combinations of values of enantiomorph dependent phases would yield the same least-squares residuals, one combination would correspond in a consistent way to one choice of enantiomorph, another to the other choice, and all possibilities (including the inconsistent combinations, if any) would normally be considered.

Four crystal structures have thus far been determined by the least-squares analysis of calculated cosine invariants as described in this section. These are:

$$\text{Estriol}(C_{18}H_{24}O_3),\ P2_1 \text{ two molecules per}$$
$$\text{asymmetric unit;}$$

Epiandrosterone($C_{19}H_{30}O_2$), $P2_1$;

5β-androstane-3α,17β-diol($C_{19}H_{32}O_2$), $P2_1$;

Silandrone($C_{22}H_{36}O_2Si$), $P2_12_12_1$.

In §§2-4, other techniques for implementing the calculated cosine invariants are given which also appear, in the light of their use in some fifteen structure determinations, to merit further application.

2.   The Use of Two-Dimensional Structure Invariants

In most of the non-centrosymmetric space groups there exist certain two-dimensional phases the values of which, because of the space group symmetries, must be 0, π, or $\pm\pi/2$. Structure invariants constructed with these phases must often, of necessity, have the values 0 or π and the corresponding cosine invariants must there-fore have the values ± 1. It is natural to expect that these cosine invariants may, in general, be calculated with greater reliability than the general cosine invariants whose values may lie anywhere between -1 and $+1$. If the origin and enantiomorph are fixed, at least in part, through the specification of two-dimensional phases, then the values (± 1) of the two-dimensional cosine invariants will lead to unique values for additional two-dimensional phases. In some space groups, e.g. $P2_12_12_1$, the values of a sufficiently large base of two-dimensional phases are sufficient, in them-selves, to yield the values of the three-dimensional phases as well via the use of the tangent techniques (Chapter VII). In other space groups, e.g. $P2_1$, the values of a sufficiently large base of two-dimensional phases, while not in themselves enough to lead to values for three-dimensional phases, when used in

conjunction with the calculated values of special cosine
invariants, often suffice, by means of the process of strong
enantiomorph discrimination (§3), to determine a base of three-
dimensional phases to serve as input to the tangent techniques.
Thus the two-dimensional structure invariants often play the
extremely important role of initiating the phase determining
process more reliably than would otherwise be possible. For
further details of their use the reader is referred to Chapters
IX, XI, XII, and XIII.

3.   Strong Enantiomorph Discrimination

3.1. Introduction. If a structure invariant L has the
value s for a crystal structure S then the value of the same
structure invariant for the enantiomorphous structure S' is -s.
Thus if s = 0 or π then L has the same value (0 or π) for both
enantiomorphs and is not suitable for enantiomorph discrimination.
If, on the other hand, s ≠ 0 or π then, since the magnitude of s
(or, equivalently, cos s) is determined by the known magnitudes
of the structure factors, the enantiomorph may be chosen by specifying arbitrarily the sign of s. This observation forms the basis
for enantiomorph selection in crystal structure determination by
direct methods.

In practice one attempts to employ a structure invariant L
whose value is approximately equal to ±π/2 in order to insure
strong enantiomorph discrimination. Even so, for complex
structures, a single structure invariant constitutes a very narrow
foundation on which to base a technique for phase determination.
For this reason one has often found, in practice, that enantiomorph specification has not been decisively made and has led to
an initial E map which contains fragments of both enantiomorphs
and presents great difficulties of interpretation.   In the present

section a technique for enantiomorph specification is described
which employs, instead of just a single invariant, a class of
several structure invariants, each member of which equals $\pm\pi/2$
approximately. In this way the base of the phase determination
process is considerably broadened with resultant unambiguous
enantiomorph discrimination.

For definiteness only the space group $P2_1$ is considered here
but the same technique may be used in most of the other noncentrosymmetric space groups. It is assumed finally that the values of
the cosine invariants, $\cos(\phi_{\vec{h}} + \phi_{\vec{k}} + \phi_{-\vec{h}-\vec{k}})$, for large A, have been
calculated from known magnitudes $|E|$, at least approximately, by
the methods of Chapter V.

3.2. Space group $P2_1$. If the integer k is fixed, then, as
h and ℓ range over those integers for which the $|E_{hk\ell}|$'s are
large, the corresponding phases $\phi_{hk\ell}$ will, in general (i.e.
provided that $|E_{0\ 2k\ 0}|$ is not too large), take on values
distributed over the range 0 to 2π. If the origin is fixed so
that $\phi_{0\ 2k\ 0} = 0$, then some of the $\phi_{hk\ell}$'s are expected to have
values in the neighborhood of 0 or π and others in the neighborhood of $\pm\pi/2$. In view of the relationship

$$\phi_{\overline{hk\ell}} = \phi_{h\overline{k}\ell}, \text{ if k is even,} \qquad (3.1)$$

or

$$\phi_{\overline{hk\ell}} = \pi + \phi_{h\overline{k}\ell}, \text{ if k is odd,} \qquad (3.2)$$

it follows that, if k is even

$$\cos(\phi_{h\overline{k}\ell} + \phi_{\overline{hk\ell}} + \phi_{0\ 2k\ 0}) = \cos 2\phi_{h\overline{k}\ell} \qquad (3.3)$$

$$\approx \pm 1 \qquad\qquad\qquad (3.4)$$

according as

$$\phi_{h\bar{k}\ell} \approx 0 \text{ or } \pi \qquad\qquad (3.5)$$

or

$$\phi_{h\bar{k}\ell} \approx \pm\pi/2, \qquad\qquad (3.6)$$

respectively. If, on the other hand, k is odd, then

$$\cos(\phi_{h\bar{k}\ell} + \phi_{\bar{h}\bar{k}\bar{\ell}} + \phi_{0\ 2k\ 0}) = -\cos 2\phi_{h\bar{k}\ell} \qquad (3.7)$$

$$\approx \mp 1 \qquad\qquad\qquad (3.8)$$

according as

$$\phi_{h\bar{k}\ell} \approx 0 \text{ or } \pi \qquad\qquad (3.9)$$

or

$$\phi_{h\bar{k}\ell} \approx \pm\pi/2 \qquad\qquad (3.10)$$

respectively. In other words the special cosine invariants (3.3),
(3.7), by doubling the phase $\phi_{h\bar{k}\ell}$, exaggerate the deviations in
the values of the $\phi_{hk\ell}$. Thus, differences of $\pi/2$ among the $\phi_{hk\ell}$
(permitting strong enantiomorph discrimination) imply, via (3.3)–
(3.10), that the values of some cosine invariants are –1 so that
the identification of these cosines by the methods of Chapter V
is feasible. This observation is the motivation for the method
of strong enantiomorph discrimination which is described here.

(It should be noted that $|E_{0\ 2k\ 0}|$ must not be too large; otherwise, as the probability distribution of cosine invariants shows, the values of most of the cosines (3.3) or (3.7) will be approximately unity, and the possibility of constructing two "orthogonal" classes of phases as described in the sequel is greatly reduced).

The basic idea is to find an integer k and two ("orthogonal") classes, I and II, of phases $\phi_{hk\ell}$ having the properties:

1. $|E_{0\ 2k\ 0}|$ is moderately large (say ≈ 2);

2. every $|E_{hk\ell}|$ corresponding to any phase $\phi_{hk\ell}$ in Class I or II is large (say > 1);

3. any two phases in Class I differ from each other by 0 or π, approximately;

4. any two phases in Class II differ from each other by 0 or π, approximately;

5. any phase in Class I differs from any phase in Class II by $\pi/2$ approximately.

It will be seen in the sequel that, while it is desirable to have Property 1, it is not essential.

In order to identify two classes, I and II, with the stated properties, determine first an integer k such that $|E_{0\ 2k\ 0}|$ is large. Place tentatively in Class I those phases $\phi_{hk\ell}$ for which the $|E_{hk\ell}|$ are large (so that Property 2 is satisfied) and the calculated values of the cosine invariants $\cos(\phi_{h\bar{k}\ell} + \phi_{\overline{hk\ell}} + \phi_{0\ 2k\ 0})$ are large:

$$\cos(\phi_{h\bar{k}\ell} + \phi_{\overline{hk\ell}} + \phi_{0\ 2k\ 0}) \approx +1. \qquad (3.11)$$

Then

$$2\phi_{hk\ell} \approx \phi_{0\ 2k\ 0} \text{ or } \pi + \phi_{0\ 2k\ 0} \qquad (3.12)$$

according as k is even or odd.   Hence

$$\phi_{hk\ell} \approx \frac{1}{2} \phi_{0\ 2k\ 0} \text{ or } \frac{1}{2} \phi_{0\ 2k\ 0} + \pi \text{ if k is even,} \qquad (3.13)$$

$$\phi_{hk\ell} \approx \frac{1}{2} \phi_{0\ 2k\ 0} \pm \pi/2 \text{ if k is odd,} \qquad (3.14)$$

and in either case Property 3 is presumably satisfied for these
members of Class I ("presumably" because calculated cosines (3.11)
are subject to error).   In order to insure that Property 3 hold
with at most minor exception for the elements of Class I, retain
only those phases $\phi_{hk\ell}$ which "interact" strongly with one or
several others, $\phi_{h'k\ell'}$, i.e. those $\phi_{hk\ell}$ such that at least one of

$$A = \frac{2}{N^{1/2}} \left| E_{hk\ell} \ E_{h'k\ell'} \ E_{h\pm h',0,\ell\pm\ell'} \right| \qquad (3.15)$$

is large and the calculated value of at least one of

$$\cos(\phi_{\overline{hk\ell}} + \phi_{\overline{h'k\ell'}} + \phi_{h+h',0,\ell+\ell'}), \qquad (3.16)$$

$$\cos(\phi_{\overline{hk\ell}} + \phi_{h'\overline{k\ell'}} + \phi_{h-h',0,\ell-\ell'}), \qquad (3.17)$$

is also large, i.e. $\approx$ +1.   It is clear from (3.16) and (3.17) that
in this case, since $\phi_{h\pm h',0,\ell\pm\ell'} = 0$ or π, only those phases will
be retained in Class I for which Property 3 must almost surely
hold.

Next, Class II consists tentatively of those phases $\phi_{hk\ell}$
for which the $\left| E_{hk\ell} \right|$ are large (so that Property 2 is satisfied)
and the calculated values of the cosine invariants $\cos(\phi_{hk\ell} + \phi_{\overline{hk\ell}} + \phi_{0\ 2k\ 0})$ are small:

$$\cos(\phi_{hk\bar{\ell}} + \phi_{\overline{hk\ell}} + \phi_{0\ 2k\ 0}) \approx -1. \tag{3.18}$$

Then

$$2\phi_{hk\ell} \approx \pi + \phi_{0\ 2k\ 0} \text{ or } \phi_{0\ 2k\ 0} \tag{3.19}$$

according as k is even or odd.   Hence

$$\phi_{hk\ell} \approx \frac{1}{2}\ \phi_{0\ 2k\ 0} \pm \frac{\pi}{2} \text{ if k is even,} \tag{3.20}$$

$$\phi_{hk\ell} \approx \frac{1}{2}\ \phi_{0\ 2k\ 0} \text{ or } \frac{1}{2}\ \phi_{0\ 2k\ 0} + \pi \text{ if k is odd,} \tag{3.21}$$

and, in either case, Property 4 is presumably satisfied.   As before, in order to insure that Property 4 hold with negligible exception, retain only those phases $\phi_{hk\ell}$ in Class II which interact strongly with one another in the sense defined earlier.

Next, if k is even, compare (3.13) with (3.20) in order to verify that Property 5 holds.   If k is odd, compare (3.14) and (3.21).

In order to secure Property 5 with at most minor exceptions, only those phases are retained such that no phase in Class II interacts strongly with any phase in Class I, i.e. if $\phi_{h_1 k \ell_1}$ is in Class I and $\phi_{h_2 k \ell_2}$ is in Class II then either

$$A = \frac{2}{N^{1/2}}\ |E_{h_1 k \ell_1}\ E_{h_2 k \ell_2}\ E_{h_1 \pm h_2,0,\ell_1 \pm \ell_2}| \approx 0 \tag{3.22}$$

or, if A of (3.22) is large, then the following calculated cosine invariants satisfy

$$\cos(\phi_{h_1 k \ell_1} + \phi_{h_2 \bar{k} \ell_2} + \phi_{-h_1 -h_2,0,-\ell_1 -\ell_2}) \approx 0 \tag{3.23}$$

or

$$\cos(\phi_{h_1 k \ell_1} + \phi_{\bar{h}_2 \bar{k} \bar{\ell}_2} + \phi_{-h_1 + h_2, 0, -\ell_1 + \ell_2}) \approx 0. \qquad (3.24)$$

Clearly, (3.22)–(3.24) are consistent with the previously derived

$$\phi_{h_1 k \ell_1} + \phi_{h_2 \bar{k} \ell_2} + \phi_{-h_1 - h_2, 0, -\ell_1 - \ell_2} \approx \pm\pi/2 \qquad (3.25)$$

and

$$\phi_{h_1 k \ell_1} + \phi_{\bar{h}_2 \bar{k} \bar{\ell}_2} + \phi_{-h_1 + h_2, 0, -\ell_1 + \ell_2} \approx \pm\pi/2 \qquad (3.26)$$

for all phases $\phi_{h_1 k \ell_1}$ in Class I and for all phases $\phi_{h_2 k \ell_2}$ in Class II, i.e. Property 5.

Since each structure invariant (3.25) and (3.26) is approximately $\pm\pi/2$, it is clear that the class consisting of all of them constitutes a collection of structure invariants each approximately $\pm\pi/2$ and therefore appropriate for decisive enantiomorph discrimination. Clearly too, owing to the built-in redundancy in defining Classes I and II, a modified procedure suggests itself. Thus Property I, while highly desirable, may possibly be dispensed with and only the notions of strongly and weakly interacting classes of phases employed. It should, however, be emphasized that in the three applications actually made thus far, Property 1 has in fact been satisfied. In any event the method is seen to be strongly dependent on the ability to calculate cosine invariants, at least approximately. The application of this method to the solution of the valinomycin structure is described in Chapter XII.

4. Space Group $P2_12_12_1$

4.1. <u>The double-angle (or squared-tangent) formula</u>. The
double-angle (or squared-tangent) formula of §5, Chapter II, valid
for $P2_12_12_1$ and other selected noncentrosymmetric space groups,
permits the evaluation ab initio of $\cos 2\phi_{\vec{h}}$, where $\vec{h}$ is an
arbitrary three-dimensional vector such that $|E_{\vec{h}}|$ is large (say >
1.5). In this way four possible values, in general, of the phase
$\phi_{\vec{h}}$ are determined. In the important case that

$$\cos 2\phi_{\vec{h}} \approx \pm 1, \qquad\qquad (4.1)$$

then only two values for $\phi_{\vec{h}}$ are possible, 0 or π if the cosine is
positive, $\pm\pi/2$ if the cosine is negative. In any event the
ambiguities are usually resolved either by arbitrary selection,
in order to fix the origin and enantiomorph as previously
described, or by use of calculated cosine invariants in conjunction
with phases previously determined. In those cases where ambiguities cannot be resolved it may be necessary to carry the several
values through the entire phase determination process unless
certain ambiguities can be rejected by means of appropriate
figures of merit (Chapter VII).

4.2. <u>One-dimensional phases</u>. In the space group $P2_12_12_1$
(and other selected space groups) the double-angle formula permits
an independent evaluation of the one-dimensional phases. This
result is important because one-dimensional phases are occasionally
not decisively determined by a straightforward application of the
Σ_1 formula.

It is a general result (Chapter VII) that, if the reciprocal
vector $\vec{h}$ is fixed, then

$$\cos \phi_{\vec{h}} = K \left\langle |E_{\vec{k}} E_{\vec{h}-\vec{k}}| \cos(\phi_{\vec{k}} + \phi_{\vec{h}-\vec{k}}) \right\rangle_{\vec{k}}, \qquad (4.2)$$

in which the average is taken over an arbitrary set of vectors $\vec{k}$ and the scaling parameter K is positive but, for the present purpose, not further defined. In most applications, including the present one, $\vec{k}$ ranges over those vectors in reciprocal space for which the magnitudes $|E_{\vec{k}}|$ and $|E_{\vec{h}-\vec{k}}|$ are large, say greater than some specified number t (e.g. t = 1.5). For the present purpose a modification of (4.2) is needed which is derived in the same way, i.e. via the conditional probability distribution of the pair of phases $\phi_{\vec{k}}$, $\phi_{\vec{h}-\vec{k}}$ For $P2_12_12_1$ the required formula is

$$E_{2h\ 0\ 0} = K \left\langle |E_{hk\ell}\ E_{h\bar{k}\bar{\ell}}|\cos(\phi_{hk\ell} + \phi_{h\bar{k}\bar{\ell}}) \right\rangle_{k,\ell} \qquad (4.3)$$

in which the integer h is fixed and the average is taken over those integers k,ℓ such that $|E_{hk\ell}|^2$ is large. Again K is a positive scaling parameter not further defined. Owing to the space group dependent relationships among the phases, (4.3) implies

$$\cos \phi_{2h\ 0\ 0} = K \left\langle (-1)^{h+k}|E_{hk\ell}|^2 \cos 2\phi_{hk\ell} \right\rangle_{k,\ell} \qquad (4.4)$$

from which, employing the double-angle formula of §5, Chapter II, the value (±1) of $\cos \phi_{2h\ 0\ 0}$ may often be inferred. In a similar way, for fixed integer k,

$$\cos \phi_{0\ 2k\ 0} = K \left\langle (-1)^{k+\ell}|E_{hk\ell}|^2 \cos 2\phi_{hk\ell} \right\rangle_{\ell,h} \qquad (4.5)$$

and, for fixed integer ℓ,

$$\cos \phi_{0\ 0\ 2\ell} = K \left\langle (-1)^{\ell+h}|E_{hk\ell}|^2 \cos 2\phi_{hk\ell} \right\rangle_{h,k}. \qquad (4.6)$$

Eq. (4.4)-(4.6) have been found to be useful in confirming Σ_1 indications or, occasionally, in the independent evaluation of one-dimensional phases when the Σ_1 indication was not definitive.

CHAPTER VII

THE TANGENT TECHNIQUES

Perhaps the most important single technique of phase determination is the tangent formula which was discovered over fifteen years ago. Simply stated, the tangent formula enables one to calculate unknown phases once a sufficiently broad base of phases have been determined. Thus its chief limitation, in practice, is the requirement that the values of a number of phases be known first. Subject to this restriction then, the tangent formula has proven to be an extremely powerful and versatile tool and it seems fair to say that, given the present state of knowledge, the solution of any complex noncentrosymmetric structure by direct methods must be strongly dependent on it or some variant of it. For this reason it is important to formulate this formula precisely and to give a rigorous derivation of it. This is the purpose of §1 of this chapter. The analysis shows that the formula depends rather weakly on the conditional distribution of the pair of phases $\phi_{\vec{k}}$, $\phi_{\vec{h}-\vec{k}}$ and suggests a method of using this distribution more effectively. In this way one is led to the modified tangent technique (§2) which, in the three or four cases that comparisons have been made, has yielded phases somewhat more accurately than the earlier tangent formula (e.g. Chapter XIII).

1. The Tangent Formula

<u>1.1. First derivation</u>. Let $\vec{h}$ be a fixed vector in recip-
rocal space and suppose that the reciprocal vector $\vec{k}$ is the
primitive random variable. Define A by

$$A = \frac{2}{N^{1/2}} \; \left| E_{\vec{h}} E_{\vec{k}} E_{\vec{h}-\vec{k}} \right| \tag{1.1}$$

where N is the number of atoms, assumed identical, in the unit
cell. Then, referring to (13.2) and (14.3) of Chapter III
(after replacing $\vec{h}$ by $-\vec{h}$), one finds

$$\left\langle \sin(\phi_{\vec{h}} - \phi_{\vec{k}} - \phi_{\vec{h}-\vec{k}}) \,\middle|\, A \right\rangle_{\vec{k}} = 0, \tag{1.2}$$

$$\left\langle \cos(\phi_{\vec{h}} - \phi_{\vec{k}} - \phi_{\vec{h}-\vec{k}}) \,\middle|\, A \right\rangle_{\vec{k}} = \frac{I_1(A)}{I_0(A)} \; , \tag{1.3}$$

in which the averages are extended over those vectors $\vec{k}$ in
reciprocal space such that A has a specified, fixed value. Hence
$\left| E_{\vec{k}} E_{\vec{h}-\vec{k}} \right|$ has a constant value with respect to this averaging
process so that (1.2) and (1.3) may be replaced by

$$\left\langle \left| E_{\vec{k}} E_{\vec{h}-\vec{k}} \right| \sin(\phi_{\vec{h}} - \phi_{\vec{k}} - \phi_{\vec{h}-\vec{k}}) \,\middle|\, A \right\rangle_{\vec{k}} = 0, \tag{1.4}$$

$$\left\langle \left| E_{\vec{k}} E_{\vec{h}-\vec{k}} \right| \cos(\phi_{\vec{h}} - \phi_{\vec{k}} - \phi_{\vec{h}-\vec{k}}) \,\middle|\, A \right\rangle_{\vec{k}} = \left| E_{\vec{k}} E_{\vec{h}-\vec{k}} \right| \frac{I_1(A)}{I_0(A)} \; . \tag{1.5}$$

Next, let $\vec{k}$ range over an arbitrary subset S of reciprocal
space. Imagine that S is decomposed into subsets each with the
property that for all reciprocal vectors $\vec{k}$ in the subset the value

of A is constant. Hence equations (1.4) and (1.5) hold for each
of these subsets, but A changes from one subset to another.
Averaging these equations over the subsets yields the following
generalizations of (1.4) and (1.5)

$$\left\langle |E_{\vec{k}}E_{\vec{h}-\vec{k}}| \sin(\phi_{\vec{h}}-\phi_{\vec{k}}-\phi_{\vec{h}-\vec{k}}) \right\rangle_{\vec{k}} = 0, \qquad (1.6)$$

$$\left\langle |E_{\vec{k}}E_{\vec{h}-\vec{k}}| \cos(\phi_{\vec{h}}-\phi_{\vec{k}}-\phi_{\vec{h}-\vec{k}}) \right\rangle_{\vec{k}} = \left\langle |E_{\vec{k}}E_{\vec{h}-\vec{k}}| \frac{I_1(A)}{I_0(A)} \right\rangle_{\vec{k}}, \qquad (1.7)$$

in which the averages are now extended over an arbitrary set of
reciprocal vectors $\vec{k}$ so that A is no longer constant with respect
to this averaging process. Making the definitions

$$C = \left\langle |E_{\vec{k}}E_{\vec{h}-\vec{k}}| \cos(\phi_{\vec{k}}+\phi_{\vec{h}-\vec{k}}) \right\rangle_{\vec{k}}, \qquad (1.8)$$

$$S = \left\langle |E_{\vec{k}}E_{\vec{h}-\vec{k}}| \sin(\phi_{\vec{k}}+\phi_{\vec{h}-\vec{k}}) \right\rangle_{\vec{k}}, \qquad (1.9)$$

$$B = \left\langle |E_{\vec{k}}E_{\vec{h}-\vec{k}}| \frac{I_1(A)}{I_0(A)} \right\rangle_{\vec{k}} >0, \qquad (1.10)$$

(1.6) and (1.7) become

$$C \sin \phi_{\vec{h}} - S \cos \phi_{\vec{h}} = 0, \qquad (1.11)$$

$$S \sin \phi_{\vec{h}} + C \cos \phi_{\vec{h}} = B, \qquad (1.12)$$

the solution of which is evidently

$$\sin \phi_{\vec{h}} = \frac{BS}{C^2+S^2} \ , \tag{1.13}$$

$$\cos \phi_{\vec{h}} = \frac{BC}{C^2+S^2} \ , \tag{1.14}$$

from which

$$\tan \phi_{\vec{h}} = \frac{S}{C} = \frac{\left\langle \left| E_{\vec{k}} E_{\vec{h}-\vec{k}} \right| \sin(\phi_{\vec{k}}+\phi_{\vec{h}-\vec{k}}) \right\rangle_{\vec{k}}}{\left\langle \left| E_{\vec{k}} E_{\vec{h}-\vec{k}} \right| \cos(\phi_{\vec{k}}+\phi_{\vec{h}-\vec{k}}) \right\rangle_{\vec{k}}} \ , \tag{1.15}$$

the so-called tangent formula. Clearly (1.15) is useful in determining the unknown phase $\phi_{\vec{h}}$ if it is assumed that the several phases $\phi_{\vec{k}}$, $\phi_{\vec{h}-\vec{k}}$ are known. It is to be emphasized that in (1.15) it is not necessary that $\vec{k}$ range over all vectors in reciprocal space or even over a random sample of reciprocal vectors. Instead, $\vec{k}$ may range over an arbitrary set of vectors in reciprocal space, in particular vectors for which $\left| E_{\vec{k}} \right|$ and $\left| E_{\vec{h}-\vec{k}} \right|$ are large. This observation is important because this is just the way in which the tangent formula is used (and in fact must be used) in practice since initial values can be found for only those phases corresponding to large values of $|E|$. Many derivations of the tangent formula are to be found which justify the use of (1.15) only in the special case that $\vec{k}$ ranges over a representative sample of reciprocal vectors, not a particularly useful case for the practicing crystallographer. It is precisely for this reason, i.e. to derive the tangent formula in the more general form given here, that it is necessary to base the analysis on the conditional distribution of the pair of phases $\phi_{\vec{k}}$, $\phi_{\vec{h}-\vec{k}}$, given A, as is done here. It is to be noted finally that (1.13) and (1.14) are used

merely to fix the quadrant of $\phi_{\vec{h}}$, (1.15) to determine its value
more precisely.

    1.2. Second derivation. Eq. (1.15) (but not (1.13) or
(1.14)) is an immediate consequence of (1.11) alone so that the
tangent formula (1.15) depends ultimately only on (1.2), i.e.
the relatively weak property of the conditional probability
distribution of the random variable $\phi_{\vec{h}} - \phi_{\vec{k}} - \phi_{\vec{h}-\vec{k}}$ that it is an even
function. No essential use is made of (1.3) except, via (1.13)
and (1.14), to fix the quadrant of $\phi_{\vec{h}}$, and the ratio of Bessel
Functions $\dfrac{I_1(A)}{I_0(A)}$ does not appear, explicitly or implicitly, in
(1.15). These remarks suggest that more effective use of (1.3),
as well as (1.2), may well yield an improved procedure for
determining the unknown phase $\phi_{\vec{h}}$ when the phases $\phi_{\vec{k}}$, $\phi_{\vec{h}-\vec{k}}$ are
assumed to be known. This analysis is carried out in §2 and
leads to the so-called modified tangent procedure.

    1.3. Cosine figure of merit. If a base of phases is assumed
to be known, the tangent formula (1.15) leads to the values of a
number of unknown phases $\phi_{\vec{h}}$. In view of §1.2, the phases $\phi_{\vec{h}}$ so
determined must satisfy (1.6) exactly, but (1.7) is not necessarily
satisfied. One is thus led to define an individual figure of
merit by means of

$$\frac{\displaystyle\sum_{\vec{k}} A\left\{\cos(\phi_{\vec{h}} - \phi_{\vec{k}} - \phi_{\vec{h}-\vec{k}}) - \frac{I_1(A)}{I_0(A)}\right\}}{\displaystyle\sum_{\vec{k}} A} \qquad (1.16)$$

which, it is anticipated, will be numerically small if the
calculated value of $\phi_{\vec{h}}$ does in fact correspond to its true value.
More generally, when a number of phases $\phi_{\vec{h}}$ are calculated simultaneously from the same base of known vectors, one defines the
cycle (or cosine) figure of merit by

$$\left\langle \frac{\sum\limits_{\vec{k}} A \left\{ \cos(\phi_{\vec{h}}-\phi_{\vec{k}}-\phi_{\vec{h}-\vec{k}}) - \frac{I_1(A)}{I_0(A)} \right\}}{\sum\limits_{\vec{k}} A} \right\rangle_{\vec{h}} \qquad (1.17)$$

which will be small when the process is converging to the correct answer. Figures of merit in excess of 0.20 usually indicate an incorrect phase determination while values numerically less than 0.10 are often (but not **always**) consistent with the correct solution. The figure of merit associated with the modified tangent procedure to be discussed next is usually a more reliable indicator of correct phasing and is therefore preferred.

2.   Modified Tangent Procedure

2.1. <u>Derivation</u>. It has been seen that the derivation of the tangent formula (1.15) depends only on (1.2) and that no essential use is made of (1.3). If the attempt is made to exploit the probability distribution of the pair $\phi_{\vec{k}}$, $\phi_{\vec{h}-\vec{k}}$ more effectively by using (1.3) as well as (1.2), one is led to a least squares formulation of the problem. First suitable weights w_c and w_s are defined by means of

$$\frac{1}{w_c} = \text{Var } \cos(\phi_{\vec{h}}-\phi_{\vec{k}}-\phi_{\vec{h}-\vec{k}}) \qquad (2.1)$$

$$\frac{1}{w_s} = \text{Var } \sin(\phi_{\vec{h}}-\phi_{\vec{k}}-\phi_{\vec{h}-\vec{k}}) \qquad (2.2)$$

so that, in view of Chapter III, w_c and w_s are known as functions of A (which is in turn, for fixed $\vec{h}$, a function of $\vec{k}$). If it is assumed that certain phases $\phi_{\vec{k}}$, $\phi_{\vec{h}-\vec{k}}$ are known, one may then attempt to determine the unknown phase $\phi_{\vec{h}}$ by minimizing the following function of $\phi_{\vec{h}}$:

$$\bar{\Phi} = \left\langle w_s \sin^2(\phi_{\vec{h}}-\phi_{\vec{k}}-\phi_{\vec{h}-\vec{k}}) + w_c \left[\cos(\phi_{\vec{h}}-\phi_{\vec{k}}-\phi_{\vec{h}-\vec{k}}) - \frac{I_1(A)}{I_0(A)} \right]^2 \right\rangle_{\vec{k}}$$

(2.3)

which, after some simplification, reduces to

$$\bar{\Phi} = \frac{1}{2} \left\langle (w_c-w_s)\cos 2(\phi_{\vec{k}}+\phi_{\vec{h}-\vec{k}}) \right\rangle_{\vec{k}} \cos 2\phi_{\vec{h}}$$

$$+ \frac{1}{2} \left\langle (w_c-w_s)\sin 2(\phi_{\vec{k}}+\phi_{\vec{h}-\vec{k}}) \right\rangle_{\vec{k}} \sin 2\phi_{\vec{h}}$$

$$- 2 \left\langle w_c \frac{I_1(A)}{I_0(A)} \cos(\phi_{\vec{k}}+\phi_{\vec{h}-\vec{k}}) \right\rangle_{\vec{k}} \cos \phi_{\vec{h}}$$

$$- 2 \left\langle w_c \frac{I_1(A)}{I_0(A)} \sin(\phi_{\vec{k}}+\phi_{\vec{h}-\vec{k}}) \right\rangle_{\vec{k}} \sin \phi_{\vec{h}}$$

$$+ \frac{1}{2} \left\langle w_c \left[1 + \left(\frac{I_1(A)}{I_0(A)} \right)^2 \right] + w_s \right\rangle_{\vec{k}} .$$

(2.4)

This procedure, to determine $\phi_{\vec{h}}$ by minimizing (2.3) or (2.4), is a substitute for the tangent formula and, since it makes stronger use of the probability distribution of the structure invariants, is expected to be an improvement over the tangent formula.

Comparison between the two techniques does confirm, in fact, that
the modified tangent procedure is superior to the older tangent
formula. Usually the improvement is only marginal (§3). However,
in at least one instance (Chapter XIII) use of the modified tangent technique resulted in a very substantial improvement.
Further comparison of the two techniques would be desirable.

   2.2. Modified tangent figures of merit. Clearly the depth
of the minimum of the function $\bar{\Phi}$ (2.3) is a measure of the
accuracy of the phase determination. Hence an individual figure
of merit (one for each $\phi_{\vec{h}}$) is defined by

$$R_{\vec{h}} = \bar{\Phi}^{1/2}_{\min} \tag{2.5}$$

where $\bar{\Phi}_{\min}$ is the minimum value of $\bar{\Phi}$. It is expected that, in
general, the smaller the value of $R_{\vec{h}}$ the better the phase determination is. Next, if a set of phases $\phi_{\vec{h}}$ is determined simultaneously from the same basic set of known phases, then one defines
first

$$A_{\vec{h}} = \sum_{\vec{k}} A, \tag{2.6}$$

where the sum is taken over all the vectors $\vec{k}$ contributing to the
value of $\bar{\Phi}$ in (2.3), recalling of course, that for fixed $\vec{h}$, A is
a function of $\vec{k}$. With this definition for $A_{\vec{h}}$, the (cycle)
modified tangent figure of merit is defined by

$$R = \frac{\displaystyle\sum_{\vec{h}} A_{\vec{h}} R_{\vec{h}}}{\displaystyle\sum_{\vec{h}} A_{\vec{h}}} \tag{2.7}$$

to be a weighted average of the $R_{\vec{h}}$'s (2.5). Clearly, when phases
are being correctly calculated, the modified tangent figure of
merit is expected to be small. In the solution of the 9α-fluoro-
cortisol structure (Chapter XI), R was found to be about 0.8 in
the early cycles when A values were running around 3, but towards
the end of the phase determination R was approximately 0.9 and
A values averaged about 1. These figures are typical for
structures of this complexity; values of R in excess of 0.95
would be indicative of incorrect phasing. However, for structures
of great complexity, with A values in the range of 0.6 and 0.8,
larger values of R would be expected in view of (2.3) and the
fact that the variances defined by (2.1) and (2.2) are known to
be decreasing functions of A.

Another figure of merit may be derived by comparing the
depth of the minimum of $\overline{\Phi}$ with the average value of $\overline{\Phi}$. Thus if
one defines an individual figure of merit (one for each $\phi_{\vec{h}}$) by
means of

$$R'_{\vec{h}} = \frac{\overline{\Phi}_{avg}}{\overline{\Phi}_{min}} - 1, \qquad (2.8)$$

where $\overline{\Phi}_{avg}$ is the average value of $\overline{\Phi}$, it would be expected that,
in general, the larger the values of $R'_{\vec{h}}$ the better the phase
determination is. Values of R'_{h} in the neighborhood of 0.75 or
greater for structures with 20-30 independent nonhydrogen atoms
appear to be acceptable, but values less than 0.25 are evidence
of incorrect phasing.

In analogy with (2.7) one defines the cycle figure of merit
R' by means of

$$R' = \frac{\sum_{\vec{h}} A_{\vec{h}} R'_{\vec{h}}}{\sum_{\vec{h}} A_{\vec{h}}}, \qquad (2.9)$$

but now the larger values for R' are associated with correct
phasing, the smaller values with incorrect phasing.

3. Comparison of the Two Tangent Techniques

The efficacy of the modified tangent procedure (2.3), and
especially its utility in comparison with the simple tangent
formula (1.15), were first tested on the data for three known
steroid crystal structures, each of which has 20-30 nonhydrogen
atoms in the asymmetric unit. These structures are epiandrosterone (Chapter VIII), the estradiol urea complex (1:1), and
6α-fluorocortisol (Chapter IX). The first of these structures
crystallizes in space group $P2_1$, and the other two crystallize in
space group $P2_12_12_1$.

In the first test of the modified tangent procedure, the
input consisted of a large set of correct phases which were refined using both tangent techniques. Since it was desired, in
these initial applications, to give approximately equal weights to
the sine and cosine contributions to the function $\bar{\phi}$ and since
the variance of the cosine is much less than that of the sine, the
weights were made proportional to the reciprocals of the corresponding standard deviations rather than the variances. Table 1
gives a comparison of the results for four cycles of refinement
of the true phases for reflections with $|E| \geq 1.3$, for each of the
three structures by means of both the simple and modified tangent
formulas. Although large sets of accurate phases are not available at the point at which the tangent procedures are applied to
an unknown structure, the results of this sort of experiment
indicate the maximum accuracy which can be expected. It was
found that, for all three structures, the average deviation of
the predicted phases from their true values ($\phi_{true} - \phi_{predicted}$)
was approximately zero, and the average absolute value of the

Table 1. Comparison of simple (top figure) and modified tangent formulas following 4 cycles of refinement of the true phases of all vectors with $|E| \geq 1.3$ using all triples with $A \gtrsim 1.0$. The deviations are $(\phi_{true} - \phi_{predicted})$.

Structure	Number of phases	$\langle$ Deviation Radians $\rangle$	$\langle$ |Deviation| Radians $\rangle$
epiandrosterone	289	0.01 0.01	0.22 0.23
estradiol·urea complex	359	-0.03 -0.03	0.20 0.21
6α-fluorocortisol	367	0.02 0.03	0.29 0.28

deviation was in the range 0.2-0.3 radians for both formulas.
Thus, the formulas are about equally good under these circumstances, and the simple tangent formula would be preferred because
it requires less computing time.

Once it had been established that the modified tangent
procedure was at least as accurate as the simple tangent formula
in an idealized case, attention was directed to the type of
situation encountered in the solution of an unknown structure,
and the structures of the estradiol·urea complex and 6α-fluoro-
cortisol were re-solved with the modified tangent procedure using
the same basic sets of phases which had previously been used to
solve the structures by means of the simple tangent formula. In
both cases, after selection of the origin and enantiomorph,
additional phases were determined through the two-dimensional
structure invariants $X = \cos(\phi_{\vec{h}} + \phi_{\vec{k}} + \phi_{-\vec{h}-\vec{k}})$ before beginning the
tangent formula (Chapter IX), and a total of 300 reflections were
then phased in the tangent formula. Table 2 summarizes the
results of these studies. As in the cases involving a large set
of correct input phases, the average deviations of the phases
from their correct values are near zero regardless of which
formula is used, but the average absolute value of the deviation
for the phases computed by the simple tangent formula was 1.5
times as great as the average absolute deviation for the modified
tangent phases in the case of 6α-fluorocortisol and twice as great
in the case of the estradiol·urea complex. Although the phases
derived by both tangent techniques resulted in intelligible
initial E-maps for these structures, the more correct modified
tangent phases produced cleaner maps as shown by the data in
Table 3. Four more atoms were discernible on the 6α-fluorocortisol
E map phased by the modified tangent formula than on the corresponding map phased by the simple tangent formula. Thus, although
the modified tangent procedure was superior in both of these
applications, the differences were not decisive. However, a case

Table 2.   Comparison of the simple (top figures) and modified
           tangent formulas for the phase buildups used to solve
           the estradiol·urea complex and 6α-fluorocortisol.   The
           deviations for the final tangent cycles, in radians, are
           $(\phi_{true}-\phi_{predicted})$.

Structure	Estradiol·urea complex (1:1)	6α-Fluorocortisol
No. input phases	62	51
No. incorrect input phases	4	5
$\langle$Deviation$\rangle$	0.08 0.00	0.06 0.03
$\langle\vert$Deviation$\vert\rangle$	0.59 0.28	0.50 0.33

Table 3.   Comparison of initial E-maps for the estradiol·urea com-
           plex and 6α-fluorocortisol which were computed using
           phases calculated by the simple (top figure) and
           modified tangent formulas.

Structure	Estradiol·urea complex	6α-Fluorocortisol
No. nonhydrogen atoms in asymmetric unit	24	27
No. atoms found	22 23	22 26
Average distance from correct position (Å)	0.23 0.15	0.12 0.10

in which the difference was decisive is described in Chapter XIII.

It is instructive to examine some typical examples of the modified tangent minimization function, $\bar{\Phi}$, and in Figures 1 and 2, $\bar{\Phi}^{1/2}$ has been plotted as a function of phase angle for the two reflections 0,1,1 and 0,2,2 respectively from the estradiol·urea complex. These curves were obtained by dividing the total possible range of the phases into 100 equal intervals (i.e. intervals of 3.6°) and evaluating $\bar{\Phi}$ in each interval. The curves are plotted for each of the seven cycles performed in the modified tangent run used to re-solve this structure.

Reflection 0,1,1 has a well-determined phase. The curves for this reflection are sinusoidal, and they remain approximately constant throughout all seven tangent cycles. Each curve has a single well-defined minimum occurring at one of the two phase values ($\pi/2$ radians) allowed for this reflection by space group symmetry. The minima are low and give good figures of merit (i.e. $R_{\vec{h}} \simeq 0.85$).

In contrast, reflection 0,2,2 has a poorly determined phase. The curves are nearly flat, and they give no clear indication for the value of this phase. Although it is difficult to see, there are actually two minima in each of the curves, these minima occur at widely separated values of the phase, and they have the same poor high figures of merit ($R_{\vec{h}}$ in the range 1.06-1.09 for the various cycles). The true value of $\phi_{0,2,2}$ is 0. The simple tangent formula calculates 0 in cycles 1 and 3 and π in all other cycles. The modified tangent formula has minima at $\pi/2$ and $-\pi/2$ in cycle 1, and the positions of the minima move closer to π in subsequent cycles. In cycle 7, the minima corresponded to phase values of -2.03 and 2.08 radians.

The reason for the disparate behavior of the two phases is clear from a _post mortem_ examination of the average values of the cosine seminvariants, $\cos(\phi_{\vec{h}} + \phi_{\vec{k}} + \phi_{-\vec{h}-\vec{k}})$, for the Σ_2 triples which were used in the tangent procedures and which involved these

Fig. 1.

Estradiol·Urea Complex
Reflection 0,1,1

$\overline{\Phi}^{\,1/2}$ as a function of $\phi_{\vec{h}}$, $\vec{h} = 0,1,1$

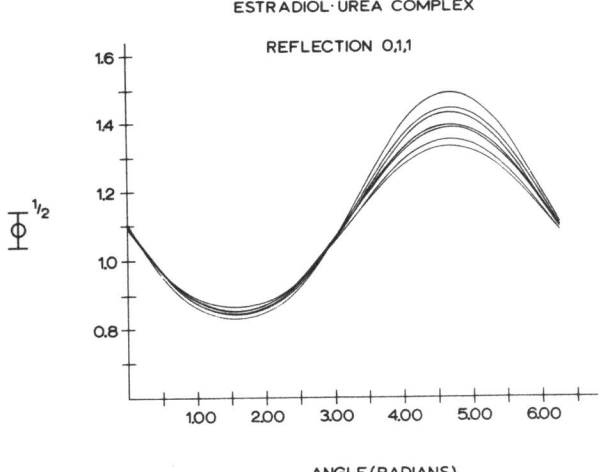

ESTRADIOL·UREA COMPLEX

REFLECTION 0,1,1

ANGLE (RADIANS)

Fig. 2.

Estradiol·Urea Complex
Reflection 0,2,2

$\bar{\Phi}^{1/2}$ as a function of $\phi_{\vec{h}}$, $\vec{h} = 0,2,2$

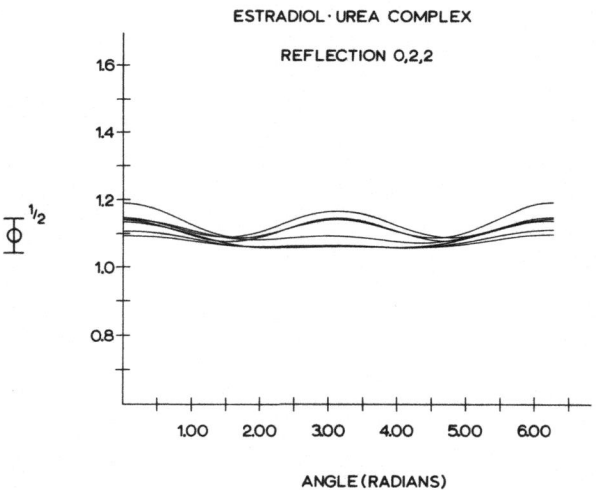

reflections. The average A value for the triples involving 0,1,1
was 1.95 and the average cosine value was 0.57, in acceptable
agreement with the theoretical expected value of 0.69. On the
other hand, the triples involving 0,2,2 (for which the average
A was 1.33) had an average cosine value of 0.10, in disagreement
with the theoretical expected value of 0.55. This low average
cosine value for 0,2,2 reflects the fact that this phase is, in
a sense, 'out of phase' with the majority of estradiol·urea
reflections, and this explains why the phase was so poorly
determined.

A number of attempts to expand a large set of known phases
for very complex molecules, mostly protein structures, by means
of the simple tangent formula have been made without conspicuous
success. In view of the comparisons made here and in Chapter
XIII, it is suggested that similar attempts employing the
modified tangent procedure instead of the older tangent formula
may meet with greater success.

4. Enantiomorph Selection by the Modified Tangent Procedure

One of the difficulties in the application of direct methods
to structures in noncentrosymmetric space groups like $P2_1$ where
all restricted phases have the same two possible values (0 or π)
lies in the selection of the enantiomorph. On the other hand, in
space group $P2_12_12_1$, enantiomorph selection is simplified because
some phases are restricted to values of 0 or π and other phases
are restricted to $\pm\pi/2$. In space groups of the $P2_1$ type, enantiomorph discrimination depends on the accurate identification of
some cosine seminvariants, $\cos(\phi_{\vec{h}}+\phi_{\vec{k}}+\phi_{-\vec{h}-\vec{k}})$, whose values are far
from +1 or −1 and will follow from the least-squares analysis of
such seminvariants (Chapter IV). Alternatively, this enantiomorph

discrimination may be accomplished by the modified tangent
procedure. When the first enantiomorph sensitive phase is
determined by this procedure, there will be two equal minima in
the minimization function $\bar{\Phi}$. The phase corresponding to one
minimum will be the phase for one enantiomorph whereas the other
minimum will occur at the phase for the other enantiomorph. In
contrast, in cases where all starting phases have values equal to
one of the two cardinal points (0 or π), no values other than
these may be calculated for any phase using the simple tangent
formula (1.15).

The enantiomorph distinguishing ability of the modified tangent procedure may be demonstrated by considering what would have
happened had the tangent techniques been applied to the starting
set of six phases (three phases determined by the origin specifying
procedure and three additional phases determined by the Σ_1 formula)
for epiandrosterone described in Chapter VIII. In the actual
structure solution, least-squares analysis of the cosine seminvariants was used to derive a larger set of phases, including
phases sensitive to the enantiomorph, before application of the
simple tangent formula. Table 4 shows the progress of the
enantiomorph selection by the modified tangent procedure. During
the first three cycles, five three-dimensional phases ($51\bar{3}$, $11\bar{2}$,
$51\bar{2}$, 310, and $62\bar{4}$) were determined, but all were calculated to
have cardinal point values by both the simple and modified tangent
formulas. In the fourth cycle, six additional phases ($41\bar{1}$ through
$62\bar{5}$) were determined, and one of these, $\phi_{41\bar{1}}$, was calculated by
the modified tangent formula to have two minima, and these minima
did not correspond to cardinal point values for the phase. It
may be seen from the second column in Table 4 that, of all the
three-dimensional reflections for which phase values had been
determined up to this point, reflection $41\bar{1}$ was the most sensitive
to the enantiomorph. The absolute value of the deviation of the
true phase of $41\bar{1}$ from the nearest cardinal point was 1.48 radians.

Table 4. Enantiomorph selection for epiandrosterone (space group $P2_1$) by the modified tangent procedure.

Vector	|Dev.| true phase from nearest cardinal pt.	|Dev.| predicted phase from cardinal pt. nearest true phase		|Dev.| predicted phase from true phase		
		Simple	Modified	A = Simple	B = Modified	A − B
4 1 $\bar{1}$ *†	1.48	3.14	1.79	1.66	0.31	1.35
5 1 $\bar{3}$ **	0.06	0.	0.18	0.06	0.12	−0.06
1 1 $\bar{2}$ **	0.98	0.	0.16	0.98	0.82	0.16
5 1 2	0.17	0.	0.	0.17	0.17	0.
3.1 0	0.25	0.	0.	0.25	0.25	0.
6 2 $\bar{4}$	0.05	0.	0.	0.05	0.05	0.
4 1 $\bar{1}$ *	1.48	3.14	2.08	1.66	0.60	1.06
2 2 3	1.06	0.	0.	1.06	1.06	0.
2 2 2	0.11	0.	0.	0.11	0.11	0.
3 1 $\bar{1}$	0.87	0.	0.	0.87	0.87	0.
3 1 $\bar{1}$ **	1.04	0.	0.20	1.04	0.84	0.20
6 2 5	0.62	0.	0.	0.62	0.62	0.
1 1 0	1.29	0.	0.	1.29	1.29	0.
1 2 $\bar{1}$ **	0.80	0.	0.21	0.80	0.59	0.21
2 2 1	0.95	0.	0.	0.95	0.95	0.
3 2 0 **	1.09	0.	0.67	1.09	0.42	0.67
4 1 $\bar{2}$ **	0.41	0.	0.14	0.41	0.27	0.14
4 2 $\bar{2}$	1.41	0.	0.	1.41	1.41	0.
7 1 5	0.10	0.	0.	0.10	0.10	0.

* Enantiomorph selecting phase.

** Phase assignment dependent on $\phi_{4,1,\bar{1}}$.

† Calculations for cycle 4. All other rows pertain to cycle 5 of the tangent procedures as described in the text.

The maximum value of this deviation is 1.57 radians, and the closer the value of the deviation to 1.57, the greater the difference between the values of the phase in question for the two enantiomorphs. For example, the difference between the values of $\phi_{41\bar{1}}$ for the two enantiomorphs is 2.96 radians, but this difference is only 0.12 radians in the case of the enantiomorph insensitive reflection $51\bar{3}$.

All phases were calculated, by the simple tangent formula, to have cardinal point values in the fourth cycle, and all phases except $\phi_{41\bar{1}}$ were calculated to be equal to the cardinal point nearest to the true phase. The same results were obtained for the simple tangent formula in the fifth cycle. However, the phases of reflections which depended on $\phi_{41\bar{1}}$ were not calculated to have cardinal point values by the modified tangent formula. Only one $(51\bar{3})$ of the seven phases which were calculated off cardinal points by the modified tangent formula in cycle 5 was less accurate than the phase calculated by the simple tangent formula.

5.   Related Techniques

Although the major emphasis in this book has been on the use of calculated values of cosine invariants as the prime tool of phase determination, other methods in common use have proved very effective and it is appropriate to describe them briefly. Only the symbolic addition and multiple solution techniques will be discussed here.

5.1. <u>Symbolic addition</u>. If A is large, say greater than 3, the conditional distribution of the cosine invariant $\cos(\phi_{\vec{h}}+\phi_{\vec{k}}+\phi_{-\vec{h}-\vec{k}})$ shows that, with high probability, the value of the cosine is approximately unity,

$$\cos(\phi_{\vec{h}} + \phi_{\vec{k}} + \Phi_{-\vec{h}-\vec{k}}) \approx +1, \qquad (5.1)$$

or, equivalently,

$$\phi_{\vec{h}+\vec{k}} \approx \phi_{\vec{h}} + \phi_{\vec{k}}. \qquad (5.2)$$

Once the origin and enantiomorph have been fixed by suitable
specification of a small number of phases, then (5.2) or the
tangent formula may usually be used to generate additional phases.
However, except for the simplest structures, this process usually
is self-terminating before a sufficient number of phases have
been determined because of the need to use small A values for
which (5.2) does not hold. If one introduces letter symbols for
a small number of well chosen phases, then (5.2) is well suited
to generate the values of additional phases in terms of the
unknown symbols, provided that attention is restricted to large A
values only. Using relationships among the phases so determined,
it is often possible to reduce greatly the number of independent
letter symbols required to determine a sufficient number of phases
to be used as input to the tangent formula. Substituting different
values for the letter symbols usually leads to a small or moderate
number of different solution sets and the correct one may be found
by using a suitable reliability index (figure of merit), or
calculating several initial E-maps, or a combination of the two.
Often the method leads to only a partial structure because of the
breakdown of (5.2) which, after all, has only statistical validity.
If this fragment is correctly placed in the unit cell, then tangent
refinement, using as input to the tangent formula the values of
the phases based on the partial structure, is usually sufficient
to yield the correct set of phases and the true structure. The
power of the method is enhanced if one uses calculated values of
the cosine invariants to restrict attention to those invariants
for which (5.1) holds. Even so, for very complex structures,

large A values are few in number and it is to be anticipated that structure solution will be possible only with the introduction of many symbols, leading to a large number of solution sets and excessive computing time.

5.2. Multiple solution. This technique is strongly dependent on the tangent formula. In addition to the origin and enantiomorph fixing phases, a small set of phases is selected and each is allowed to range through a small set of values. Each combination of values is used as input to the tangent formula and that input set which is closest to the true solution leads to an accurate extension of the basis set and to the correct crystal structure. Considerable effort has been expended in determining the best choice for the basic set of phases and the procedure devised to achieve this goal (the convergence method) is both efficient and effective. An advantage of the method is not only that it permits the immediate use of the tangent formula but a weighting scheme for use with the tangent formula is also possible whereby a weight is attached to each phase which is a function of the reliability with which the phase is determined. Thus the usual tangent formula is replaced by

$$\tan \phi_{\vec{h}} = \frac{\displaystyle\sum_{\vec{k}} w_{\vec{k}} w_{\vec{h}-\vec{k}} |E_{\vec{k}} E_{\vec{h}-\vec{k}}| \sin(\phi_{\vec{k}} + \phi_{\vec{h}-\vec{k}})}{\displaystyle\sum_{\vec{k}} w_{\vec{k}} w_{\vec{h}-\vec{k}} |E_{\vec{k}} E_{\vec{h}-\vec{k}}| \cos(\phi_{\vec{k}} + \phi_{\vec{h}-\vec{k}})} = \frac{T_{\vec{h}}}{B_{\vec{h}}}, \qquad (5.3)$$

$$w_{\vec{h}} = \tanh \left\{ \frac{\sigma_3}{\sigma_2^{3/2}} |E_{\vec{h}}| (T_{\vec{h}}^2 + B_{\vec{h}}^2)^{1/2} \right\}, \qquad (5.4)$$

$$\sigma_n = \sum_{j=1}^{N} z_j^n . \qquad (5.5)$$

Initial weights are chosen to be unity for the input phases with the exception of those determined by means of Σ_1. Initial weights for the latter depend upon the reliability of the phase determination as given by appropriate probability measures (§1.1, Chapter V). In this way the whole process of tangent refinement is considerably shortened and made more reliable. Figures of merit are used in order to determine the best solutions prior to calculation of initial E-maps. Although this method has had a good measure of success, it is again to be anticipated that, with very complex structures, successful structure determination will require a large basic set of phases to be used as input to the tangent formula resulting in a correspondingly large number of solution sets.

THE APPLICATIONS

In tackling an unknown crystal structure by direct methods, one naturally uses whatever tools appear to be applicable. In the applications described in Part C, the solutions were, for the most part, carried out by a combination of the techniques described in Part B. However, the particular examples selected here were chosen because they provided a particularly transparent illustration of a single technique and, in each case, it is this aspect of the structure determination which is emphasized. In all cases, however, there was a strong dependence on the calculated values of the cosine invariants, and this feature of the phase determination is also to be stressed.

CHAPTER VIII

TWO ANDROSTANE DERIVATIVES;
LEAST-SQUARES ANALYSIS OF
THE COSINE INVARIANTS

1. Introduction

The structures of 5α-androstan-3β-ol-17-one ($C_{19}H_{30}O_2$) and
5β-androstane-3α,17β-diol ($C_{19}H_{32}O_2$) were solved via the least-
squares analysis of the calculated cosine invariants, cos
($\phi_1 + \phi_2 + \phi_3$). Both these substances crystallize in space group $P2_1$
with two molecules in the unit cell. The computational procedure
whereby phases were derived from the invariants is discussed in
detail. The computed cosine invariants were compared with the
observed values, and it was found that invariants for which the
true value is relatively large are computed more accurately than
invariants having smaller values.

The most difficult part of solving acentric crystal structures by probability methods lies in the determination of a basic
set of phases. Depending on the space group, one to three phases
may be arbitrarily specified to select the origin and a few additional phases are usually determined by Σ_1 relationships. After
approximately 50 phases have been found, the tangent formula may
be employed to evaluate the remaining phases needed to solve the
structure.

It is the aim here to show that the structure invariants,
$\cos(\phi_{\vec{h}_1} + \phi_{\vec{h}_2} + \phi_{\vec{h}_3})$, may be used to compute additional phases when

only a few initial phases are known. These invariants were first
used experimentally to solve the structure of the female sex
hormone estriol. Subsequently this method has been used successfully to solve the structure of 17β-trimethylsiloxy-4-androsten-
3-one as well as the structures of 5α-androstan-3β-ol-17-one
(epiandrosterone, $C_{19}H_{30}O_2$) and 5β-androstane-3α,17β-diol
($C_{19}H_{32}O_2$). The purposes of this chapter are to describe the
process of phase determination by least-squares analysis of
structure invariants (with reference being made to the phase-
determination procedure for epiandrosterone), to point out where
difficulties are likely to be encountered, and to compare the
merits of various methods of computing the structure invariants
based on the results for epiandrosterone and 5β-androstane-
3α,17β-diol.

2. Phase Determination

In space group $P2_1$, the phases of reflections of the type
h0ℓ, where both h and ℓ are even, are structure invariants whose
values depend only on the arrangement of atoms within the unit
cell, but the values of all other phases depend on the location
of the origin of the unit cell as well as the atomic positions.
The location of the origin may be specified by arbitrarily
assigning three phases that are linearly independent. The linear
dependence of phases is a function of space-group symmetry and,
for space group $P2_1$, the origin may be uniquely determined by
assigning phases to one reflection with k=1, and to two reflections a0b and c0d which satisfy the following conditions: (1) a, b
not both even, (2) c,d not both even, and (3) the sums a+c and
b+d not both even. Phases were assigned to 40$\bar{1}$, 50$\bar{3}$, and 11$\bar{1}$ to
specify the origin for epiandrosterone. This set was chosen from
among the many possible sets of linearly independent phases
because, in each case, both the observed structure-factor

amplitude, $|F_{obs}|$, and the normalized structure-factor amplitude, $|E_{obs}|$, were large, and each vector was found to occur in many vector triples $(\vec{h}_1, \vec{h}_2, \vec{h}_3)$* having the property that

$$\vec{h}_1 + \vec{h}_2 + \vec{h}_3 = 0, \qquad (2.1)$$

and which also satisfy the condition that $|E_1 E_2 E_3|$ be large. It is demonstrated below that the probability that the structure invariants, $\cos(\phi_1 + \phi_2 + \phi_3)$, can be accurately computed increases as the product $|E_1 E_2 E_3|$ increases. Therefore, reflections with large $|E|$ are introduced into the set of reflections with known phases whenever possible.

In space group $P2_1$, the Σ_1 formula

$$E_{2h\ 0\ 2\ell} \simeq \frac{\sigma_2^{3/2}}{\sigma_3} \left\langle (-1)^k (|E_{hk\ell}|^2 - 1) \right\rangle_k \qquad (2.2)$$

may be used to obtain, with calculable probability, phases that are structure invariants; σ_n is defined by the relationship

$$\sigma_n = \sum_{j=1}^{N} z_j^n \qquad (2.3)$$

where N is the number of atoms in the unit cell, and Z_j is the

---

* The abbreviations $\phi_1 = \phi_{\vec{h}_1}$, $\phi_2 = \phi_{\vec{h}_2}$, $\phi_3 = \phi_{\vec{h}_3}$, $E_1 = E_{\vec{h}_1}$, $E_2 = E_{\vec{h}_2}$, and $E_3 = E_{\vec{h}_3}$ will be used throughout the remainder of this chapter.

atomic number of the jth atom. The probability that the
normalized structure factor is positive (phase = 0) is given by
the relationship

$$P_+(E_{2h\ 0\ 2\ell}) \cong$$

$$\frac{1}{2} + \frac{1}{2}\ \tanh\left[\frac{\sigma_3}{\sigma_2^{3/2}}\ |E_{2h\ 0\ 2\ell}|\sum_k (-1)^k (E_{hk\ell}^2 - 1)\right].\quad (2.4)$$

The results of the application of the Σ_1 formula to those
reflections in the epiandrosterone data for which $|E_{obs}|$ was
greater than unity are presented in Table 1. A sign was
considered to be determined if the probability that it had the
indicated value was greater than 95% and using this criterion
the reflections $20\bar{4}$, $60\bar{4}$, and $40\bar{2}$ were added to the set of vectors
with known phases.

Table 1. Σ_1 results for epiandrosterone

Reflection	$|E_o|$	E_c (by Σ_1)	Probability phase positive	True sign
$20\bar{6}$	1.06	0.56	0.58	+
$20\bar{4}$	1.23	−4.54	0.03	−
$60\bar{4}$	1.75	−2.92	0.05	−
$80\bar{4}$	1.42	0.79	0.64	+
$40\bar{2}$	1.83	2.81	0.95	+
$80\bar{2}$	1.05	0.76	0.60	+

The phases of reflections whose indices satisfy equation (2.1) are linearly dependent, and it is possible to relate all reflections to the origin-determining-reflections and those determined by Σ_1 by means of this equation. It is to be noted that, since the value of the determinant

$$\begin{vmatrix} 4 & 0 & -1 \\ 5 & 0 & -3 \\ 1 & 1 & -1 \end{vmatrix},$$

consisting of the origin determining triple alone, is 7 (not ±1), the corresponding matrix H is not primitive. Therefore not all phases are accessible, via (2.1), modulo the origin fixing triple. Nevertheless, the values of all phases are uniquely determined because the matrix H is primitive modulo $\vec{\omega}$, where $\vec{\omega}$ is the invariant modulus (2 0 2), whence the origin is uniquely specified. However, if one adjoins the symmetry related phase $\phi_{1\,\bar{1}\,\bar{1}}$ (which surely may be done) then, since the greatest common division of the four 3 x 3 subdeterminants of the matrix

$$\begin{pmatrix} 4 & 0 & -1 \\ 5 & 0 & -3 \\ 1 & 1 & -1 \\ 1 & -1 & -1 \end{pmatrix}$$

is unity, then this augmented matrix is primitive and all phases are accessible, via (2.1), modulo the origin fixing set. In this instance, since the three phases determined by Σ_1 are to be added to the basic set, the primitivity property is retained and every phase is surely accessible via (2.1). This use of formulas relating the phases of such linearly dependent reflections allows the set of known phases to be expanded to such an extent that calculation of a Fourier synthesis may reveal the crystal structure. The simplest of the mathematical formulas of this type is the Σ_2 type relationship

$$E_{-\vec{h}} = E^*_{\vec{h}} \simeq \frac{\sigma_2^{3/2}}{\sigma_3} \left\langle E_{\vec{k}} E_{-\vec{h}-\vec{k}} \right\rangle_{\vec{k}} \tag{2.5}$$

which is valid for all space groups. In this equation $\vec{h}$ is a fixed vector, and $\vec{k}$ ranges over all vectors for which $-\vec{h}-\vec{k}$ exists. It is obvious that the indices of the triple $(\vec{h},\vec{k},-\vec{h}-\vec{k})$ sum to zero, and consequently the equivalence

$$\vec{h} = \vec{h}_1, \quad \vec{k} = \vec{h}_2, \quad -\vec{h}-\vec{k} = \vec{h}_3 \tag{2.6}$$

exists for individual terms contributing to the average in equation (2.5). In the case of noncentrosymmetric space groups, other phase relations which have found considerable use are

$$\phi_{\vec{h}} \simeq -\left\langle (\phi_{\vec{k}} + \phi_{-\vec{h}-\vec{k}}) \right\rangle_{\vec{k}} \tag{2.7}$$

and the tangent formula,

$$\tan \phi_{\vec{h}} = \frac{-\sum_{k} |E_{\vec{k}} E_{-\vec{h}-\vec{k}}| \sin(\phi_{\vec{k}} + \phi_{-\vec{h}-\vec{k}})}{\sum_{k} |E_{\vec{k}} E_{-\vec{h}-\vec{k}}| \cos(\phi_{\vec{k}} + \phi_{-\vec{h}-\vec{k}})} \tag{2.8}$$

The tangent formula is an extremely powerful tool. It cannot be used effectively, however, until several triples exist involving vector $\vec{h}$ in conjunction with pairs $(\vec{k},-\vec{h}-\vec{k})$ for which $\phi_{\vec{k}}$ and $\phi_{-\vec{h}-\vec{k}}$ are known.

The difficulty with using any of the equations (2.5), (2.7) and (2.8) when only a few phases are known is that they all involve summations, but only a very few terms in these summations can be computed, and the calculated $\phi_{\vec{h}}$ may be in error as a

result. The probability that a single term will accurately yield
the unknown phase increases as the magnitude of $|E_1 E_2 E_3|$ increases.
Equation (2.7) has found use in connection with the symbolic
addition procedure, in which symbols are assigned to a few phases
and equation (2.7) applied to triples having two known phases and
a large value of $|E_1 E_2 E_3|$. The third phase is then either known or
else it can be related to one of the assigned symbols, and it is
placed in the set of known phases. Considerable care must be
exercised in the use of equation (2.7) when more than one triple
contributes to a new phase determination because of the ambiguity
arising from the multiple-valued nature of the phases. It is
possible to replace ϕ with $(\phi + 2\pi k)$ where k is any integer, and
there is nothing in equation (2.7) to indicate which of these is the
proper choice. One method of circumventing this problem is to
introduce arbitrary values for the phases of a few (usually one
to four) reflections having large $|E|$, and to vary the values of
phases systematically and use each set of phases so obtained as
input for the tangent formula (multiple solution technique).
Often one or two of the sets of phases output by the tangent
formula can be selected as more likely to be correct based on such
criteria as the α index,

$$\alpha_{\vec{h}} = \frac{2\sigma_3}{\sigma_2^{3/2}} \, |E_{\vec{h}}| \left[\left\{ \sum_k |E_{\vec{k}} E_{-\vec{h}-\vec{k}}| \, \sin(\phi_{\vec{k}} + \phi_{-\vec{h}-\vec{k}}) \right\}^2 + \right.$$

$$\left. \left\{ \sum_k |E_{\vec{k}} E_{-\vec{h}-\vec{k}}| \, \cos(\phi_{\vec{k}} + \phi_{-\vec{h}-\vec{k}}) \right\}^2 \right]^{1/2} \qquad (2.9)$$

which is relatively large for the correct solution. However, it
may be necessary to compute Fourier syntheses from several sets of
phases before the structure is solved. The obvious drawback to

this approach is that even if only two arbitrary phases have been
introduced, it may be necessary to assign several values to each
before the correct solution is found. Even if only 4 values are
assigned to each of 2 arbitrary phases, it may still be necessary
to compute 16 Fourier syntheses and, in unfavorable cases, a
considerably larger number may be required.

The need for a procedure to eliminate the introduction of
deliberate ambiguities during the early stages of phase determination has led to the renewed investigation of the structure
invariants $\cos(\phi_1 + \phi_2 + \phi_3)$ described in §2 of Chapter V. The
modified triple product procedure described in §2.1 of Chapter V
was used in the present structure determinations.

In the case of epiandrosterone, the phases of six independent
spectra and the phases of their eight symmetry-related reflections
were initially known. Using this set of reflections as vectors
$\vec{k}$ and $-\vec{h}-\vec{k}$ all possible vectors $\vec{h}$ were generated, and for each of
these vectors, the summation $\sum_{k} A_k$, over all triples $(\vec{h}, \vec{k}, -\vec{h}-\vec{k})$ was
constructed. The reflection $51\bar{3}$, for which this summation was
the largest ($\Sigma A = 5.98$), was selected as the first reflection
whose phase was to be determined from the structure invariants by
minimizing the function $\bar{\phi}$ defined in equation (1.2) of Chapter
VI. This reflection and its three symmetry-related reflections
were added to the set of vectors with known phases, additional
triples were generated, summations were incremented, and the 3
vectors ($11\bar{2}$, $51\bar{2}$, and $80\bar{3}$) with the largest summations were
added to the set with known phases. This process was continued
for 21 additional cycles in which 5, 7, 9, 11, ... vectors were
successively selected until the order in which the phases of all
533 reflections with $|E|>1$ were to be found, was specified.
Twelve cycles of this procedure were sufficient to select 150
reflections among which there were 1971 independent vector triples
whose indices were related by equation (2.1). For each of these
triples, the average in equation (2.2), Chapter V, was computed

using an exponent p = 1/2 and allowing $\vec{\ell}$ to vary over all
measurable reflections, and the corresponding R_3 terms were
computed by means of equation (2.6), Chapter V.

To find the scaling parameter K, the triples with their
associated averages and R_3 terms were sorted on increasing value
of A and divided into 17 groups with 120 triples in each of the
first 15 groups and 86 and 85 triples in the last 2 groups; the
average A was found for each group. Within each group, the
triples were sorted so that the terms $\psi/|E_1E_2E_3|$ [equation (2.2),
Chapter V] were in decreasing order, and inspection of equation
(2.1), Chapter V, shows that this amounts to sorting according to
the value of the invariant since A and R_3 are approximately constant
within a group.

Some intermediate calculations which were performed to find
the value of K for the second group of triples (which had <A> =
0.53 and $R_3/|E_1E_2E_3| \simeq 0.23$) are presented in Table 2. An
extensive table of the conditional probability that $\cos(\phi_1+\phi_2+\phi_3)$
is greater than X for several values of A and X has been published
(see Chapter III for an abbreviated version), and the probability
that a given invariant with A $\simeq$ 0.53 is greater than X = 0.071,
0.170, 0.267, ..., 0.995 was read from the more extensive table.
The value of $\psi/|E_1E_2E_3|$ for that invariant which should be
closest to X in value was then found. For example, the probability
that $\cos(\phi_1+\phi_2+\phi_3)$ > 0.622 is 42% if A = 0.53, and 42% of the
invariants in the experimental group will be greater than 0.622 if
the invariant for the 50th term in the sorted list, which had
$\psi/|E_1E_2E_3|$ = 0.000371 and $R_3/|E_1E_2E_3|$ = 0.217, is set equal to

0.622. This will be the case if K = 1092.*

Values of K were also computed in analogous fashion for the other values of X, and they are listed in Table 2. If this sample of averages were large and random, and if the accuracy of averages computed by equation (2.2), Chapter V, were independent of the true value of the invariant, all the values for K so obtained should be nearly equal, but it is observed that this is not the case. K cannot be negative because, if this were true, invariants for which the averages were smallest (negative) would give the largest values of $\cos(\phi_1+\phi_2+\phi_3)$. The instability seen in the values of K results from division by a number which is very close to zero (i.e. $\Psi/|E_1E_2E_3|$), and if, because of experimental error, a term with a positive average is used when a negative one should be, K will have the wrong sign. Consequently, it is necessary to find that region in which the calculated values of K are positive and reasonably constant, and to find the average K in this region. In the case of these data, K was found to be reasonably stable for all groups of triples in the range X = 0.622 to 0.955, regardless of the average value of A for that group. The values of K for the various ranges of A are plotted in Fig. 1 to show the dependence of K on A, and it was found from a least-squares analysis that K equals 1015 + 267 A for the epiandrosterone data.

---

* From equation (2.1), Chapter V,

$$K = \frac{0.622 - R_3/|E_1E_2E_3|}{\Psi/|E_1E_2E_3|}$$

and

$$K = (0.622 - 0.217)/0.000371 = 1092.$$

Table 2. K values for the group of 120 epiandrosterone
triples having an average A=0.53

The term $R_3/|E_1E_2E_3|$ is approximately constant for these
triples and is in the range 0.215–0.255.

X	Probability $\cos(\phi_1+\phi_2+\phi_3)>X$	$\dfrac{\Psi}{|E_1E_2E_3|}$ x 10^4	K
0.071	0.64	1.21	−1277
0.170	0.61	1.56	−386
0.267	0.57	1.98	233
0.362	0.54	2.45	535
0.454	0.50	3.07	742
0.540	0.46	3.34	864
0.622	0.42	3.71	1092
0.698	0.38	3.94	1142
0.765	0.34	4.12	1309
0.825	0.29	4.61	1290
0.878	0.25	5.30	1229
0.921	0.20	6.18	1110
0.955	0.15	7.44	946
0.980	0.10	8.27	884
0.995	0.05	9.53	801

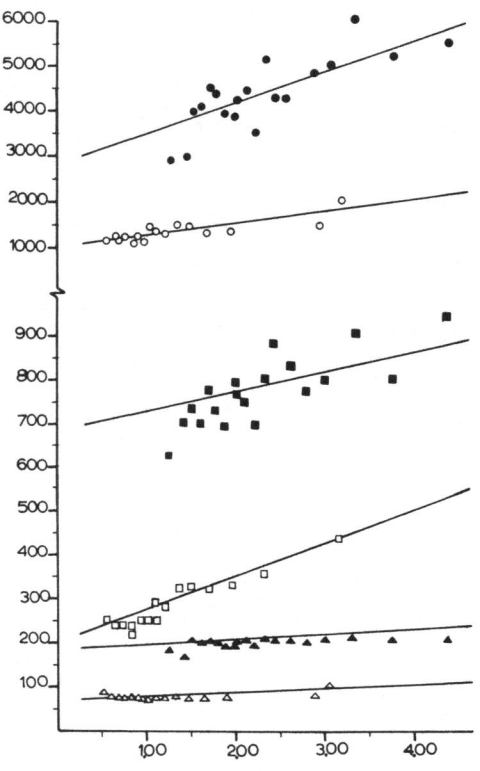

Fig. 1. The dependence of K on A. Epiandrosterone, $|E_{\vec{\ell}}|$ greater
than: 0.0, $\bigcirc$; 1.0, $\square$; 2.0, $\triangle$. 5β-androstane-3α,17β-
diol, $|E_{\vec{\ell}}|$ greater than: 0.0, $\bullet$; 1.0, $\blacksquare$; 2.0, $\blacktriangle$.

The 1971 structure invariants, $\cos(\phi_1 + \phi_2 + \phi_3)$, involving
the first 150 reflections were then calculated, and the phases
of these reflections were found by 12 cycles of the least-squares
procedure outlined above. The invariants used during the first
three cycles are listed in Table 3. Several of these invariants
were calculated to be greater than 1 and a few less than -1, but
before calculation of the function $\bar{\Phi}$ [equation (1.2), Chapter VI],
such invariants were set equal to +1 or -1, respectively. In
noncentrosymmetric space groups, the enantiomorph is selected
the first time a phase is assigned that has substantially
different values for the two enantiomorphs. This requires the
use of a structure invariant whose value is different from ±1.
The first calculated invariant to be used for epiandrosterone
involved the reflections $11\bar{1}$, $60\bar{4}$, and $51\bar{3}$; since the computed
value 0.46 was significantly different from ±1, the function $\bar{\Phi}$
had two equal minima in the first cycle, and the value 0.73 was
arbitrarily chosen for $\phi_{51\bar{3}}$ to specify the enantiomorph. The
results of the least-squares process for the first three cycles
are presented in Table 4. In the second cycle, two additional
reflections ($11\bar{2}$ and $51\bar{2}$) were encountered each of which had two
equal minima in the least-squares function. These phases are
ambiguous because the last arbitrary choice of phase was made in
specifying the enantiomorph, and these ambiguities arise because
there is only one invariant contributing to each phase determination.
If a calculated phase is not 0 or π and there is only one contributor, there will always be two solutions because, for all
angles except 0 and π, there is a second angle having the same
value of the cosine. A second contributor is needed to resolve
such ambiguities, and, unfortunately, it is not always possible
to avoid them at the beginning of phase determination since there
may not be an alternate path for building up a set of phases.

In the third cycle, other double minima occur but there are
no additional ambiguities of the type encountered with the

Table 3. Some structure invariants used during
the least-squares calculations for epiandrosterone

	$\vec{h}_1$	$\vec{h}_2$	$\vec{n}_3$	A	Predicted $\cos(\phi_1+\phi_2+\phi_3)$ *	Observed $\cos(\phi_1+\phi_2+\phi_3)$
Cycle 1	1 1 $\bar{1}$	$\bar{6}$ 0 4	5 $\bar{1}$ $\bar{3}$	2.92	0.46	1.00
	1 1 $\bar{1}$	4 0 $\bar{2}$	5 $\bar{1}$ 3	3.06	1.09	1.00
Cycle 2	4 0 $\bar{1}$	$\bar{5}$ $\bar{1}$ 3	1 1 $\bar{2}$	4.45	0.81	0.50
	4 0 $\bar{1}$	1 1 $\bar{1}$	$\bar{5}$ $\bar{1}$ 2	3.64	0.92	0.99
	4 0 $\bar{1}$	4 0 $\bar{2}$	$\bar{8}$ 0 3	3.76	1.32	1.0
	6 0 $\bar{4}$	$\bar{1}$ $\bar{1}$ 2	$\bar{5}$ 1 2	2.44	0.39	0.41
	4 0 $\bar{1}$	$\bar{4}$ 0 2	0 0 1	3.50	1.32	1.0
	5 1 $\bar{3}$	$\bar{5}$ $\bar{1}$ 2	0 0 1	3.60	0.97	1.0
	1 1 $\bar{1}$	$\bar{1}$ $\bar{1}$ 2	0 0 1	3.70	0.81	0.56
	4 0 $\bar{1}$	$\bar{5}$ 0 3	1 0 $\bar{2}$	3.51	0.95	1.0
	4 0 $\bar{1}$	$\bar{6}$ 0 4	2 0 $\bar{3}$	1.73	0.48	1.0
Cycle 3	1 1 $\bar{1}$	1 $\bar{1}$ $\bar{2}$	$\bar{2}$ 0 3	1.92	0.24	0.56
	4 0 $\bar{2}$	$\bar{1}$ $\bar{1}$ 2	$\bar{3}$ 1 0	1.40	1.01	0.33
	4 0 $\bar{1}$	$\bar{1}$ $\bar{1}$ 1	$\bar{3}$ 1 0	2.00	0.98	0.97
	5 1 $\bar{3}$	$\bar{8}$ 0 3	3 $\bar{1}$ 0	2.12	1.07	0.98
	1 1 $\bar{2}$	5 1 $\bar{2}$	$\bar{6}$ $\bar{2}$ 4	2.78	1.08	0.71
	1 1 $\bar{1}$	5 1 $\bar{3}$	$\bar{6}$ $\bar{2}$ 4	3.33	1.21	0.99
	2 0 $\bar{4}$	$\bar{1}$ $\bar{1}$ 0	$\bar{1}$ 1 4	0.40	-0.03	0.90
	2 0 $\bar{4}$	$\bar{1}$ $\bar{2}$ 0	$\bar{1}$ 2 4	0.43	-0.23	-0.32
	2 0 $\bar{4}$	$\bar{6}$ $\bar{1}$ 5	4 1 $\bar{1}$	0.64	-0.19	0.24
	2 0 $\bar{4}$	3 2 2	$\bar{5}$ $\bar{2}$ 2	0.71	0.18	0.99
	2 0 $\bar{4}$	$\bar{8}$ 1 2	6 $\bar{1}$ 2	0.72	0.37	0.76
	2 0 $\bar{4}$	3 $\bar{2}$ $\bar{1}$	1 2 5	0.73	0.19	0.89
Invariants	2 0 $\bar{4}$	1 1 $\bar{3}$	$\bar{3}$ $\bar{1}$ 7	0.79	0.09	0.23
involving 20$\bar{4}$	2 0 $\bar{4}$	1 0 $\bar{2}$	$\bar{3}$ 0 6	0.79	0.47	1.00
	2 0 $\bar{4}$	$\bar{1}$ $\bar{1}$ 3	$\bar{1}$ 1 1	0.84	-0.07	0.38
	2 0 $\bar{4}$	$\bar{4}$ $\bar{1}$ 0	2 1 4	0.88	0.19	0.98
	2 0 $\bar{4}$	$\bar{5}$ $\bar{1}$ 3	3 1 1	0.96	-0.18	0.45
	2 0 $\bar{4}$	$\bar{1}$ $\bar{5}$ $\bar{1}$	$\bar{1}$ 5 5	0.96	0.38	0.67
	2 0 $\bar{4}$	$\bar{6}$ $\bar{6}$ 1	4 6 3	0.97	0.12	0.94
	2 0 $\bar{4}$	$\bar{4}$ 0 1	2 0 3	1.00	0.37	1.00
	2 0 $\bar{4}$	$\bar{2}$ 0 3	0 0 1	1.10	-0.06	1.00

* The exponent p [(2.1) of Chapter V] was 1/2, and ℓ ranged over all measurable
reflections ($|E_\ell|>0$). The scaling parameter used was K = 1015 + 267 A.

Table 4. Some results of the least-squares
phase determination cycles for epiandrosterone

Cycle	$\vec{h}$	Number of invariants contributing to phase determination	1st (lowest) minimum		2nd minimum	
			ϕ_h (rad)	R_h	ϕ_h (rad)	R_h
Input (origin and Σ_1)	$20\bar{4}$		3.14			
	$60\bar{4}$		3.14			
	$50\bar{3}$		0.00			
	$40\bar{2}$		0.00			
	$40\bar{1}$		0.00			
	$11\bar{1}$		0.00			
1	$51\bar{3}$*	2	0.73	0.27	-0.73	0.27
	$60\bar{4}$	1	3.14	0.29		
	$80\bar{3}$*	1	0.00	0.00		
	$51\bar{3}$	2	0.73	0.27	0.00	0.27
2	$40\bar{2}$	1	0.00	0.25		
	$11\bar{2}$*	1	1.35	0.00	0.11	0.00
	$51\bar{2}$*	1	0.41	0.00	-0.41	0.00
	$11\bar{1}$	2	0.73	0.27	0.00	0.27
	$60\bar{4}$	2	3.14	0.25		
	$62\bar{4}$*	2	1.23	0.13		
	$20\bar{3}$*	2	3.14	0.27	0.00	0.86
	$80\bar{3}$	1	0.00	0.00		
	$51\bar{3}$	3	0.73	0.20		
	$10\bar{2}$*	1	0.32	0.00	-0.32	0.00
3	$40\bar{2}$	2	0.00	0.21		
	$11\bar{2}$	2	1.40	0.09	0.61	0.30
	$51\bar{2}$	2	0.31	0.05		
	$40\bar{1}$	5	0.00	0.00		
	$11\bar{1}$	3	0.00	0.21	0.56	0.27
	310*	3	-2.50	0.17		
	001*	3	0.00	0.40		
4	$20\bar{4}$	1	-2.32	0.00	2.32	0.00
5	$20\bar{4}$	2	0.00	0.63		
6	$20\bar{4}$	2	0.00	0.63		
7	$20\bar{4}$	3	0.00	0.59	-3.10	0.70
8	$20\bar{4}$	4	0.00	0.46	3.14	0.61
9	$20\bar{4}$	5	0.00	0.45	3.14	0.67
10	$20\bar{4}$	5	0.00	0.51	-3.12	0.74
11	$20\bar{4}$	10	0.00	0.68	-3.10	0.76
12	$20\bar{4}$	12	0.00	0.67		

* This phase was first determined in this cycle.

reflections $11\bar{2}$ and $51\bar{2}$. The two minima for $20\bar{3}$ do not create
an ambiguity because the second residual is much larger than the
first. While the two residuals for $10\bar{2}$ are equal, this phase is
still determined because it is known to be a centrosymmetric
phase with a true value of either 0 or π, and both of the
calculated minima lie close to one of these values. In no instance
were more than two minima observed.

A disturbing feature in the results of the least-squares
phase determination procedure is that the calculated phase for
$20\bar{4}$ was 0 (see Table 4) whereas the Σ_1 formula had strongly
determined this phase to be π. Input phases that have been
determined by origin specification or through use of the Σ_1 formula
are normally used at their original value in all phase-determina-
tion cycles regardless of the calculated value in the previous
cycle. Consequently, the least-squares procedure was repeated with
the alternative phase for $20\bar{4}$; little difference was observed in
the two sets of cycle residuals (Table 5) and $\phi_{20\bar{4}}$ was still
calculated to be 0. Three additional runs, in which $\phi_{20\bar{4}} = 0$, and
which differed from each other in the initial minimum slected for
$11\bar{2}$ and $51\bar{2}$ were then performed. Again, the cycle and centrosymmetric residuals were all about equally good, and it was also
impossible to select the set of phases most likely to be correct
based on the criterion that there be little fluctuation of
calculated values of a single phase in successive cycles. Because
of the ambiguities with the reflections $20\bar{4}$, $11\bar{2}$, and $51\bar{2}$, it was
necessary to perform 8 runs in the tangent formula. The set of 150
phases resulting from the first least-squares run in which $\phi_{20\bar{4}}$
equaled 0 was arbitrarily chosen, and 11 cycles in which no phases
were forced to the input values, were performed using the tangent
formula to determine phases for all 533 independent X-ray spectra
with $|E| > 1$. All phases were then refined for five additional
cycles. The phase of the $20\bar{4}$ reflection fluctuated from 0 to π

Table 5. Average residuals for the least-squares
phase determination cycles for epiandrosterone

	Cycle residual			Centrosymmetric residual		
Cycle	No. reflection	$\phi_{20\bar{4}}=\pi$	$\phi_{20\bar{4}}=0$	No. reflection	$\phi_{20\bar{4}}=\pi$	$\phi_{20\bar{4}}=0$
1	1	0.27	0.27	0	—	—
2	7	0.22	0.22	3	0.00	0.00
3	13	0.21	0.22	6	0.07	0.07
4	22	0.29	0.29	10	0.34	0.30
5	31	0.30	0.30	12	0.46	0.50
6	42	0.34	0.34	12	0.42	0.36
7	55	0.36	0.37	14	0.49	0.44
8	70	0.35	0.36	16	0.45	0.43
9	87	0.37	0.42	16	0.42	0.43
10	106	0.41	0.42	18	0.37	0.33
11	127	0.40	0.39	22	0.43	0.41
12	150	0.37	0.37	23	0.33	0.37

during the tangent formula cycles, and in the final refinement
cycles, it was consistently calculated to be π.

A Fourier map was prepared using all 533 phases resulting
from the final tangent formula refinement cycle, and peaks were
found that appeared to fit the expected model of epiandrosterone.
Positive electron density was found at the positions of all the
nonhydrogen atoms, but the axial carbons, C(18) and C(19), as well
as the two oxygen atoms, O(3) and O(17), were not so well resolved
as the atoms in the steroid nucleus. Four cycles of least-squares
refinement of the positional and isotropic thermal parameters of
the 21 atoms were performed using all observed reflections and a
block-diagonal approximation to the normal equations. The R
indices in successive cycles were 47.8, 38.3, 36.9, and 36.7%,
and the phase of the $20\bar{4}$ reflection, as calculated using the
original atomic positions, was equal to 0. Since this refinement
appeared to converge after two or three cycles, and since there
were no individual isotropic temperature factors which became
either very large or very small and the overall geometry of the
steroid molecule appeared to be quite satisfactory, it was
hypothesized that the entire molecule had been translated into a
false minimum as the result of an incorrect phase assignment
during the early stages of the phase buildup.

Examination of the unit-cell dimensions, and of the Patterson
synthesis, showed that the molecule was located in the unit cell
with its long axis almost parallel to the twofold screw axis. One
conspicuously large peak, which indicated the position of the
center of mass of the molecule, was observed on the Harker section,
and it was found that this position was related to the location of
the molecule on the Fourier map by a translation perpendicular to
the $20\bar{4}$ planes over a distance approximately equal to one and one
half times the spacing of these planes. The molecule was
translated so that the position of its center of mass coincided
with the location of this large peak, and following refinement

of the atomic positional and anisotropic thermal parameters the R index fell to a final value of 6.4%. The final phase of the reflection $20\bar{4}$ was π.

To try to understand the behavior of the $20\bar{4}$ reflection, the true invariants, $\cos(\phi_1+\phi_2+\phi_3)$, were computed using the phases for the refined structure and were compared to the predicted invariants. The 15 invariants involving the $20\bar{4}$ reflection which were used in the least-squares phase determining procedure are listed in Table 3. In all but one case, the value of the predicted invariant is less than the observed value for the refined structure. This distinctly nonrandom deviation of the predicted invariants from their true values is not the normal situation as can be seen from Table 6 where the root-mean-square deviation and average deviation of all predicted invariants involving a common vector are listed for several centrosymmetric reflections. Although there are other cases (201 and 603) where the r.m.s. deviation is relatively large, there is no other case where the average deviation is as large as it is in the case of $20\bar{4}$, and a large average deviation is indicative of nonrandom error. The fact that a large nonrandom error occurred in the computation of the invariants involving the $20\bar{4}$ reflection is, of course, consistent with the facts that its phase was wrongly indicated in the least-squares phase calculations, and that it had a comparatively high residual (see Table 4). The 201 reflection is unusual in that it is involved in several invariants whose observed values are small, and the observed errors for such invariants are discussed in greater detail below.

Examination of the observed invariants presented in Table 3 also shows that the invariant involving $11\bar{1}$, $60\bar{4}$, and $51\bar{3}$, which was predicted to be significantly different from ±1 and which was presumed to cause selection of the enantiomorph, has an observed value of +1. Consequently, the enantiomorph was not really specified until $\phi_{11\bar{2}}$ was selected in the second cycle. This example points out the critical nature of the invariant(s) which actually

Table 6.   R.m.s. and average deviations* of epiandrosterone
invariants involving a common reflection

Reflection	Number of invariants	R.m.s. deviation	Average deviation
$20\bar{6}$	21	0.48	0.19
$30\bar{6}$	22	0.50	0.21
$20\bar{4}$	15	0.64	−0.56
$60\bar{4}$	26	0.55	−0.19
$20\bar{3}$	32	0.53	−0.19
$50\bar{3}$	43	0.32	0.08
$80\bar{3}$	22	0.38	0.16
$10\bar{2}$	45	0.30	−0.20
$40\bar{2}$	44	0.37	0.17
$40\bar{1}$	49	0.34	0.03
100	30	0.42	−0.04
001	48	0.41	0.04
201	26	0.71	0.00
203	23	0.44	−0.02
603	20	0.60	0.31
105	35	0.37	−0.01
205	24	0.37	0.09

*The deviation is the difference between the predicted and the
observed values of cos $(\phi_1+\phi_2+\phi_3)$.

result in enantiomorph selection. It would be desirable to try
several different ways of selecting the origin until a case could
be found where there are at least two collaborating invariants not
equal to ±1 that contribute to the phase assignment which specifies
the enantiomorph. In the present situation, enantiomorph selection
actually took place at the later, second cycle, not the first. As
it turned out, one and only one phase had substantially different
values for the two enantiomorphs in the second cycle and so the
enantiomorph was unambiguously selected, unwittingly, in this
cycle. (The comments at end of §1, Chapter VI, are relevant here.)

To determine if the epiandrosterone molecule could be found
at the correct position on a Fourier map resulting from phases
determined by the least-squares analysis of the structure invariants without making any assumptions concerning molecular packing,
four additional syntheses were computed. In each case, $\phi_{20\bar{4}}$ was
taken to be π, and the phasing differed on the basis of the initial
phase used for $11\bar{2}$ and $51\bar{2}$, two of the reflections for which two
equal least-squares minima had been calculated. Also in each case,
150 phases were determined by the least-squares analysis, but only
8 cycles (two of which were refinement cycles) were performed with
the tangent formula, and a total of only 330 phases were determined,
which is approximately 15 phases per nonhydrogen atom in the asymmetric unit. The epiandrosterone molecule was found in the correct
position on each of the resultant four maps although the definition
of the molecule was considerably better in one case (map 2) as
shown in Table 7 by the values of the average r.m.s. deviation of
the atoms from their refined positions, the number of real atoms
among the highest thirty peaks on the map, and the rank of the
largest spurious peak after the peaks had been sorted according to
their relative heights. As expected, choice of the second least-
squares phase for $11\bar{2}$ resulted in selection of the second enantiomorph. The large individual deviations on map 4 make it doubtful whether the molecule could have been recognized had the correct

Table 7. Comparison of four E maps resulting from different initial choice of phase for 11̄2 and 51̄2

Map	Phase(s) forced to second minimum in cycles 2 and 3	Number of real atoms among top 30 peaks	Rank of largest spurious peak	R.m.s. deviation*
1	‾‾‾	16	1	0.29Å
2	11̄2	20	19	0.13
3	51̄2	17	7	0.27
4	11̄2,51̄2	16	6	0.39

*R.m.s. deviation (in Å) of original map positions from refined atomic positions.

position not been known, but nevertheless, it was possible to re-
fine this molecule to the correct solution. Since the deviations
for the map of the computed synthesis using the second phase for
$11\bar{2}$ were markedly smaller than those found on the other maps, and
since this map had many fewer spurious peaks with a height com-
parable to the height of the real atoms, it was expected that the
corresponding residuals for the least-squares phase determination
cycles might be significantly less than in the other cases, but
this was not observed to be true. Furthermore, the observed residuals were about the same as those encountered in the least-squares
run which eventually resulted in a Fourier map on which the molecule was translated perpendicular to the $20\bar{4}$ plane.

To determine the overall accuracy of the invariants used to
solve the structure, and to see if some groups of invariants are
calculated more accurately than others, the true invariants were
computed, and the r.m.s. deviation of the predicted invariants
from their observed values was used as an indicator of the preci-
sion of invariants calculated using equation (2.1), Chapter V.
For purposes of this comparison, invariants predicted to be greater
than +1 or less than -1 were not forced to the nearest allowed
value of the cosine. If the predicted invariants were distributed
at random, then the expected value of the r.m.s. deviations of the
invariants from their observed values ranges from 1.15 for very
large values of A to 0.82 for very small values of A if all invariants are calculated to be within the allowed range of the cosine.
The observed values of 0.48 for 1971 epiandrosterone invariants and
0.43 for 1903 invariants for 5β-androstane-3α,17β-diol are substantially smaller than this. The invariants were then grouped according to the value of A, and the r.m.s. deviation was computed for
each range of A values. The results for epiandrosterone are presented in Table 8(a) and show that predicted invariants with high
values of A (1.5-5.0) are computed much more accurately than are
those with low A (0.0-1.0). Similar data for 5β-androstane-3α,17β-

Table 8. R.m.s. deviations of the predicted structure invariants,
$\cos(\phi_1+\phi_2+\phi_3)$, from the observed values for the refined structure

The column headings are threshold values for $|E_\ell|$. When $|E_\ell|>1.0$ or 1.5, only the permutation $\vec{k}_1=\vec{h}_1$, $\vec{k}_3=\vec{h}_3$ was used to compute the averages in equation (2.1), Chapter V. When $|E_\ell|>2.0$, 2.25 or 2.5, the three even permutations of $\vec{k}_1$, $\vec{k}_2$ and $\vec{k}_3$ were used.

(a) R.m.s. deviations for the epiandrosterone invariants.

A range	No. of invari-ants	p=1/2						p=2					
		0.0	1.0	1.5	2.0	2.25	2.5	0.0	1.0	1.5	2.0	2.25	2.5
0.2-0.4	36	0.78	0.68	0.71	0.67	0.71	0.92	0.84	0.75	0.78	0.79	0.84	1.09
0.4-0.6	254	0.61	0.62	0.61	0.58	0.58	0.64	0.60	0.60	0.60	0.60	0.61	0.75
0.6-0.8	362	0.53	0.58	0.56	0.54	0.54	0.58	0.59	0.60	0.60	0.60	0.61	0.71
0.8-1.0	369	0.50	0.53	0.52	0.51	0.52	0.56	0.57	0.56	0.57	0.60	0.62	0.76
1.0-1.2	244	0.49	0.49	0.48	0.47	0.47	0.51	0.57	0.54	0.55	0.58	0.60	0.81
1.2-1.4	170	0.40	0.38	0.38	0.38	0.38	0.42	0.48	0.45	0.46	0.48	0.50	0.64
1.4-1.6	138	0.42	0.44	0.43	0.43	0.44	0.47	0.52	0.49	0.50	0.54	0.56	0.75
1.6-1.8	119	0.31	0.32	0.32	0.32	0.32	0.35	0.41	0.38	0.39	0.40	0.42	0.57
1.8-2.0	80	0.33	0.32	0.33	0.32	0.33	0.37	0.47	0.42	0.44	0.47	0.51	0.68
2.0-2.2	57	0.35	0.39	0.38	0.36	0.37	0.39	0.46	0.45	0.46	0.49	0.52	0.67
2.2-2.4	37	0.30	0.34	0.33	0.32	0.32	0.33	0.38	0.36	0.37	0.37	0.39	0.47
2.4-2.6	29	0.30	0.34	0.34	0.33	0.33	0.36	0.44	0.43	0.45	0.47	0.49	0.72
2.6-2.8	21	0.25	0.22	0.21	0.21	0.22	0.25	0.40	0.36	0.35	0.36	0.41	0.52
2.8-3.0	8	0.29	0.20	0.21	0.22	0.21	0.23	0.31	0.25	0.25	0.27	0.27	0.32
3.0-5.0	47	0.24	0.28	0.27	0.26	0.26	0.28	0.46	0.45	0.45	0.49	0.55	0.72
0.2-5.0	1971	0.48	0.50	0.49	0.47	0.48	0.53	0.55	0.53	0.54	0.56	0.58	0.73

(b) R.m.s. deviations for the 5β-androstane-3α,17β-diol invariants.

A range	No. of invari-ants	The exponent p=1/2					
		0.0	1.0	1.5	2.0	2.25	2.5
1.0-1.2	23	0.47	0.64	0.57	0.57	0.61	0.67
1.2-1.4	115	0.57	0.47	0.45	0.45	0.44	0.48
1.4-1.6	187	0.53	0.43	0.43	0.41	0.41	0.47
1.6-1.8	229	0.54	0.44	0.44	0.42	0.44	0.46
1.8-2.0	239	0.45	0.38	0.38	0.36	0.37	0.41
2.0-2.2	227	0.41	0.37	0.36	0.35	0.36	0.40
2.2-2.4	173	0.38	0.24	0.24	0.26	0.27	0.33
2.4-2.6	150	0.35	0.25	0.25	0.24	0.25	0.30
2.6-2.8	108	0.36	0.30	0.29	0.28	0.28	0.33
2.8-3.0	70	0.37	0.27	0.27	0.26	0.26	0.29
3.0-9.0	382	0.29	0.20	0.19	0.19	0.20	0.24
1.0-9.0	1903	0.43	0.35	0.34	0.33	0.34	0.38

diol are given in Table 8(b) and confirm the observation that the
accuracy of predicted invariants increases as A increases. It
should be recalled that the invariants actually used to solve
these structures were computed using an exponent p [see equation
(2.1), Chapter V] equal to 1/2, and $\vec{\ell}$ ranged over all measurable
reflections (i.e. $|E_{\vec{\ell}}|>0$).

It is also a point of interest to determine if the range of $\vec{\ell}$
can be restricted since the computation of Ψ is time consuming
even on very fast computers. Consequently, Ψ was computed for the
epiandrosterone and 5β-androstane-3α,17β-diol invariants by im-
posing each of the restrictions $|E_{\vec{\ell}}|>1.0$, 1.5, 2.0, 2.25, and 2.5
in turn. The calculation of the epiandrosterone invariants was
also repeated using an exponent p=2 to test experimentally the relative reliability of exponents 2 and 1/2. The r.m.s. deviations
of the invariants computed by each of these methods are shown, as
a function of A, in Table 8, and the approximate number of contributors to Ψ computed by making the various restrictions on the
range of $\vec{\ell}$, as well as the amount of time required to calculate
1971 epiandrosterone invariants by each method, are listed in
Table 9. These data demonstrate that a substantial amount of time
can be saved, without sacrificing accuracy, by restricting the
range of $\vec{\ell}$. There is little difference in the accuracy of invariants computed with the different restrictions for either structure
except for those computed with $|E_{\vec{\ell}}|>2.5$ which have slightly larger
deviations. However, in view of the small number of contributors
to Ψ when $|E_{\vec{\ell}}|$ is required to be greater than 2, and since the
amount of computing time required is not much greater when $|E_{\vec{\ell}}|>2$,
it is suggested that the optimum procedure would be to require that
$|E_{\vec{\ell}}|$ be greater than 2. The theoretical justification for this
restricted averaging process, using conditional probability distributions, is suggested by the work in Chapter IV, but details of
the analysis are omitted. The invariants computed using exponent
p=2 seem to be less accurate than those computed using p=1/2, es-

pecially at higher values of A. Since invariants with large A are
in general the most accurate and the most useful invariants, it
seems to be advisable to use exponent p=1/2 rather than an exponent equal to 2.

Table 9.   Number of contributors to individual averages [equation
           (2.2), Chapter V] and IBM 1130[++] computing time required
           to calculate 1971 structure invariants for epiandrosterone
           using various restrictions on the range of $\vec{\ell}$.

$\lvert E_{\vec{\ell}}\rvert >$	Number of contributors	Time (hours)
0.0	2000–4000	13.25
1.0*	400–800	4.5
1.5*	150–300	1.75
2.0†	150–300	1.5
2.25†	100–200	1.0
2.5†	50–100	0.5

* Using only the permutation $\vec{k}_1 = \vec{h}_1$ and $\vec{k}_3 = \vec{h}_3$.

† All three even permutations used to generate contributors.

†† 3.6μs, 8K word core, single-disk storage.

     If the three vectors in a given triple are denoted by $\vec{k}_1, \vec{k}_2$,
and $\vec{k}_3$, three different values of Ψ may be computed by permuting
the order of the vectors in the triple if the range of $\vec{\ell}$ is restricted. For example, one value is obtained by making the equiv-
alences $\vec{k}_1 = \vec{h}_1$ and $\vec{k}_3 = \vec{h}_3$ [where $\vec{h}_1$ and $\vec{h}_3$, as used in equation (2.2),
Chapter V, are any two of the three vectors in the triple], and
other averages are formed if $\vec{k}_3 = \vec{h}_1$ and $\vec{k}_2 = \vec{h}_3$ or $\vec{k}_2 = \vec{h}_1$ and $\vec{k}_1 = \vec{h}_3$.
If $\vec{\ell}$ ranges over all measurable reflections (i.e. $\lvert E_{\vec{\ell}}\rvert > 0$), the
three even permutations of vectors in the triple $(\vec{k}_1, \vec{k}_2, \vec{k}_3)$ will

result in the same value for Ψ. No new values will be obtained by
a permutation of the type $\vec{k}_1 = \vec{h}_3$ and $\vec{k}_3 = \vec{h}_1$, because the only difference in the way in which $\vec{h}_1$ and $\vec{h}_3$ enter into equation (2.2),
Chapter V is that $\vec{h}_3$ is used in the form $-\vec{h}_3$ whereas the sign of
$\vec{h}_1$ is not changed, and the change of sign on $\vec{h}_3$ is compensated by
the fact that, for every vector $\vec{\ell}$ which is used, $-\vec{\ell}$ is also used.
When the restrictions $|E_{\vec{\ell}}| > 1.0$ or 1.5 were made, only the permuta-
tion $\vec{k}_1 = \vec{h}_1$ and $\vec{k}_3 = \vec{h}_3$ was used, whereas all three permutations were
made and entered into the averages when the range of $\vec{\ell}$ was more
severely restricted. The dependence of K on A, as a function of
the restriction imposed on $\vec{\ell}$, was also studied, and the results
are seen in Fig. 1. K appears to become largely independent of A
as the restriction on the range of $\vec{\ell}$ becomes more severe, and when
$|E_{\vec{\ell}}|$ is required to be greater than 2 (or larger), it seems to be
a constant.

A further study of the errors entering into the calculation
of the structure invariants revealed that the magnitude of such
errors depends on the observed values of the invariants. The
average and r.m.s. deviations for two of the sets of invariants
computed using p=1/2 and a scaling parameter having a linear dependence on A, are shown in Table 10 for various ranges of the observed values of the invariants. The major conclusion to be drawn
from these data is that invariants whose observed values are small
are the most poorly computed. Not only is the r.m.s deviation for
these invariants much larger than the deviations found when the
observed value is larger, but the average deviation is also very
large and indicates that there is a substantial nonrandom error
entering into the computation of such invariants which appear, on
the average, to be calculated substantially larger than they really
are. Furthermore, this trend is independent of A. It does seem
to be true that invariants whose observed values are near unity are
computed more accurately if A is large, but the dependence of accuracy on A is not as striking as the dependence of accuracy on

Table 10. R.m.s. and average deviations of the computed structure invariants grouped according to the true values of the invariants

The deviations are $(\cos_{predicted} - \cos_{observed})$.

(a) Deviations for the epiandrosterone invariants:

			Observed $\cos(\phi_1+\phi_2+\phi_3)$				
			-1.00 to -0.00	0.00 to 0.25	0.25 to 0.50	0.50 to 0.75	0.75 to 1.00
	No. of invariants		204	68	112	163	474
A 0.0-1.0	$|E_{\vec{h}}|>0$	avg. dev.	0.64	0.18	0.15	-0.04	-0.16
		r.m.s. dev.	0.79	0.41	0.49	0.47	0.48
	$|E_{\vec{h}}|>2$	avg. dev.	0.90	0.36	0.26	0.04	-0.16
		r.m.s. dev.	0.97	0.44	0.40	0.35	0.36
	No. of invariants		54	21	50	74	290
A 1.0-1.5	$|E_{\vec{h}}|>0$	avg. dev.	0.69	0.35	0.25	0.00	-0.13
		r.m.s. dev.	0.77	0.45	0.44	0.40	0.37
	$|E_{\vec{h}}|>2$	avg. dev.	0.94	0.48	0.32	0.08	-0.10
		r.m.s. dev.	0.98	0.53	0.41	0.26	0.26
	No. of invariants		25	10	18	64	344
A 1.5-5.0	$|E_{\vec{h}}|>0$	avg. dev.	0.83	0.16	0.23	0.12	-0.04
		r.m.s. dev.	0.89	0.30	0.39	0.26	0.24
	$|E_{\vec{h}}|>2$	avg. dev.	1.09	0.44	0.36	0.14	-0.07
		r.m.s. dev.	1.14	0.48	0.45	0.20	0.17

(b) Deviations for the 5β-androstane-3α,17β-diol invariants:

			Observed $\cos(\phi_1+\phi_2+\phi_3)$				
			-1.00 to 0.00	0.00 to 0.25	0.25 to 0.50	0.50 to 0.75	0.75 to 1.00
	No. of invariants		26	12	26	44	122
A 1.0-1.5	$|E_{\vec{h}}|>0$	avg. dev.	0.51	-0.02	-0.11	0.05	-0.08
		r.m.s. dev.	0.76	0.33	0.60	0.54	0.52
	$|E_{\vec{h}}|>2$	avg. dev.	0.94	0.22	0.17	0.15	-0.10
		r.m.s. dev.	0.99	0.33	0.37	0.32	0.32
	No. of invariants		76	42	86	213	1256
A 1.5-90	$|E_{\vec{h}}|>0$	avg. dev.	0.28	0.18	0.06	-0.02	-0.04
		r.m.s. dev.	0.55	0.44	0.41	0.39	0.40
	$|E_{\vec{h}}|>2$	avg. dev.	0.78	0.50	0.28	0.11	-0.04
		r.m.s. dev.	0.84	0.55	0.37	0.26	0.22

the observed value of the invariant. Consequently, the large r.m.s.
deviations seen when all invariants with low A are grouped together
result primarily from the fact that invariants whose observed
values are relatively small are much more frequent at low A.

3. Summary

The following conclusions are based on the analysis of the
epiandrosterone and 5β-androstane-3α,17β-diol data.

1. If an incorrect phase assignment is made in the early
stages of a phase buildup, the molecule may still appear on a
resulting Fourier map but be translated perpendicular to the
planes having the same indices as the reflection in question.

2. Invariants involving certain reflections may be computed
less accurately than a general set of invariants, and they may be
subject to a nonrandom error.

3. The residuals calculated during the least-squares phase
determining procedure do not distinguish all false minima from the
true minimum.

4. An exponent p=1/2 in equation (2.2), Chapter V, seems to
be somewhat more reliable than an exponent p=2.0.

5. Invariants computed after imposing restrictions on $|E_{\vec{\ell}}|$
are, on the average, as accurate as those computed when $\vec{\ell}$ is allowed to range over all reflections, but computing time is substan-
tially less. The condition $|E_{\vec{\ell}}|>2$ is suggested for general use.

6. The scaling parameter K is largely independent of A when
the range of $\vec{\ell}$ is severely restricted. When the range of $\vec{\ell}$ is un-
restricted, K is a function of A, and in the case of these struc-
tures, the dependence was linear.

7. Invariants whose observed values are large are computed
more accurately than those which are, in reality, small. Invariants which are small show a relatively large nonrandom error since
their predicted values are usually larger than their observed values.

8.   Invariants with large A appear to be computed much more
accurately than invariants with small A, but this is largely a reflection of lower accuracy of the computation of invariants whose
observed values are relatively small because such invariants are
more frequent at low A.

CHAPTER IX

THE ESTRADIOL·UREA COMPLEX AND 6α-FLUOROCORTISOL;
TWO-DIMENSIONAL INVARIANTS

1. Introduction

Experience has shown that a sufficiently accurate and large
set of phases used as input to the tangent formula usually results
in extension of the set of known phases. For moderately complex
structures (20 to 40 independent non-hydrogen atoms) which have
appreciable overlap in the Patterson synthesis it is desirable to
determine some 40 to 60 phases correctly before applying the tangent formula. (In contrast, 6 - 8 phases often suffice in the
multiple solution technique because these are non-interacting among
themselves but interact strongly with others.) In the structures
of estriol and two androstane derivatives the correct phasing of
this basis set was achieved through the least-squares analysis of
the cosine invariants $\cos(\phi_1+\phi_2+\phi_3)$ which were evaluated by the
modified triple product formula (§2.1, Chapter V).

In the case of the two androstane derivatives, a statistical
analysis comparing the cosine invariants calculated by the modified triple product formula with the true values calculated from
the solved structure showed that calculated cosines are subject
to error. Consequently, an initial phasing procedure based upon
cosine invariants for triples involving only vectors in the centrosymmetric projections was suggested because these "two-dimen-

283

sional" invariants have only two possible values (±1)* whereas
the general (3D) invariants can have any value in the range −1 to
+1. The phasing problem would therefore be reduced to one of
selecting which invariants are +1.

For many space groups of orthorhombic or higher symmetry,
correct phasing of a large set of two-dimensional reflections can
be successfully extended to a three-dimensional solution through
the tangent formula. This method was applied to the structures
of estradiol·urea (1:1) and 6α-fluorocortisol** both of which
crystallize in the space group $P2_1 2_1 2_1$. In the case of the
estradiol·urea complex, it was not possible to identify unambiguously those invariants which were +1 on the basis of the results
of the modified triple product procedure. The intensity statistics for estradiol·urea indicated greater overlap in the Patterson
than is generally observed, and it was indeed found that the
fraction of invariants whose value was −1 was much greater than
that predicted by theory in which such overlap was assumed not to
exist. Application of the MDKS formula (§2.2 of Chapter V)

---

* In space group $P2_1 2_1 2_1$, there are two-dimensional triples of
the type $(h\bar{k}0, \bar{h}0\ell, 0k\bar{\ell})$, where $h+k+\ell=2n+1$, for which the cosine value must be 0. These triples are of no use in the phas-
ing procedure described here because any combination of the
permitted values of the three phases is consistent with the
zero value of the cosine.

** Essentially the same technique has been more recently applied
with success to the structure determination of methyl-phenyl
glyoxylate.

afforded a better evaluation of the cosine invariants for this
structure. The structure of 6α-fluorocortisol did not have extensive overlap in the Patterson, and the modified triple product
evaluation of the two-dimensional invariants was adequate.

Following the solutions of the structures, the observed
values of the cosine invariants for the solved structures were
compared with the triple product and MDKS predicted values in an
attempt to define the optimal procedure for the use of these
formulas to calculate the values of the two-dimensional invariants.

2. Diffraction Measurements

All crystallographic measurements were made on a General
Electric single crystal orienter, and the intensity data were
collected by the stationary counter-stationary crystal technique.
The systematic absences in the diffraction patterns were consistent with the orthorhombic space group $P2_12_12_1$ for both the
estradiol·urea complex ($C_{18}H_{24}O_2 \cdot CON_2H_4$) and 6α-fluorocortisol
($C_{21}H_{29}O_5F$), and the cell constants were a=24.631, b=7.951, and
c=9.302Å for the former and a=13.568, b=11.447 and c=12.247Å for
the latter. The final R factors $[R=\Sigma(\left||F_o|-|F_c|\right|)/\Sigma|F_o|]$ for the
observed data were 6% for the estradiol·urea complex and 7% for
6α-fluorocortisol.

A comparison of the observed distributions of the normalized
structure factor magnitudes, $|E|$, for these structures with the
theoretical distribution for noncentrosymmetric structures is
presented in Table 1. The extremely large values of the averages
$<(|E|^2-1)^2>$ and $<(|E|^2-1)^3>$ in the case of the estradiol·urea complex were the first indications that extensive overlap in the
Patterson function existed and that the statistics of this
structure would be exceptional.

Table 1. Intensity statistics.

	Theoretical	Calculated Estradiol·urea	Calculated 6α-fluorocortisol
Noncentric (3D) reflections			
$\langle(|E|^2-1)^2\rangle$	1.00	1.44	1.05
$\langle(|E|^2-1)^3\rangle$	2.00	4.86	3.92
Centric (2D) reflections			
$\langle(|E|^2-1)^2\rangle$	2.00	4.12	2.32
$\langle(|E|^2-1)^3\rangle$	8.00	35.85	17.29

3. Calculation of Structure Invariants

The two-dimensional cosine invariants for the two structures were evaluated using the modified triple product formula

$$\cos\ (\phi_1 + \phi_2 + \phi_3) \approx \frac{K\psi}{|E_1 E_2 E_3|}\ +\ \frac{R_3}{\lceil E_1 E_2 E_3 \rceil} \tag{3.1}$$

(§2.1, Chapter V)*. The quantity R_3,

$$R_3 = \frac{\sigma_3}{4\sigma_2^{3/2}}\left[\frac{3}{2}(|E_1 E_2|^2 + |E_2 E_3|^2 + |E_3 E_1|^2) + |E_1|^2 + |E_2|^2 + |E_3|^2 - \frac{7}{2}\right], \tag{3.2}$$

is a function only of the normalized structure factor magnitudes of the reflections forming the Σ_2 triple. The function ψ, where

$$\psi = \left\langle (|E_{\vec{k}}|^{1/2} - \overline{|E|^{1/2}})(|E_{\vec{h}_1 + \vec{k}}|^{1/2} - \overline{|E|^{1/2}})(|E_{-\vec{h}_3 + \vec{k}}|^{1/2} - \right.$$

$$\left. \overline{|E|^{1/2}})\Big| |E_{\vec{k}}| > t \right\rangle_{\vec{k}}, \tag{3.3}$$

---

* The usual abbreviations will be employed in this chapter: $|E_i| = |E_{\vec{h}_i}|$, $\phi_i = \phi_{\vec{h}_i}$, $i = 1,2,3$, in which it is assumed that $\vec{h}_1 + \vec{h}_2 + \vec{h}_3 = 0$, so that $\cos\ (\phi_1 + \phi_2 + \phi_3)$ is a structure invariant. A is defined as

$$A = \frac{2\sigma_3}{\sigma_2^{3/2}}\ |E_1 E_2 E_3|$$

where $\sigma_n = \sum_{j=1}^{N} Z_j^n$, Z_j is the atomic number of the atom labeled j, and there are N atoms in the unit cell.

is a restricted average over all reflections $\vec{k}$ such that $|E_{\vec{k}}|$ is
greater than some threshold t, and $|E|^{1/2}$ is the average value of
the square root of the normalized structure factor magnitudes for
all reflections in reciprocal space. Analysis of the evaluation
of cosines by the triple product formula for two androstane
structures (Chapter VIII) indicated that 2.0 was a suitable value
for the constant t.   The K values of 305.2 for estradiol·urea and
954.7 for 6α-fluorocortisol were chosen in such a way as to make
the empirical distribution of predicted invariants agree as
closely as possible with the theoretical distribution (§2.1,
Chapter V).

The same cosine invariants were evaluated using the (D-S)/S
formula (cf. the related (2.9) of Chapter V)

$$\cos(\phi_1+\phi_2+\phi_3) \simeq (D-S)/S, \qquad (3.4)$$

where

$$D = \left\langle (|E_{-\vec{h}_3+\vec{k}}|^2-1) \ \Big| \ |E_{\vec{k}}|>t, |E_{\vec{h}_1+\vec{k}}|>t \right\rangle_{\vec{k}} \qquad (3.5)$$

and

$$S = \left\langle (|E_{-\vec{h}_3+\vec{k}}|^2-1) \ \Big| \ |E_{\vec{k}}|>t \right\rangle_{\vec{k}} +$$

$$\left\langle (|E_{-\vec{h}_3+\vec{k}}|^2-1) \ \Big| \ |E_{\vec{h}_1+\vec{k}}|>t \right\rangle_{\vec{k}}. \qquad (3.6)$$

The threshold t is an arbitrary fixed number exceeding unity and
D, for example, is the average of $(|E_{-\vec{h}_3+\vec{k}}|^2-1)$ taken over all
reflections $\vec{k}$ such that $|E_{\vec{k}}|>t$ and $|E_{-\vec{h}_1+\vec{k}}|>t$. The optimum value

of t is unknown. On the one hand, it is desirable that t be large
for, if t is large, then S is large and errors which occur in the
numerator of (3.4) are not unduly exaggerated by division by S.
However, if t is large, then the numbers of contributors to the
averages in (3.5) and (3.6) are small and errors arising from the
finite sampling are large. If, on the other hand, t is chosen to
be small so as to increase the numbers of contributors to these
averages and thus to reduce the errors arising from the finite
sampling, then S is small and whatever errors occur in the numerator of (3.4) are exaggerated by division by the small number S.
As a compromise, the D and S terms were computed using t = 1.0
and t = 1.3. For structures of this complexity (20-30 nonhydrogen atoms in the asymmetric unit), it has been found that the
predicted cosine values are erratic if there are fewer than 200
contributors to D. A threshold value of t = 1.3 results in a
minimum of approximately 400 contributors to D. As the complexity
of the structure increases, the minimum number of contributors to
D increases but t remains approximately constant.

Because of the extensive overlap in the Patterson function of
real crystals, the (D-S)/S formula yields a greater percentage of
negative cosine invariants than is actually observed when the
structure is solved and the true values of the invariants are computed. Therefore, equation (3.4) has been altered in the manner
described in §2.2, Chapter V by the introduction of scaling parameters, K and M, to give the relationship

$$\cos (\phi_1 + \phi_2 + \phi_3) \simeq M(D-KS), \qquad (3.7)$$

the MDKS formula. The constant K of the MDKS formula is evaluated
so that within each group of triples having approximately the same
value of A, the proper proportion of invariants, as predicted by
theory, will be negative. The value for M is then found such that

the empirical distribution of cosine invariants agrees with the
theoretical distribution (§2.2, Chapter V and Chapter VIII). In
these structures, M and K were found to be independent of A. The
actual values of M and K are presented in Table 2.

4. Phasing the Basic Set

Initial attempts to phase the estradiol·urea complex using
two-dimensional cosine invariants computed by the modified triple
product formula proved unsuccessful. The same cosine invariants
were then reevaluated using the (D-S)/S formula (3.4), and only
those invariants for which (D-S)/S>0 were retained for use in
determining phases. These cosine invariants were assumed to have
the value unity and are presented in Table 3. The criterion that
the calculated value of (D-S)/S be positive, in order for the invariant to be accepted as having a value of +1, was a stringent
restriction since it resulted in the selection of many fewer invariants than the theoretical number which should have values of
+1. For example, among the invariants with A>3, more than 98%
should have values of +1, but the (D-S)/S calculation resulted in
the acceptance of only 65% of the invariants with A values in this
range, and these were invariants for which the (D-S)/S values were
highest. Among the groups of invariants having lower A values,
slightly less than 50% of the invariants were accepted as positive,
although even in the group of invariants with 1.0<A<1.5, more than
75% should have values of +1. Initial phasing based upon this set
of cosine invariants led to the solution of the estradiol·urea
structure, and it is this procedure that will be described in detail. The values of the cosine invariants calculated by the modified triple product formula were sufficiently good to allow solution of the 6α-fluorocortisol structure by a similar phasing procedure. Only those aspects of the phasing of 6α-fluorocortisol
that differ materially from the procedure followed for

Table 2. Values of the scaling constants M and K
          used in the MDKS formula (3.7).

	Threshold (t=1.0)		Threshold (t=1.3)	
Structure	M	K	M	K
Estradiol·urea	8.72	0.78	3.30	0.65
6α–Fluorocortisol	14.57	0.64	6.04	0.58

Table 3.   2D Triples for estradiol·urea having A>1.0 for
which the calculated value of the quantity
(D-S)/S was positive. The terms D and S are
defined in equations (3.5) and (3.6) respectively.

No.	VECTOR TRIPLE									A	COS*
1	20	1	0	-14	5	0	-6	-6	0	7.02	1
2	20	1	0	-20	0	-1	0	-1	1	6.59	1
3	20	1	0	-13	5	0	-7	-6	0	6.16	1
4	0	1	2	0	-4	3	0	3	-5	6.11	1
5	0	1	2	0	4	3	0	-5	-5	5.67	1
6	21	2	0	-14	5	0	-7	-7	0	5.08	1
7	20	1	0	-20	0	-3	0	-1	3	4.96	1
8	0	1	2	0	-3	3	0	2	-5	4.75	1
9	30	1	0	-30	0	-1	0	-1	1	4.71	1
10	0	1	2	0	1	3	0	-2	-5	4.68	1
11	20	1	0	-30	1	0	10	-2	0	4.46	1
12	21	2	0	-13	5	0	-8	-7	0	4.18	1
13	0	4	3	0	-1	4	0	-3	-7	4.01	1
14	0	1	1	0	1	4	0	-2	-5	3.95	1
15	0	1	2	0	3	7	0	-4	-9	3.91	1
16	20	1	0	-10	2	0	-10	-3	0	3.74	1
17	20	1	0	-11	5	0	-9	-6	0	3.69	1
18	0	4	3	0	1	4	0	-5	-7	3.67	1
19	0	1	2	0	-3	5	0	2	-7	3.60	1
20	0	1	2	0	-5	7	0	4	-9	3.58	1
21	20	1	0	-12	6	0	-8	-7	0	3.42	1
22	20	1	0	-13	6	0	-7	-7	0	3.34	1
23	21	2	0	-21	0	-5	0	-2	5	3.26	1
24	0	1	2	0	-4	5	0	3	-7	3.21	1
25	0	1	2	0	-3	7	0	2	-9	3.20	1
26	29	3	0	-14	5	0	-15	-8	0	3.18	1
27	0	1	1	0	3	2	0	-4	-3	3.06	1
28	0	1	1	0	-4	3	0	3	-4	3.00	1
29	1	3	0	14	5	0	-15	-8	0	2.99	1
30	21	1	0	-14	5	0	-7	-6	0	2.99	1
31	8	1	0	7	7	0	-15	-8	0	2.85	1
32	0	1	4	0	3	5	0	-4	-9	2.84	1
33	20	0	1	1	0	4	-21	0	-5	2.83	1
34	20	0	3	1	0	4	-21	0	-7	2.76	1
35	20	1	0	-9	5	0	-11	-6	0	2.66	1
36	14	5	0	-14	0	-5	0	-5	5	2.63	1
37	8	1	0	-14	5	0	6	-6	0	2.61	1
38	21	2	0	-21	0	-7	0	-2	7	2.53	1
39	0	1	2	0	1	5	0	-2	-7	2.53	1
40	21	2	0	-7	3	0	-14	-5	0	2.52	1

* True value of the cosine invariant.

Table 3 (continued)

No.	VECTOR TRIPLE									A	COS*
41	21	2	0	-10	3	0	-11	-5	0	2.42	1
42	0	3	3	0	-1	4	0	-2	-7	2.41	1
43	0	1	1	0	-3	4	0	2	-5	2.40	1
44	0	1	3	0	1	4	0	-2	-7	2.37	1
45	0	1	3	0	-4	3	0	3	-6	2.36	1
46	0	1	4	0	-3	5	0	2	-9	2.32	1
47	0	1	1	0	2	5	0	-3	-6	2.32	1
48	20	3	0	-20	0	-5	0	-3	5	2.30	1
49	10	3	0	-14	5	0	4	-8	0	2.27	1
50	30	0	1	-13	0	3	-17	0	-4	2.27	1
51	28	0	0	-8	-1	0	-20	1	0	2.23	1
52	7	1	0	8	7	0	-15	-8	0	2.16	1
53	0	1	2	0	-1	4	0	0	-6	2.15	1
54	0	4	1	0	-1	2	0	-3	-3	2.09	1
55	8	1	0	20	1	0	-28	-2	0	2.03	1
56	0	4	3	0	-2	5	0	-2	-8	2.00	1
57	0	3	2	0	-1	3	0	-2	-5	2.00	1
58	7	1	0	-13	5	0	6	-6	0	1.96	1
59	0	1	1	0	2	2	0	-3	-3	1.94	1
60	10	2	0	-6	6	0	-4	-8	0	1.90	1
61	9	1	0	-15	8	0	6	-9	0	1.88	-1
62	21	2	0	-14	4	0	-7	-6	0	1.85	1
63	0	3	2	0	-4	3	0	1	-5	1.84	-1
64	20	0	3	-21	0	7	1	0	-10	1.83	-1
65	18	0	1	-20	0	1	2	0	-2	1.76	1
66	18	0	1	-30	0	1	12	0	-2	1.74	-1
67	21	2	0	-15	4	0	-6	-6	0	1.74	1
68	10	2	0	4	3	0	-14	-5	0	1.74	1
69	4	2	0	11	6	0	-15	-8	0	1.70	1
70	20	3	0	-5	5	0	-15	-8	0	1.69	1
71	4	3	0	11	5	0	-15	-8	0	1.69	1
72	10	3	0	5	5	0	-15	-8	0	1.67	1
73	18	1	0	-14	5	0	-4	-6	0	1.66	1
74	20	0	5	-21	0	5	1	0	-10	1.64	-1
75	20	1	0	0	2	0	-20	-3	0	1.63	1
76	0	1	4	0	1	5	0	-2	-9	1.63	1
77	21	2	0	-4	6	0	-17	-8	0	1.63	1
78	18	0	1	2	0	2	-20	0	-3	1.62	1
79	9	1	0	14	5	0	-23	-6	0	1.61	1
80	20	0	1	-20	0	3	0	0	-4	1.60	1
81	18	1	0	-11	6	0	-7	-7	0	1.59	1
82	30	1	0	-17	4	0	-13	-5	0	1.55	1
83	30	0	1	-17	0	8	-13	0	-9	1.53	1

* True value of the cosine invariant.

Table 3 (continued)

No.	VECTOR TRIPLE			A	COS*
84	20 0 1	0 0 4	-20 0 -5	1.52	1
85	18 1 0	-11 5 0	-7 -6 0	1.52	1
86	7 1 0	21 2 0	-28 -3 0	1.51	1
87	0 2 2	0 -5 5	0 3 -7	1.51	1
88	5 1 0	-12 6 0	7 -7 0	1.51	1
89	20 0 1	1 0 6	-21 0 -7	1.49	1
90	0 2 2	0 3 5	0 -5 -7	1.49	1
91	0 4 1	0 -1 4	0 -3 -5	1.48	1
92	18 0 1	-1 0 3	-17 0 -4	1.47	1
93	8 1 0	13 1 0	-21 -2 0	1.46	-1
94	30 0 1	-15 0 3	-15 0 -4	1.45	1
95	0 1 3	0 -3 4	0 2 -7	1.44	1
96	18 1 0	-13 5 0	-5 -6 0	1.43	1
97	0 3 2	0 -5 5	0 2 -7	1.43	-1
98	18 0 1	-20 0 3	2 0 -4	1.42	1
99	30 0 1	-22 0 7	-8 0 -8	1.42	1
100	20 0 1	-21 0 5	1 0 -6	1.41	1
101	18 1 0	-15 8 0	-3 -9 0	1.39	1
102	0 3 2	0 1 3	0 -4 -5	1.38	1
103	8 1 0	-13 5 0	5 -6 0	1.38	1
104	0 4 1	0 1 4	0 -5 -5	1.37	1
105	0 4 2	0 -1 4	0 -3 -6	1.36	1
106	22 0 0	-9 -5 0	-13 5 0	1.36	1
107	18 0 1	-19 0 3	1 0 -4	1.34	1
108	18 0 1	2 0 4	-20 0 -5	1.34	1
109	18 1 0	10 2 0	-28 -3 0	1.33	1
110	1 0 3	20 0 5	-21 0 -8	1.32	1
111	9 1 0	-14 5 0	5 -6 0	1.32	1
112	20 0 1	-21 0 2	1 0 -3	1.31	-1
113	0 3 4	0 -5 5	0 2 -9	1.31	-1
114	6 3 0	9 5 0	-15 -8 0	1.31	1
115	0 1 3	0 -3 6	0 2 -9	1.30	1
116	18 0 1	-1 0 4	-17 0 -5	1.30	1
117	7 3 0	-22 5 0	15 -8 0	1.30	1
118	20 1 0	-14 4 0	- 6 -5 0	1.29	1
119	21 0 2	-1 0 4	-20 0 -6	1.29	1
120	10 0 0	20 0 1	-30 0 -1	1.28	-1
121	0 4 3	0 0 6	0 -4 -9	1.27	1
122	3 0 2	17 0 4	-20 0 -6	1.26	1
123	3 1 0	-11 6 0	8 -7 0	1.26	1
124	6 0 0	14 0 5	-20 0 -5	1.25	1
125	9 1 0	-13 5 0	4 -6 0	1.24	-1
126	6 2 0	-20 3 0	14 -5 0	1.23	1

* True value of the cosine invariant.

Table 3 (continued)

No.	VECTOR TRIPLE			A	COS*
127	0 3 4	0 1 5	0 -4 -9	1.21	-1
128	0 1 4	0 -3 4	0 2 -8	1.20	1
129	0 2 2	0 -3 3	0 1 -5	1.17	1
130	0 3 3	0 0 4	0 -3 -7	1.16	1
131	0 4 2	0 -3 3	0 -1 -5	1.13	1
132	4 3 0	-10 3 0	6 -6 0	1.10	1
133	0 0 4	1 0 4	-1 0 -8	1.02	-1
134	0 2 0	0 1 3	0 -3 -3	1.01	1
135	0 2 0	0 3 5	0 -5 -5	1.01	1
136	22 0 0	-2 0 -5	-20 0 5	1.00	1

* True value of the cosine invariant.

estradiol·urea will be discussed.

The list of triples (Table 3) presumed to have cosine invariant values of +1 based on the (D-S)/S calculations was inspected
in order to find a set of reflections which interacted with many
other reflections and which would consequently be suitable for
selection of an origin and enantiomorph. The reflections (0,4,3),
(20,1,0), and (1,0,4) interacted well and also satisfied the
parity restrictions placed on the origin defining reflections in
space group $P2_12_12_1$, and reflection (0,1,1) proved to be suitable
for specifying the enantiomorph. These reflections, $|E|$ values,
and assigned phases, as well as the reason for the phase assignment,
are displayed in Table 4 along with similar information for the
other two-dimensional reflections in the basis set.

From invariants #1 and #37 (Table 3), it is apparent that

$$\phi_{8,1,0} = \phi_{20,1,0} = 0.$$

(Table 5 may be consulted to aid in transforming the phases of
reflections to correspond to sign changes in the indices). From

Table 4. Basis set for the estradiol·urea complex as determined
          from the two-dimensional invariants predicted by the
          MDKS formula. The serial number(s) of the invariant(s)
          in Table 3 used for each phase assignment are indicated.

h	k	ℓ	$|E|$	ϕ	Invariant number	h	k	ℓ	$|E|$	ϕ	Invariant number
0	1	1	2.02	$\pi/2$	enantiomorph	0	2	8	1.21	0	56
0	4	3	3.31	0	origin	0	0	4	1.69	0	80
1	0	4	2.69	0	origin	20	0	5	2.34	$-\pi/2$	84
20	1	0	4.15	0	origin	20	3	0	1.93	0	48
8	1	0	1.72	0	1,37	1	0	10	1.60	0	64,74*
20	0	1	2.74	$-\pi/2$	2	0	2	0	1.51	0	75
0	3	2	1.87	$-\pi/2$	27	1	0	6	1.30	0	89,100
0	3	4	1.80	$-\pi/2$	28	13	1	0	1.54	$-\pi/2$	93*
0	2	5	2.72	π	43	10	3	0	2.11	π	70,72
0	1	4	3.01	$\pi/2$	14	10	2	0	2.40	π	16
0	3	7	2.20	$\pi/2$	13	30	1	0	2.24	0	11
0	5	7	2.01	$\pi/2$	18	11	5	0	1.93	$-\pi/2$	41
0	3	6	1.65	$-\pi/2$	47	9	6	0	2.18	$\pi/2$	16
0	1	3	2.47	$\pi/2$	45,57	0	4	2	1.55	π	106
0	1	2	2.80	$\pi/2$	10	30	0	1	3.47	$-\pi/2$	9
0	3	3	2.45	$\pi/2$	8	10	0	0	0.97	0	120*
0	3	5	2.52	$\pi/2$	4	1	0	8	1.78	0	133*
0	5	5	2.15	$\pi/2$	5	14	5	0	3.28	π	
20	0	3	2.50	$-\pi/2$	7	6	6	0	2.42	0	1
0	4	9	2.05	π	15,20	7	7	0	2.30	$-\pi/2$	6
0	2	7	1.84	π	19,42	13	6	0	1.60	$-\pi/2$	22
0	4	5	1.76	0	24	15	8	0	3.73	$-\pi/2$	31
0	2	9	1.69	π	25	29	3	0	1.21	$-\pi/2$	26
21	0	7	2.08	$-\pi/2$	34	1	3	0	1.30	$\pi/2$	29
21	2	0	3.07	$-\pi/2$	38	14	0	5	1.75	$\pi/2$	36
0	1	5	1.67	$\pi/2$	39,76	7	3	0	1.27	$\pi/2$	40
21	0	5	1.96	$-\pi/2$	23,33	4	8	0	1.67	π	49
28	0	0	2.27	π	51	4	3	0	1.27	0	68
28	2	0	1.48	0	55	5	5	0	1.16	$\pi/2$	72
0	4	1	1.18	0	54,91	6	0	0	2.26	π	124
0	0	6	1.39	π	53						
0	2	2	1.66	0	59,87						

*Phased incorrectly due to undetected negative invariants.

Table 5. Phase transformations corresponding
          to sign changes of indices ($\alpha=\pm\pi/2$,
          $\beta=0$ or π).

	hkℓ	h$\bar{k}\bar{\ell}$	$\bar{h}$k$\bar{\ell}$	$\bar{h}\bar{k}\ell$
ouu	α	$-\alpha$	α	$-\alpha$
oug	α	$-\alpha$	$-\alpha$	α
gou	α	α	$-\alpha$	$-\alpha$
uou	α	$-\alpha$	$-\alpha$	α
ugo	α	$-\alpha$	α	$-\alpha$
uuo	α	α	$-\alpha$	$-\alpha$
ogu	β	β	$\pi+\beta$	$\pi+\beta$
uog	β	$\pi+\beta$	β	$\pi+\beta$
guo	β	$\pi+\beta$	$\pi+\beta$	β
ggg	β	β	β	β

invariant #2, the phases of (0,1,1) and (20,1,0) determine $\phi_{20,0,1}$ to be $-\pi/2$. The systematic analysis of those invariants in which the phases of two reflections were known, thereby determining the phase of the third reflection, yielded a set of 49 phases. Table 4 lists the order in which the phasing proceeded together with the serial number(s) of the invariant(s) determining the phase. The only conflicts in phasing detected during this process concerned (0,1,5) and (0,5,5). Decisions made regarding these conflicts based upon A values and calculated cosine values proved to be correct. It was observed that a block of 13 strongly interacting reflections could be introduced into the basic set by assigning a phase to (14,5,0). When $\phi_{14,5,0}$ was assumed to be π, three resultant phase assignments (6,6,0; 4,8,0; and 6,0,0) agreed with Σ_1 indications, and this phasing was therefore considered to be correct.

The triples in Table 3 which have true cosine invariant values of -1 are so indicated and constitute a 10% error in invariant evaluation. Fortunately they caused incorrect phase assignment to only 4 of the 62 reflections used as tangent formula input (a 6% error). Starting from this basis set, the tangent formula was used to determine 238 additional phases, and 5 or more contributors were required for each new phase. The 62 input phases were held constant during all cycles of tangent formula refinement.

During the attempts to determine phases using cosine invariants computed by the modified triple product formula, the space group invariants (0,2,0) and (0,2,2) were involved in numerous conflicts between cosine calculations and Σ_1 indications. Furthermore, when these reflections were among those whose phases were input to the tangent formula, the refinement invariably altered the phase assignment by π. When phasing was based upon invariants computed by the (D-S)/S formula, the ambiguities regarding the phases of (0,2,0) and (0,2,2) were removed and both phases were determined to be 0, in agreement with weak Σ_1 indi-

cations. The tangent formula calculated the phase of (0,2,0) to
be π through all 7 cycles of refinement, and, although $\phi_{0,2,2}$ was
calculating 0 in early cycles, it eventually became π also. An
E map constructed from this tangent formula output revealed the
24 non-hydrogen atoms of the complex among the highest peaks in
the map. The true values of the phases of the (0,2,0) and (0,2,2)
reflections proved to be 0 in agreement with their initially
determined values, but in conflict with the tangent formula calculations of π which were based upon 113 and 134 contributors
respectively. This miscalculation of the tangent formula is a
result of the unusual statistical distribution of negative cosine
invariants and illustrates clearly how a too-early and too-heavy
dependence on the tangent formula may lead to occasional incorrect
phasing.

Less interaction of the two-dimensional reflections having
large $|E|$ values was observed in the structure of 6α-fluorocortisol.
Consequently, the working set of triples having A<1.00 which could
be considered to have values of +1 based upon triple product or
MDKS calculations was composed of approximately 150 invariants.
In addition to the origin and enantiomorph defining reflections,
four of the Σ_1 predictions that were considered most reliable were
required to build successfully a basic set of 51 input phases.
Three of these phases were for three-dimensional reflections each
of which occurred in invariants with two-dimensional reflections
for which A was greater than 2.5 and the computed cosine was
nearly 1.0, so that they could be phased with relative certainty.
Twenty-two of the non-hydrogen atoms were located in the first
E-map and the remaining five atoms of the steroid A-ring were located in a Fourier synthesis based upon the E map positions. Four
of the 51 input phases were determined to be incorrect.

5. Statistical Distribution of Invariant Values

In Table 6, the true distributions of negative cosine invariants, as a function of A, for triples composed of two-dimensional
(centric) reflections in the estradiol·urea and the 6α-fluoro-
cortisol structures, are compared with the theoretical probability
distribution of negative invariants. At all A values examined,
the number of negative invariants for estradiol·urea is greater
than that predicted by theory, and the excess of negative invariants increases as A increases. This observation explains the
numerous conflicts encountered in the early attempts to solve this
structure. In comparison, 6α-fluorocortisol has fewer negative
invariants than are theoretically predicted except in the range
2<A<3, and even in this range the number of excess negative invariants is small. This difference in percentages of negative
invariants in the two structures is directly related to the difference in intensity statistics seen in Table 1. Structures
having large values for the averages $\left\langle (|E|^2-1)^2 \right\rangle$ and $\left\langle (|E|^2-1)^3 \right\rangle$
may be expected to have high percentages of negative invariants,
especially for large A.

In order to compare the relative accuracy with which the
cosine invariants, cos $(\phi_1+\phi_2+\phi_3)$, are calculated by different
methods, the sets of invariants calculated by the modified triple
product formula (3.1) and the MDKS formula (3.7) for estradiol·urea
were each arranged in increasing order of their calculated values
after first breaking the triples into groups having similar A
values. Each set of invariants was then broken into quarters, and
the percentage of actual negative invariants occurring in each
quarter was computed and is displayed in Table 7. For example,
30% of the invariants, with A values in the range 1.0-1.5, whose
values as computed by MDKS (t = 1.0) ranked in the lowest quarter
for all invariants in this A range, actually had values of -1.
It is encouraging to note that for both formulas, and at nearly

Table 6. Percentages of negative two-dimensional cosine invariants.

A Range	Estradiol·urea No. invariants	%Neg.	6α-Fluorocortisol No. invariants	%Neg.	%Neg. (theoretical)*
1.0-1.5	282	24%	154	17%	22.3%
1.5-2.0	87	20	34	9	14.8
2.0-3.0	64	11	22	9	7.6
3.0-7.0	48	10	9	0	<2

*Percentage negative 2D invariants = $100/(1+\exp(A))$.

Table 7. Percentages of actual negative cosine invariants for estradiol·urea occurring in each quarter of the predicted invariants sorted in increasing order of calculated cosine. The 1st quarter consists of those invariants which had the lowest computed values, and the 4th quarter consists of those invariants with the highest computed values.

Formula	A Range	1st Quarter	2nd Quarter	3rd Quarter	4th Quarter
MDKS	1.0-1.5	30%	35%	23%	7%
(t=1.0)	1.5-2.0	35	26	9	18
	2.0-3.0	37	6	0	0
	3.0-7.0	38	0	0	0
MDKS	1.0-1.5	32%	32%	21%	10%
(t=1.3)	1.5-2.0	47	13	9	18
	2.0-3.0	25	6	12	0
	3.0-7.0	38	0	0	0
TPROD	1.0-1.5	36%	26%	22%	10%
(t=2.0)	1.5-2.0	43	26	18	0
	2.0-3.0	19	19	6	0
	3.0-7.0	23	0	8	8

all A values, the greatest percentages of negative invariants are
found among the quarter with lowest calculated invariant values.
Also, in general, the percentage of negative invariants in each
higher ranking quarter is less than the percentage in the next
lower quarter. The deviations from this trend probably result
from the smallness of the sample size. A comparison of the results for the two formulas shows that MDKS, especially with a
threshold value of 1.0, is more successful than the triple product
formula in identifying the negative invariants in the most useful,
higher A ranges; that is, the negative invariants occur more
frequently in the lower ranking quarters.

An important question which now presents itself is the matter
of what restrictions should be placed on the invariants which are
to be accepted as forming the basis of a phase determination so
that a maximum number of triples are available, but a minimum
number of triples with negative cosine invariants are included in
the working set. In order to answer this question, the invariants
were again divided into four groups having similar A values
(1.0-1.5, 1.5-2.0, 2.0-3.0, and 3.0-7.0) and sorted according to
their values as predicted by the triple product and MDKS formulas.
Within each group of triples with similar A values, certain percentages of the highest ranking invariants were accepted as having
values of unity, and the total number of invariants available for
use, as well as the percentage error (i.e. the percentage of invariants accepted as having values of +1 whose true values were
-1) in several cases are presented in Table 8. If 100 per cent of
the invariants which should be positive based on the MDKS (t=1.0)
calculations are accepted, then 400 estradiol·urea invariants are
available for use, and 17% of these invariants have true cosine
values of -1. Since 77% of the invariants with A values in the
range 1.0-1.5 should, in theory, have values of +1 as should 85%
of the invariants having 1.5<A<2.0, then accepting 50% of the invariants which "should be positive" as calculated by MDKS means

Table 8. Comparison of criteria for acceptance of predicted invariants.

Formula	Percentage of presumed positive invariants accepted	Estradiol·urea		6α-Fluorocortisol	
		No. invariants accepted	% error	No. invariants accepted	% error
MDKS (t=1.0)	100%	400	17%	179	13%
	75	303	12	134	13
	50	197	12	89	9
Triple product (t=2.0)	100%	400	16%	179	11%
	75	303	16	134	10
	50	197	13	89	8
MDKS and Triple product	100%	364	16%	161	12%
	75	247	12	107	10
	50	130	8	66	5

that the highest ranking 38.5% of the invariants with A values in
the range 1.0-1.5 were accepted as were 42.5% of the highest
ranking invariants having A values in the range 1.5-2.0. The
results shown in Table 8 demonstrate that, for both structures, if
all invariants with 1.0<A<7.0 are considered, approximately the
same amount of error is made by using the results of either formula. If only 50% of the highest ranking invariants which should
be positive are accepted, the error is reduced by 3-5% but the
number of available triples is only about half the number available if 100% of the invariants which should be positive are accepted. If it is required that a given triple must pass restrictions placed on its calculated value from both formulas simultaneously, the percentage error is reduced an additional 3-4%, but
only in the case where 50% of the invariants which should be positive are accepted.

The data presented in Table 9 show a further breakdown of
the percentage of negative invariants for estradiol·urea accepted
as positive. If A<2, the requirement that the computed value be
among the upper 50% of the invariants expected to be positive as
predicted by both formulas results in substantial reduction in
error. If A>2.0, restriction to the highest 75% of the invariants
predicted to be positive by MDKS greatly reduces the errors encountered. Further restriction only serves to reduce the number
of available triples, and the need for corroboration between the
formulas is not evident.

6. Conclusion

The necessity of acquiring a strong base of correctly phased
reflections prior to beginning tangent formula refinement is
demonstrated by the ambiguous behavior of the highly interacting
(0,2,0) and (0,2,2) reflections of estradiol·urea. The simplicity
and effectiveness of consistently phasing a basic set of two-di-

Table 9. Percentage of negative invariants accepted as positive for estradiol·urea.

A Range	Percentage of presumed positive invariants accepted	MDKS (t=1.0)	Triple product (t=2.0)	MDKS and Triple product
1.0–2.0	100%	21%	19%	19%
	75	15	19	16
	50	16	16	10
2.0–7.0	100%	9%	8%	8%
	75	1	5	2
	50	0	6	0

mensional reflections from accurately computed cosines have been
demonstrated. While the triple product formula was satisfactory
and perhaps preferable for computing cosine invariants for 6α-
fluorocortisol, it is apparent that the extensive overlap in the
estradiol·urea Patterson introduced elements not incorporated in
the derivation of the triple product formula. In the MDKS formula,
the problem of overlap is less troublesome and the accuracy of the
computed cosines for estradiol·urea illustrates the success of
this formula. It is hoped that the unambiguous phasing procedure
and the analysis of the comparison between the predicted and the
observed true values of the structure invariants described here
will provide useful guidelines for future structure determinations.

CHAPTER X[*]

HEXESTROL; QUADRUPLES

It frequently happens that the cosine seminvariant for an
important Σ_2 triple is calculated ambiguously by the triple
product and MDKS formulas. In such cases, consideration of the
relationship of the triple in question to triples which do have
well-determined cosines may allow its cosine to be evaluated.
Several types of relationships or identities among structure
invariants were discussed in §3 of Chapter V, and the present
chapter deals with the application of one of these relationships,
the quadruple relationship, to the structure of d,l-hexestrol.
The use of the auxilary phase determining formulas Σ_1 and pairs
to isolate groups of related Σ_2 triples is described elsewhere
in this monograph.

If $\bar{\Phi}_1$, $\bar{\Phi}_2$, $\bar{\Phi}_3$, and $\bar{\Phi}_4$ are structure invariants defined by
equations (3.1)-(3.4) of Chapter V, then the identity

$$\bar{\Phi}_1 + \bar{\Phi}_2 + \bar{\Phi}_3 + \bar{\Phi}_4 = 0$$

depends only on Friedel's law and is valid regardless of the space
group. If the space group symmetry restricts each of the invariants
$\bar{\Phi}_1$, $\bar{\Phi}_2$, $\bar{\Phi}_3$, and $\bar{\Phi}_4$ to be equal to 0 or π, as is true for all such

---

* This Chapter was written by Dr. Charles Weeks

invariants in centrosymmetric space groups and most invariants
for which all three component phases correspond to reflections in
the centrosymmetric projections of noncentrosymmetric space groups,
then all the cosine invariants cos $\bar{\Phi}_1$, cos $\bar{\Phi}_2$, cos $\bar{\Phi}_3$, and cos $\bar{\Phi}_4$
must have values of ±1, and the number of negative cosine invari-
ants in each such quadruple must be zero, two, or four. Most
quadruples for which the A values of all triples are large will
have four positive cosine invariants, and quadruples of four
negative cosines will be rare. The purpose of the present chapter
is to show how consideration of the quadruple relationships in a
set of triples having restricted cosine values may aid the analysis
of an unknown structure.

One of the first steps in a structure analysis is to calculate
the cosine values using the triple product and MDKS formulas. If
the cosines have restricted values, as they do when the phases
of all reflections comprising the triple are restricted, the
problem is greatly simplified. The triples can then be divided
into groups having similar A values, and the triples having
calculated cosines among the top T% or among the bottom B% in
each group as calculated by either or both formulas are then used
as a basis for determining phases by least-squares analysis of
structure invariants after the origin has been selected. Triples
among the top T% are given cosine values of unity, and those
among the bottom B% are given values of -1. The value of T is
chosen so that fewer triples than theoretically predicted will be
accepted as having cosine values of unity in each A range.
Similarly, if 10% of the triples with a given A value are expected
to have cosine values of -1, B is chosen to be a number less than
10. This procedure, with small variations, was used to obtain
initial phase sets for the estradiol·urea complex (Chapter IX)
and 9-t-butyl-9,10-dihydroanthracene (Chapter XIII).

It often happens, however, that the cosine for an important
triple which makes a useful phase accessible is calculated

ambiguously (i.e. is not among the top T% or among the bottom
B%). The quadruple analysis provides a means whereby it may be
possible to determine such cosines through their relationships
with triples whose cosines were among the top T% or among the
bottom B% (cf Chapter V, Tables 1-5). The quadruple presented
in Table 1 illustrates the situation in which three of the cosine
invariants, (cos $\overline{\Phi}_1$, cos $\overline{\Phi}_2$, and cos $\overline{\Phi}_3$), are strongly indicated
to have values of unity, but the fourth invariant is undetermined.

Table 1. A quadruple from the d,l-hexestrol
structure (space group $P2_1/c$).

| | | | | | cos $\overline{\Phi}_n^*$ | |
$\vec{h}$	$\vec{k}$	$\vec{h}-\vec{k}$	A	TPROD	MDKS
8 5 $\overline{12}$	$\overline{6}$ $\overline{1}$ 11	$\overline{2}$ $\overline{4}$ 1	3.01	1.06	1.08
$\overline{8}$ $\overline{5}$ 12	2 11 $\overline{9}$	6 $\overline{6}$ $\overline{3}$	3.14	1.09	1.80
6 1 $\overline{11}$	$\overline{2}$ $\overline{11}$ 9	$\overline{4}$ 10 2	2.31	1.16	1.13
2 4 $\overline{1}$	$\overline{6}$ 6 3	4 $\overline{10}$ $\overline{2}$	1.66	0.82	0.62

* Cos $\overline{\Phi}_1$ = cos($\phi_{8,5,\overline{12}}$ + $\phi_{\overline{6},\overline{1},11}$ + $\phi_{\overline{2},\overline{4},1}$), etc.

If $\phi_{\overline{6},6,3}$ and $\phi_{4,\overline{10},\overline{2}}$ were known, it would still not be possible
to evaluate $\phi_{2,4,\overline{1}}$ from the fourth triple by itself. Inspection
of the quadruple, however, reveals that cos $\overline{\Phi}_4$ must equal unity
if the accepted values for the other three invariants are correct.
Thus, the cosine invariants for any three of these triples contain
information concerning the value of the fourth cosine. This is
another manifestation of the over-determined nature of the phase
problem. Since each triple will, on the average, occur in
many quadruples, a triple whose cosine is determined from the
analysis of one quadruple may be used in a cyclic procedure to
determine the value of another cosine from a second quadruple.

This type of quadruple analysis of the cosine invariants for two-dimensional triples (triples with restricted cosine values) is therefore a convenient device for obtaining more information from the invariants than is obtained if the triples are only considered as isolated entities, and it is a procedure which has the advantage that it is amenable to automation.

The utility of the cyclic quadruple analysis described above has been tested by application to the unknown structure of d,l-hexestrol. Racemic hexestrol, $C_{18}H_{22}O_2$, crystallizes in space group $P2_1/c$ with one molecule in the asymmetric unit. A centrosymmetric structure was chosen for this experiment because it is then possible to obtain a much larger set of two-dimensional triples. The triples involving the 167 reflections with $|E_{obs}|$ greater than 1.8 were exhaustively generated (lowest possible $A \approx 1.3$), and the cosine invariants were computed by the modified triple product (equation 2.1 of Chapter V) and MDKS (equation 2.15 of Chapter V) formulas using thresholds of 2.0 and 1.3 respectively. These 806 triples were then broken into ten groups such that the A values were approximately constant within a group, and the highest ranking T% of the invariants in each group were accepted as having cosine values of unity. The T percentages used for each group are defined in Table 2. In addition, negative values were assigned to 34 cosines resulting in a total of 422 accepted cosines. Negative cosine values were chosen for the two lowest ranking triples with A greater than 3.0 because these cosines were calculated much lower than the cosines for other triples in this A range. All triples with A values between 1.5 and 3.0 for which the cosine was computed to be negative by both formulas were accepted, and this resulted in an acceptance rate for negative cosines of about 5% whereas the theoretical percentage of negative cosines is 7.5% if $2<A<3$ and 15% if $1.5<A<2$.

The 806 triples were input to a computer program which generated 712 nonredundant quadruples. These quadruples were then

Table 2. Acceptance criteria for the 806 hexestrol triples involving reflections with $|E| > 1.8$.

Total No. Invariants	A Range	% Positive Cosines			Required Minimum Number Quadruples Indicating Positive Cosine (cycles 1-10)									
		Theoretical	Observed*	Accepted(T%)	1	2	3	4	5	6	7	8	9	10
21	1.30-1.50	80%	76%	35%	5	5	5	4	4	3	3	2	2	2
92	1.50-1.80	84	87	40	5	5	5	4	4	3	3	2	2	2
98	1.80-2.00	87	89	45	5	5	5	4	4	3	3	2	2	2
72	2.00-2.15	89	86	50	3	3	2	2	2	2	2	2	2	2
80	2.15-2.30	90	94	55	3	3	2	2	2	2	2	2	2	2
95	2.30-2.50	91	93	60	3	3	2	2	2	2	2	2	2	2
81	2.50-2.75	93	93	65	2	2	2	2	2	2	2	2	2	2
81	2.75-3.00	94	98	70	2	2	2	2	2	2	2	2	2	2
93	3.00-3.50	96	98	75	2	2	2	2	2	2	2	2	2	2
93	3.50-8.00	99	99	80	2	2	2	2	2	2	2	2	2	2

* Calculated using the phases for the refined structure.

used as the basis for a cyclic analysis of the triples as
described above. The number of quadruple interactions relating
each triple having an unknown cosine with three triples having
known cosines was counted, and it was noted whether each quadruple
indicated a positive or negative value for the unknown cosine.
In each cycle, additional cosines were accepted as determined if
they had a specified minimum number of positive or negative contributors. The minimum numbers of contributors were functions
of the cycle and of the A values, and the numbers of contributors
required for cosines to be accepted as +1 are recorded in Table 2.
One negative contributor was required for all triples in the
first cycle, and two negative contributors were required for
acceptance in all later cycles. In general, fewer positive contributors were required for triples having high A values, and
more contributors were required in the early cycles because it is
desirable that the probability of incorrect cosine evaluation be
very low in these cycles. Since the program is capable of detect-
ing conflicts among known cosines, it was anticipated that such
conflicts would begin to arise as the acceptance criteria were
gradually lowered. In this application, however, no such conflicts were observed. The program also recognizes ambiguous
indications for unknown cosines. One such example (triple 282)
was found in the hexestrol data, and the cosine in question was
not accepted. A search for the cause of this conflict suggested
that one of the input negative cosines was incorrect, and this
deduction was verified after the structure was solved. Fewer
contributors were required for the negative cosines because the
input set of negative cosines was much smaller than the set of
positive cosines, and it was unlikely that any negative cosine
would have very many contributors. The total number of cosines
which were known following each cycle is recorded in Table 3.

Table 3. Total number of determined cosines
following each cycle of two-dimensional quadruple analysis.

Cycle	1	2	3	4	5
Total cosines determined	461	474	490	495	497

Cycle	6	7	8	9	10
Total cosines determined	503	504	512	514	514

In all, 92 additional cosines were known at the end of the
quadruple analysis, and 17 of these cosines were determined to
be negative.

The hexestrol triples which were predicted to have negative
invariants or which were found to have negative invariants after
the successful solution of the structure, are presented in Table 4.
There were 61 negative cosines among the set of 806 invariants
which was examined. Of the 34 cosine invariants which were input to the quadruples program with negative values, seven were
incorrectly determined (serial numbers 465, 395, 303, 245, 223,
122, and 48), and one of the additional 17 invariants (serial
number 385) determined as negative by the quadruples program was
incorrectly assigned because it was involved in a quadruple with
triple 303. All of the invariants which were initially assigned
positive values based on the T and B percentages given in Table 2
were correct, and all invariants assigned positive values during
the course of the quadruple analysis were correct. The two
conflicting quadruples which contributed to triple 282 are given
in Table 5. The first of these quadruples, which indicated a +1
value for cosine 282, involved three triples for which the cosine
values as calculated by both formulas, were very high, and in

Table 4. Hexestrol cosine invariants for which the true or the predicted values were negative.

SERIAL	$\vec{h}$	$\vec{k}$	$-\vec{h}-\vec{k}$	A	TPROD	MDKS	QUAD* IN	OUT	TRUE	NO. QUADS
764	3 2 -15	-2 -13 8	-1 11 7	3.96	0.20	-0.33	-1	-1	-1	1
706	3 2 -11	-4 2 11	1 -4 0	3.41	0.45	-0.04	U	U	-1	0
704	6 1 -11	-6 2 1	0 -3 10	3.40	0.00	-1.00	-1	-1	-1	2
572	1 11 -6	-1 -9 -5	0 -2 11	2.82	-0.28	-0.88	-1	-1	-1	1
546	3 1 -15	-6 -2 12	3 1 3	2.76	0.05	-0.18	U	U	-1	0
531	3 3 -11	-5 9 6	2 -12 5	2.71	0.04	-0.29	U	-1	-1	0
530	6 1 -11	-5 3 11	-1 -4 0	2.71	0.11	-0.26	U	-1	-1	2
518	3 5 -5	-4 -9 -5	1 4 10	2.68	0.32	0.03	-1	-1	-1	3
500	5 6 -13	-1 4 11	-4 -10 2	2.62	-0.30	-0.53	U	-1	-1	1
489	4 7 -12	-2 -2 7	-2 -5 5	2.58	0.32	-0.53	U	-1	-1	5
476	5 8 -14	-6 2 12	1 -10 2	2.54	-0.12	-0.57	-1	-1	+1	0
465	2 4 -14	-4 -10 13	2 6 1	2.51	-0.23	-0.50	-1	-1	-1	0
429	7 4 -14	-7 -2 3	0 -2 11	2.43	-0.22	-0.12	-1	-1	-1	1
398	5 9 -13	-1 -5 11	-4 -4 2	2.36	-0.33	-0.96	-1	-1	+1	0
395	3 1 -15	-4 -11 13	1 10 2	2.36	-0.09	-0.85	-1	-1	-1	0
394	3 2 -11	-5 4 10	2 -6 1	2.36	0.32	0.09	-1	-1	+1	0
390	8 3 -13	-2 -9 10	-6 6 3	2.35	-0.27	-0.03	U	U	-1	1
385	3 14 -6	-6 -4 1	3 -10 5	2.34	0.11	-0.18	-1	-1	+1	3
381	7 8 -8	-4 3 7	-3 -11 1	2.33	0.11	-0.44	U	-1	-1	2
375	5 1 -11	-5 -3 0	0 2 11	2.32	0.44	0.08	U	-1	-1	2
371	2 1 -12	-2 2 7	0 -3 5	2.31	-0.13	-0.52	U	-1	-1	3
357	6 1 -11	-6 6 3	0 -7 8	2.29	-0.15	-0.00	-1	-1	+1	3
303	6 0 -12	-6 -4 1	0 4 11	2.18	-0.57	-0.54	-1	-1	+1	1
300	2 13 -8	-5 -11 5	3 -2 3	2.17	-0.39	-0.75	-1	-1	-1	1

Column header for COS columns: $\longleftarrow \text{COS}(\phi_{\vec{h}} + \phi_{\vec{k}} + \phi_{-\vec{h}-\vec{k}}) \longrightarrow$

Table 4 (Continued)

$$\longleftarrow \text{COS}(\phi_{\vec{h}} + \phi_{\vec{k}} + \phi_{-\vec{h}-\vec{k}}) \longrightarrow$$

SERIAL	$\vec{h}$	$\vec{k}$	$-\vec{h}-\vec{k}$	A	TPROD	MDKS	QUAD* IN	QUAD* OUT	TRUE	NO. QUADS
297	6 1 -13	-4 4 13	-2 -5 0	2.17	-0.03	-0.30	-1	-1	-1	0
293	5 9 -13	-3 -13 12	-2 4 1	2.16	0.48	0.13	U	U	-1	1†
289	3 13 -12	-2 2 7	-1 -15 5	2.16	-0.13	-0.81	-1	-1	-1	0
281	4 11 -13	-2 2 7	-2 -13 6	2.14	-0.15	-1.22	-1	-1	-1	0
275	5 9 -6	-5 3 0	0 -12 6	2.13	0.45	-0.01	U	U	-1	0
265	1 4 -11	-2 -4 1	1 0 10	2.11	-0.06	-0.58	-1	-1	-1	2
249	2 4 -14	-3 -14 6	1 10 8	2.07	-0.44	-0.76	-1	-1	-1	0
245	5 1 -11	-5 -8 3	0 7 8	2.06	-0.07	-0.15	-1	-1	-1	2
239	0 12 6	0 -12 6	0 0 -12	2.04	0.43	-0.49	U	U	-1	0
238	2 18 -1	-4 -9 -5	2 -9 6	2.04	0.48	0.36	-1	-1	+1	2
237	4 7 -12	-7 -8 8	3 1 4	2.04	0.50	-0.43	-1	-1	-1	3
234	3 2 -11	-4 -4 2	1 2 9	2.03	0.15	-0.29	U	U	-1	0
223	5 8 -14	-4 11 13	-1 -19 1	2.01	-0.26	-0.07	-1	-1	+1	1
220	2 2 -7	-3 -11 1	1 9 6	2.01	0.90	0.45	U	U	-1	3
215	1 4 0	-2 3 -9	1 -7 9	2.00	-0.03	-0.56	-1	-1	-1	0
210	7 3 -11	-5 -4 10	-2 1 1	1.99	0.21	-0.19	-1	-1	-1	0
203	1 3 -1	-1 -11 -6	0 8 7	1.97	0.04	-0.90	U	-1	-1	0
196	3 19 -3	-4 -9 -5	1 -10 8	1.97	-0.14	0.28	U	U	-1	0
191	2 1 -1	-1 3 1	-1 -4 0	1.96	-0.04	-0.47	-1	-1	-1	0
168	0 8 6	0 -8 6	0 0 -12	1.91	0.05	-0.80	U	U	-1	0
166	1 15 -5	-2 -5 3	1 -10 2	1.90	-0.51	-1.17	-1	-1	-1	0
163	2 6 0	-3 1 -5	1 -7 5	1.90	-0.31	-0.97	-1	-1	-1	0
162	1 2 -3	-2 -9 -6	1 7 9	1.90	0.21	0.15	U	-1	-1	1
136	7 3 -14	-2 -11 11	-5 8 3	1.84	-0.13	-0.10	-1	-1	-1	0
126	4 11 -13	-5 -1 11	1 -10 2	1.83	-0.28	-0.50	-1	-1	-1	0

Table 4 (Continued)

$$\longleftarrow COS(\phi_{\vec{h}} + \phi_{\vec{k}} + \phi_{-\vec{h}-\vec{k}}) \longrightarrow$$

SERIAL	$\vec{h}$	$\vec{k}$	$-\vec{h}-\vec{k}$	A	TPROD	MDKS	QUAD* IN	OUT	TRUE	NO. QUADS.
122	7 3 -14	-4 11 13	-3 -14 1	1.82	-0.08	-0.00	-1	-1	+1	1
114	2 2 -12	-2 -4 1	0 2 11	1.80	-0.13	-0.54	-1	-1	-1	1
113	2 14 -11	-3 -5 5	1 -9 6	1.79	0.51	-0.31	U	U	-1	0
101	3 5 -5	-3 2 -3	0 -7 8	1.77	0.63	0.88	U	-1	-1	3
99	1 4 -12	-4 -8 12	3 4 0	1.77	-0.08	-0.02	-1	-1	-1	0
92	2 11 -9	-2 -4 1	0 -7 8	1.76	0.46	0.73	U	-1	-1	3
91	7 1 -12	-5 3 11	-2 -4 1	1.76	0.30	-0.20	U	-1	-1	1
80	3 11 -11	-4 -4 2	1 -7 9	1.74	-1.02	-0.49	-1	-1	-1	3
78	4 7 -12	-4 -4 2	0 -3 10	1.74	0.29	-0.25	U	-1	-1	1
69	2 2 -7	-6 7 2	4 -9 5	1.72	0.26	0.17	U	-1	-1	5
52	2 2 -7	-4 3 7	2 -5 0	1.66	-0.00	-0.19	-1	-1	-1	4
48	3 13 -12	-4 -4 2	1 -9 10	1.64	-0.15	-0.19	-1	-1	+1	0
45	2 5 -3	-2 2 -5	0 -7 8	1.62	0.29	0.15	U	-1	-1	1
37	2 2 -12	-4 -7 12	2 5 0	1.57	-0.41	-0.76	-1	-1	-1	1
22	4 5 -14	-3 -8 13	-1 3 1	1.50	0.12	0.36	U	-1	-1	0
19	4 7 -12	-2 -11 11	-2 4 1	1.48	0.36	-0.27	U	-1	-1	1
16	1 3 -1	-2 -12 -5	1 9 6	1.46	0.29	-0.23	-1	-1	-1	0
14	2 5 -3	-4 -8 -6	2 3 9	1.46	-0.17	-0.70	U	U	-1	0
10	3 1 -14	-2 -3 11	-1 2 3	1.43	-0.64	-0.13	U	U	-1	0
4	2 4 -1	-2 -12 -5	0 8 6	1.39	0.71	0.10	U	U	-1	0

*The cosine values input to and output from the quadruples program. Undetermined cosines are indicated by U.

†The quadruple involving this triple indicated a positive cosine, but cosines were never accepted as positive if there was only a single contributor (see Table 2).

Table 5. Conflicting quadruple contributors to triple 282.

Serial	$\vec{h}$	$\vec{k}$	$-\vec{h}-\vec{k}$	A	TPROD	MDKS	QUAD	TRUE	No. Quads
					$\longleftarrow$	$\cos(\phi_{\vec{h}}+\phi_{\vec{k}}+\phi_{-\vec{h}-\vec{k}})$	$\longrightarrow$		
282	6 2 $\overline{12}$	$\bar{5}$ 1 11	$\bar{1}$ $\bar{3}$ 1	2.14	0.78	0.81	U	+1	3
736	$\bar{6}$ $\bar{2}$ 12	1 11 $\bar{6}$	5 $\bar{9}$ $\bar{6}$	3.67	1.14	1.74	+1	+1	5
665	5 $\bar{1}$ $\overline{11}$	$\bar{1}$ $\overline{11}$ 6	$\bar{4}$ 12 5	3.18	1.20	1.07	+1	+1	7
517	1 3 $\bar{1}$	$\bar{5}$ 9 6	4 $\overline{12}$ $\bar{5}$	2.68	1.60	1.70	+1	+1	
282	6 2 $\overline{12}$	$\bar{5}$ 1 11	$\bar{1}$ $\bar{3}$ 1	2.14	0.78	0.81	U	+1	2
587	$\bar{6}$ $\bar{2}$ 12	5 $\bar{8}$ $\bar{3}$	1 10 $\bar{9}$	2.86	0.99	1.28	+1	+1	
245	5 $\bar{1}$ $\overline{11}$	5 8 3	0 7 8	2.06	-0.08	-0.15	-1	+1	0
175	1 3 $\bar{1}$	$\bar{1}$ $\overline{10}$ 9	0 7 $\bar{8}$	1.93	1.14	0.80	+1	+1	0

addition, each of these three triples was involved in at least
three other quadruples in which all cosine values were determined
and which were consistent with the positive value of the cosine
in question.   In contrast, the second quadruple indicated that
cosine 282 was negative, but there were internal quadruple checks
on only one of the other three triples (no. 587) in this
quadruple.   One of the other triples (no. 175) was quite strongly
indicated to have a positive cosine although there were no
quadruple conformations of this value.   The remaining triple
(no. 245) was one of the seven triples which was later found to
have been input to the quadruples program with an incorrect
value.   However, the evidence provided by this conflict strongly
suggested that cosine 245 was actually positive since a change
in the value of this cosine would be the easiest way to resolve
the conflict.   Unfortunately, it was not possible to detect any
of the other six incorrectly determined negative cosines by a
similar analysis.   However, three of these incorrect cosines were
not involved in any quadruples with other triples which all had
determined cosines, and thus there was no quadruple confirmation
of their values.   If all triples which had no quadruple contributors had been eliminated at the end of the quadruple
analysis, four (including 245) of the incorrect negative cosines
would have been eliminated, but the price which would have been
paid would have been the rejection of 14 negative and 48 positive
cosines for which the values input to the quadruples procedure
were correct, and the working set would have been reduced from
514 triples to 448 triples.   In addition, triple 303 occurred
only in the quadruple which was used to determine triple 385, and
therefore it had no quadruples which checked its internal
consistency with the remainder of the input set.   The remaining
incorrect triples (223 and 122) occurred in a quadruple together,
and they were internally consistent.   However neither occurred in
any other quadruples.

   The 513 triples (triple 245 was eliminated because of the
ambiguity it caused in triple 282) whose cosines were known at
the end of the quadruples analysis provided a foundation for the
phasing of the hexestrol structure. Phases were assigned to
$(4,2,\overline{11})$, $(5,6,\overline{1})$, and $(2,5,\overline{5})$ in order to select the origin, and
the Σ_1 formula determined the values of $\phi_{6,0,\overline{12}}$ and $\phi_{2,0,10}$ to be
0. It was observed that only a few phases would be accessible
from the 513 triples unless $(5,0,\overline{10})$ was introduced as an
ambiguity, and $\phi_{5,0,\overline{10}}$ was first chosen to be 0, a value which
later proved to be correct. The number of known phases was
increased to 142 using the calculated values (±1) of the 513
cosine invariants in nine cycles, and the 20 nonhydrogen atoms of
the structure were among the 21 highest peaks in the E map based
on these phases, all of which proved to be correct. The total
number of phases known at the end of each cycle as well as the
number of contributors required for new phases in each cycle are
recorded in Table 6.

   Table 6. Summary for the cycles of phase acquisition
             via calculated cosine invariants.

Cycle	1	2	3	4	5
Total phases	14	28	41	53	87
Minimum number of contributors	1	1	1	1	1

Cycle	6	7	8	9
Total phases	113	128	137	142
Minimum number of contributors	2	3	3	3

Ambiguous phases (phases indicated to have different values by
different contributors) were not accepted in any cycle. The

normalized structure factors used to make the E map are given
in Table 7 along with some information concerning the types of
triples which were used to determine each phase. The number of
contributing triples each phase had in the cycle in which it was
first determined is recorded as well as the number of triples in
which the phase in question was involved and was consistent with
one or two other phases first determined in the same cycle. The
number of contributing and consistent triples having cosines
which were determined by the quadruple analysis or cosines which
are negative are also indicated. Examination of this table
reveals that there were few phases which were totally dependent
on triples made available by the quadruple analysis. However,
these triples did provide many additional internal checks on the
consistency of the phasing.

The only major point of confusion in this analysis arose
when triple 303 (one of the seven triples which were incorrectly
input to the quadruples program with a negative cosine) was used
to find the phase of reflection $(0,4,11)$ in cycle 3. The phase
of $(2,9,6)$ was found in cycle 4 from a triple involving $(0,4,11)$,
and an ambiguity arose in the case of $(2,4,\bar{1})$ which was also
accessible by a route which did not depend on $(0,4,11)$. Reflections $(0,4,11)$ and $(2,9,6)$ were used to determine seven more
phases in cycle 5, and an additional ambiguity was encountered in
the case of $(1,4,\overline{11})$. In cycle six, nine additional ambiguities
were found as well as three conflicts among phases which were
already determined. At this point, the situation was analyzed,
and it was found that the smallest adjustment which would allow
the resolution of the ambiguities and conflicts would be to change
the value of the cosine for triple 303 and the related triple 385.

The conflicts which resulted from the use of the other incorrectly determined cosines were much less serious because they
occurred later in the phasing process and affected very few
phases. Reflection $(2,4,\overline{14})$ was incorrectly phased in cycle 5

Table 7. The 142 hexestrol phases on which the E map was based.

Serial	Reflection	E	No. Contributors	No. Consistencies	No. Cosines Determined by Quad	No. Negative Cosines
1	4 2 $\overline{11}$	3.28				
2	5 6 $\overline{1}$	3.13				
3	2 5 $\overline{5}$	2.73				
4	6 0 $\overline{12}$	2.83				
5	2 0 10	2.48				
6	5 0 $\overline{10}$	3.52				
7	2 7 $\overline{6}$	2.45	1	1	0	0
8	3 5 $\overline{5}$	2.19	1	2	0	0
9	6 2 $\overline{1}$	2.95	1	1	0	0
10	7 0 0	1.91	1	1	0	0
11	1 2 1	-2.51	1	3	0	0
12	3 1 4	1.81	1	1	0	0
13	0 6 9	3.18	1	2	0	0
14	1 4 10	2.93	1	1	0	0
15	3 1 $\overline{14}$	-1.80	1	1	1	0
16	3 0 $\overline{12}$	-2.61	1	1	0	0
17	7 2 $\overline{11}$	-2.51	1	2	0	0
18	5 4 $\overline{10}$	-2.28	1	2	2	0
19	1 3 $\overline{8}$	2.23	1	4	2	0
20	6 6 $\overline{3}$	2.13	1	1	0	0
21	4 4 $\overline{2}$	-1.81	2	3	1	0
22	6 4 $\overline{1}$	-1.85	1	0	0	0
23	1 4 0	2.41	2	2	1	0
24	2 6 1	-2.40	2	3	0	0
25	3 1 5	1.93	2	0	0	0
26	4 9 5	-1.90	1	0	1	1
27	1 2 9	2.60	2	3	0	0
28	0 2 11	2.48	1	1	1	0
29	7 4 $\overline{14}$	2.44	1	1	0	0
30	4 5 $\overline{14}$	-1.82	1	2	0	0
31	1 6 $\overline{13}$	-2.96	1	1	1	0
32	6 4 $\overline{12}$	-2.03	2	1	0	0
33	4 8 $\overline{10}$	-1.93	2	0	0	0
34	1 10 $\overline{9}$	-2.38	2	2	0	0
35	2 12 $\overline{8}$	2.88	2	3	1	0
36	1 11 $\overline{6}$	2.64	2	3	3	0
37	7 2 $\overline{3}$	1.83	2	1	0	0
38	4 10 $\overline{2}$	-1.94	1	1	1	0
39	0 11 6	-2.48	1	0	1	0
40	0 13 7	2.20	1	1	2	0
41	0 4 11	-1.89	1	0	0	0

Table 7. (Continued)

Serial	Reflection			E	No. Contributors	No. Consistencies	No. Cosines Determined by Quad	No. Negative Cosines
42	5	6	$\overline{13}$	-2.67	1	0	0	0
43	4	10	$\overline{13}$	-2.51	1	0	0	0
44	7	6	$\overline{10}$	-2.61	1	0	0	0
45	4	15	$\overline{8}$	-2.16	1	0	0	0
46	3	16	$\overline{8}$	1.98	3	0	0	0
47	3	4	0	1.85	1	0	0	0
48	1	8	2	2.24	1	1	1	0
49	3	2	3	1.85	1	1	1	0
50	2	14	3	1.81	1	0	0	0
51	1	17	3	2.79	4	0	0	0
52	1	9	5	1.96	1	0	0	1
53	2	9	6	-2.23	1	0	0	0
54	3	2	$\overline{15}$	-2.33	2	4	1	0
55	2	4	$\overline{14}$	-1.90	1	0	0	0
56	5	8	$\overline{14}$	2.24	1	3	1	1
57	4	4	$\overline{13}$	-2.21	1	1	0	0
58	6	2	$\overline{12}$	-2.47	1	4	0	1
59	5	3	$\overline{12}$	2.33	1	7	0	0
60	4	7	$\overline{12}$	2.04	1	6	5	1
61	4	8	$\overline{12}$	1.96	1	2	0	0
62	6	1	$\overline{11}$	2.46	1	7	0	1
63	3	2	$\overline{11}$	1.96	1	0	0	0
64	7	3	$\overline{11}$	1.99	1	3	0	0
65	1	5	$\overline{11}$	2.36	1	1	0	0
66	1	8	$\overline{9}$	1.97	1	3	1	0
67	2	13	$\overline{8}$	-2.58	1	3	2	0
68	2	2	$\overline{7}$	-2.10	2	9	4	2
69	8	2	$\overline{7}$	2.53	2	4	1	0
70	1	14	$\overline{7}$	-2.96	1	3	1	0
71	3	4	$\overline{6}$	2.99	4	11	3	0
72	5	9	$\overline{6}$	2.56	2	6	1	0
73	1	13	$\overline{4}$	-1.89	1	8	0	0
74	4	14	$\overline{3}$	2.37	1	2	0	0
75	3	19	$\overline{3}$	2.16	1	3	3	0
76	6	7	$\overline{2}$	1.96	1	6	2	1
77	2	18	$\overline{1}$	2.19	1	2	2	1
78	2	6	0	-1.96	1	3	0	1
79	1	10	2	2.09	1	3	1	1
80	4	8	4	2.41	1	2	1	0
81	0	3	5	1.86	1	5	2	0
82	3	6	5	1.88	1	2	1	0
83	1	7	5	2.29	1	10	4	1
84	0	8	7	-2.32	1	3	1	0
85	0	7	8	-1.99	1	1	1	2
86	1	0	10	-2.28	1	2	0	0
87	0	0	12	-2.59	1	0	1	0

Table 7. (Continued)

Serial	Reflection	E	No. Contributors	No. Consistencies	No. Cosines Determined by Quad	No. Negative Cosines
88	7 2 $\bar{13}$	2.35	2	0	0	0
89	3 8 $\bar{13}$	-2.03	3	2	0	0
90	2 1 $\bar{12}$	-2.69	2	2	0	1
91	8 5 $\bar{12}$	3.04	2	8	0	0
92	4 6 $\bar{12}$	-2.34	2	2	0	0
93	2 3 $\bar{11}$	1.91	2	6	5	0
94	5 3 $\bar{11}$	-2.07	3	4	4	1
95	1 4 $\bar{11}$	2.30	4	5	2	2
96	2 11 $\bar{9}$	-2.20	2	4	0	0
97	4 0 $\bar{8}$	1.85	2	1	0	0
98	7 8 $\bar{8}$	-2.51	4	3	2	1
99	4 16 $\bar{8}$	1.84	4	1	1	0
100	6 10 $\bar{7}$	2.27	4	5	1	0
101	5 11 $\bar{5}$	-2.06	3	0	0	1
102	5 8 $\bar{3}$	-2.21	8	3	1	0
103	1 3 $\bar{1}$	-1.85	3	8	0	0
104	2 4 $\bar{1}$	1.83	9	4	4	1
105	3 11 $\bar{1}$	-2.28	2	4	0	0
106	3 14 $\bar{1}$	2.11	3	2	0	0
107	5 3 0	-2.00	2	3	1	0
108	2 5 0	1.93	2	9	5	0
109	5 5 0	-1.92	3	8	2	0
110	2 2 5	1.94	8	7	1	0
111	4 8 6	-1.88	4	2	0	0
112	1 10 3	2.18	2	1	0	0
113	0 3 10	-2.13	8	3	4	2
114	7 3 $\bar{14}$	-2.11	4	1	3	0
115	6 1 $\bar{13}$	-2.31	7	2	0	1
116	7 1 $\bar{12}$	2.11	3	1	1	1
117	3 11 $\bar{11}$	-2.14	6	2	3	0
118	2 9 $\bar{10}$	-2.43	3	0	0	0
119	4 3 $\bar{7}$	-1.86	7	0	3	2
120	1 11 $\bar{7}$	-2.99	3	1	1	1
121	3 14 $\bar{6}$	-2.27	5	4	2	1
122	4 12 $\bar{5}$	-2.57	6	4	1	0
123	2 5 $\bar{3}$	1.91	5	0	1	1
124	4 6 4	2.29	3	1	1	0
125	3 10 5	-2.54	4	1	1	0
126	0 8 6	1.83	4	1	0	0
127	0 12 6	-1.89	6	0	3	0
128	0 12 7	1.91	3	0	1	0

Table 7. (Continued)

Serial	Reflection	E	No. Contributors	No. Consistencies	No. Cosines Determined by Quad	No. Negative Cosines
129	5 1 $\overline{11}$	2.13	3	1	2	1
130	2 11 $\overline{11}$	1.80	3	0	1	2
131	3 16 $\bar{9}$	2.34	4	0	0	0
132	2 13 $\bar{6}$	2.49	4	1	1	0
133	1 15 $\bar{5}$	-2.18	3	2	0	1
134	1 2 $\bar{3}$	-1.90	3	0	0	0
135	5 5 $\bar{2}$	2.25	5	2	1	0
136	1 19 $\bar{1}$	-2.19	6	0	0	0
137	1 9 6	-1.91	3	2	1	1
138	8 3 $\overline{13}$	-2.07	3	0	0	1
139	4 0 $\overline{12}$	-2.56	1*	0	0	0
140	3 3 $\overline{11}$	2.56	3	0	0	0
141	2 6 $\overline{11}$	1.88	3	0	0	0
142	2 7 9	-2.46	3	0	0	0

* Confirms moderately strong Σ_1 prediction.

because of triple 465, and when $\phi_{3,14,\bar{6}}$ was introduced in cycle 7 despite an ambiguity (four contributors indicating this phase to be π opposed an indication of 0 from triple 249), a conflict arose which could be resolved by making the cosine for either triple 249 or triple 465 positive. Both the triple product and MDKS values were lower for cosine 249. It was decided that this cosine must be the true negative, and the error in cosine 465 was detected. The only other incorrect negatives which entered into the phasing were triples 122 and 223. These triples, which are related to each other by the quadruple shown in Table 8, indicated in cycle 9 that the phase of reflection $(4,11,\overline{13})$ was 0 whereas triples 126 and 281 gave the opposite value. Triples 126 and 281 both have negative cosines, and they were input to the quadruples program with negative cosines but did not enter into any quadruples. The triple product and MDKS values calculated for these cosines were lower than the values for triples 122 and 223, and they indicated that the former had the true negative cosines. The phase of $(4,11,\overline{13})$ was not, however, included in the set used to make the E map.

Table 8. A quadruple relating triples 122 and 223.

Serial	$\vec{h}$	$\vec{k}$	$-\vec{h}-\vec{k}$	A	$\cos(\phi_{\vec{h}}+\phi_{\vec{k}}+\phi_{-\vec{h}-\vec{k}})$		QUAD	No. Quads
					TPROD	MDKS		
219	7 3 $\overline{14}$	$\overline{5}$ $\overline{8}$ 14	$\overline{2}$ 5 0	2.01	0.63	0.37	+1	3
122	$\overline{7}$ $\overline{3}$ 14	4 $\overline{11}$ $\overline{13}$	3 14 $\overline{1}$	1.82	-0.08	-0.00	-1	1
223	5 8 $\overline{14}$	$\overline{4}$ 11 13	$\overline{1}$ $\overline{19}$ 1	2.01	-0.26	-0.07	-1	1
192	2 $\overline{5}$ 0	$\overline{3}$ $\overline{14}$ 1	1 19 $\overline{1}$	1.96	1.47	1.23	+1	4

CHAPTER XI[*]

9α-FLUOROCORTISOL; PAIRS

9α-Fluorocortisol is a steroid containing 27 nonhydrogen
atoms which crystallizes in space group $P2_1 2_1 2_1$ with one molecule
per asymmetric unit in a cell having the dimensions a = 10.087,
b = 23.710, and c = 7.660Å. The procedure used to obtain an
initial set of 49 phases for this structure will be described in
detail because it illustrates the use of many different phase
determining formulas which have been discussed in this monograph.
It is an especially pleasing example because all of the deductions
based on the predicted cosine seminvariants proved to be correct,
and this resulted in a set of 49 phases for two-dimensional reflections which were all correct. This set of 49 phases was input to
the tangent formula which was used to determine 201 additional
phases, and 26 of the nonhydrogen atoms were found among the
highest 35 peaks on the E map based on all 250 phases. The
strength of the particular set of base phases which were used is
demonstrated by the fact that, throughout the tangent cycles,
multiple minima were encountered in the modified tangent minimization function only twice. Wherever possible, enough data are
given that alternative methods of phasing this structure can be
investigated, and although it is undoubtedly true that many such

----

*   This chapter was written by Charles Weeks.

approaches would be ultimately successful, the merit of an approach which requires inspection of a single Fourier map, in which nearly the entire structure can be found among the strongest peaks, is obvious.

In space group $P2_12_12_1$, the reflections in the three centrosymmetric projections are especially easy to work with, not only because their phases have restricted values, but also because several formulas are known which may allow the determination of many such phases. Consequently, when faced with an unknown $P2_12_12_1$ structure, it is often advantageous to concentrate on the two-dimensional reflections first. This approach was profitably pursued with the estradiol·urea complex as described elsewhere in this monograph (Chapter IX), and the problem of phasing 9α-fluorocortisol was first attacked in this fashion also. In those cases where Σ_2 interactions among the two-dimensional reflections are sparse, or where the computed cosine seminvariants are ambiguous or lead to many conflicting phase assignments, it is, of course, necessary to introduce three-dimensional phases at an earlier stage. However, under favorable conditions such as those encountered with the 9α-fluorocortisol data, it is possible to build up a large set of reflections whose phases are known, with perfect accuracy, before entering the tangent formula.

The two-dimensional reflections for which $|E|_{obs}$ is greater than 1.3 are presented in Table 1, and the Σ_2 triples, which involve these reflections and which also have A values greater than 1.3, are presented in Table 2. The cosine seminvariants, $\cos(\phi_1+\phi_2+\phi_3)$, corresponding to each of these triples were computed by means of both the modified triple product and MDKS formulas using thresholds of 2.0 and 1.3 respectively. The triple product scaling parameter K had a value of 450, and the MDKS scaling parameters were defined as M = 6.19-0.91*A and K = 0.41 + 0.03*A. A partial selection of the origin was made by assigning phases to (6,1,0) and (7,12,0) because examination of Tables 1

Table 1. Two-dimensional reflections having |E|obs>1.3

0gg	|E|	0gu	|E|	0ug	|E|	0uu	|E|
0 16 2	3.02	0 8 5	2.17	0 15 4	3.47	0 23 3	2.67
0 2 6	2.90	0 26 3	2.12	0 1 6	2.42	0 7 3	2.19
0 18 2	2.55	0 12 1	2.11	0 25 2	2.38	0 11 1	1.93
0 14 4	2.11	0 4 3	2.08	0 19 2	2.10	0 13 5	1.76
0 16 4	1.77	0 8 7	2.03	0 3 6	2.00	0 11 5	1.71
0 4 6	1.42	0 22 1	1.94	0 1 4	1.90	0 23 5	1.68
0 24 2	1.41	0 6 3	1.91	0 17 2	1.71	0 27 1	1.63
0 12 2	1.39	0 10 7	1.60	0 23 4	1.59	0 9 1	1.44
0 0 4	1.37	0 8 3	1.40	0 15 6	1.36	0 15 1	1.36
		0 24 3	1.31				

g0g	|E|	g0u	|E|	u0g	|E|	u0u	|E|
10 0 4	2.08	8 0 5	1.46	1 0 4	1.56	5 0 7	2.39
4 0 0	1.88	4 0 1	1.36	1 0 6	1.50	1 0 1	2.24
4 0 6	1.61			5 0 6	1.41	5 0 3	2.16
8 0 0	1.44			7 0 6	1.39	5 0 1	1.75
6 0 6	1.38					7 0 1	1.61
0 0 4	1.37					9 0 1	1.55
4 0 2	1.33					3 0 3	1.53

gg0	|E|	gu0	|E|	ug0	|E|	uu0	|E|
4 6 0	2.18	8 11 0	4.25	7 12 0	3.48	1 11 0	2.60
2 10 0	1.98	6 1 0	4.05	7 22 0	2.54	1 15 0	2.13
6 10 0	1.92	2 1 0	3.12	5 6 0	1.85	1 21 0	1.98
4 0 0	1.88	8 19 0	2.49	5 10 0	1.70	5 7 0	1.91
10 10 0	1.83	6 17 0	2.47	7 20 0	1.66	1 13 0	1.79
8 4 0	1.76	4 17 0	2.12	7 4 0	1.65	7 17 0	1.49
2 18 0	1.74	6 19 0	2.07	1 12 0	1.60	1 23 0	1.43
2 8 0	1.68	8 9 0	1.73	3 22 0	1.37	3 1 0	1.37
6 20 0	1.62	6 21 0	1.68	9 10 0	1.36	5 21 0	1.33
6 14 0	1.60	10 9 0	1.64	3 6 0	1.31	11 9 0	1.31
6 12 0	1.55	4 21 0	1.54	5 24 0	1.30		
2 6 0	1.49	2 19 0	1.41				
10 12 0	1.47	2 23 0	1.30				
8 0 0	1.44						
4 20 0	1.41						
2 20 0	1.34						

and 2 revealed that not only did these reflections have large
normalized structure factor magnitudes, but they also occurred in
many Σ_2 triples which have large A values. The phases of (1,11,0)
and (1,13,0) were then readily determined from triples 1 and 7,
and the phase of (5,10,0) could be found from triple 24 once the
phase of (1,11,0) was known. A record of the phase assignments
and the reason for each assignment are given in Table 3. Throughout the procedure described here, new phases were not determined
from single Σ_2 triples unless the triples passed certain restrictions placed on the values of their cosine seminvariants as
calculated by the triple product and MDKS formulas. If A was
greater than 3, both calculated cosine values were required to be
greater than 0.5. If A was in the range 2-3, both calculated
cosines were required to be greater than 0.75, and if A was in
the range 1.5-2, both calculated cosines were required to be
greater than unity. Inspection of Table 2 shows that, with such
conservative acceptance criteria, no cosine was assigned an
incorrect value.

Although the phases of all hk0 reflections are fixed by the
assignment of phases to (6,1,0) and (7,12,0), it was not possible,
at this point, to determine any more of these phases solely on
the basis of Σ_2 triples having well-calculated cosines. However,
the pair relationships (Chapter V, sections 1.2 and 3.2, and
exercises II.2.4 and II.2.5) did provide a means for finding the
phases of several other important hk0 reflections. In space group
$P2_12_12_1$, two types of formulas exist which allow the cosine
seminvariants, $\cos(\phi_1+\phi_2)$, where the phases ϕ_1 and ϕ_2 are both
restricted to be either 0 or π or both restricted to be $\pm\pi/2$, to
be computed from normalized structure factor magnitudes alone.
The first type of formula, in the form applicable to 0kℓ reflections, is

Table 2. Cosine seminvariants for triples of two-dimensional reflections which have
A>1.3.

SERIAL	$\bar{h}$			$\bar{k}$			$-\bar{h}-\bar{k}$			SIGNS*	TPS*	A**	$\cos(\phi_{\bar{h}}+\phi_{\bar{k}}+\phi_{-\bar{h}-\bar{k}})$ TPROD	MDKS	TRUE
1	6	1	0	1	11	0	-7	-12	0	+++	P I	7.06†	0.84	1.27	1
2	6	1	0	2	10	0	-8	-11	0	+++	0	6.58†	0.70	0.77	1
3	1	11	0	-8	11	0	7	-22	0	+--	P I	5.42†	1.03	1.47	1
4	6	1	0	-6	0	-6	0	-1	6	++-	P I	5.22	0.71	0.09	0
5	2	1	0	6	10	0	-8	-11	0	+++	0	4.93†	0.72	0.94	1
6	0	16	2	0	-15	4	0	-1	-6	+-+	0	4.89†	0.80	0.98	1
7	6	1	0	-7	12	0	1	-13	0	+--	0	4.87†	0.84	0.99	1
8	2	1	0	-10	10	0	8	-11	0	+--	P I	4.68†	0.89	0.89	1
9	4	0	0	2	1	0	-6	-1	0	+++	0	4.59†	1.02	1.06	1
10	7	12	0	-7	0	-1	0	-12	1	++-	0	4.58†	0.83	1.43	1
11	1	15	0	-1	0	-4	0	-15	4	++-	P I	4.46	0.28	0.21	1
12	1	11	0	-1	0	-1	0	-11	1	++-	P I	4.34	0.57	1.01	0
13	8	11	0	-8	0	-5	0	-11	5	++-	P I	4.10	0.39	0.45	1
14	8	11	0	-7	12	0	-1	-23	0	+-+	0	4.07†	1.28	1.03	1
15	2	1	0	-8	11	0	6	-12	0	+--	P I	3.97†	1.12	1.01	1
16	6	1	0	1	21	0	-7	-22	0	+++	P I	3.92†	1.03	1.33	1
17.	4	6	0	-8	11	0	4	-17	0	+--	0	3.80	0.57	0.61	1
18	2	1	0	8	11	0	-10	-12	0	+--	0	3.75†	0.96	0.67	1
19	0	23	3	0	-15	4	0	-8	-7	+-+	P I	3.64	0.32	0.22	-1
20	0	16	2	0	-14	4	0	-2	-6	+-+	0	3.58†	0.85	1.04	1
21	4	0	0	-2	-1	0	-2	1	0	++-	P I	3.54†	1.01	0.89	1
22	8	0	0	-2	-1	0	-6	1	0	++-	P I	3.51†	1.55	1.32	1
23	5	7	0	-5	0	-3	0	-7	3	++-	P I	3.49	0.29	0.07	0
24	6	1	0	-5	10	0	-1	-11	0	+-+	0	3.45†	0.87	0.87	1
25	0	18	2	0	-15	4	0	-3	-6	+-+	0	3.41†	1.38	1.58	1
26	6	1	0	2	18	0	-8	-19	0	+++	0	3.39	0.70	0.52	1
27	0	17	2	0	-15	4	0	-2	-6	+-+	P I	3.32†	1.14	1.37	1
28	7	22	0	-7	0	-1	0	-22	1	++-	0	3.07†	1.27	2.09	1
29	2	6	0	-8	11	0	6	-17	0	+--	0	3.03	0.85	0.46	1
30	0	7	3	0	-15	4	0	8	-7	+--	0	2.98†	1.22	1.53	1
31	5	6	0	-5	0	-3	0	-6	3	++-	0	2.95	0.62	0.49	1
32	1	12	0	-1	0	-1	0	-12	1	++-	0	2.93†	0.50	0.59	1
33	7	4	0	-8	11	0	1	-15	0	+--	0	2.90	0.43	0.29	1
34	6	1	0	-4	17	0	-2	-18	0	+-+	P I	2.89	1.02	0.07	1
35	2	8	0	-8	11	0	6	-19	0	+--	0	2.86	0.50	-0.03	1
36	0	22	1	0	-7	3	0	-15	-4	+-+	P I	2.84†	1.38	2.07	1
37	6	1	0	-7	22	0	1	-23	0	+--	0	2.83†	1.29	0.74	1
38	0	11	1	0	-26	3	0	15	-4	+--	P I	2.73	0.70	0.59	-1
39	2	10	0	-8	11	0	6	-21	0	+--	0	2.73	0.89	0.49	1
40	6	1	0	-8	9	0	2	-10	0	+--	P I	2.68	0.75	0.17	1
41	0	11	1	0	4	3	0	-15	-4	+++	P I	2.68	1.09	1.77	1
42	6	1	0	-8	19	0	2	-20	0	+--	P I	2.61	0.97	0.03	1
43	0	16	2	0	-8	5	0	-8	-7	+-+	P I	2.58†	0.85	0.59	1
44	6	1	0	-7	20	0	1	-21	0	+--	0	2.57†	1.11	0.63	1
45	0	12	1	0	-16	2	0	4	-3	+--	P I	2.56†	0.79	1.63	1
46	0	18	2	0	-16	4	0	-2	-6	+-+	0	2.53†	1.24	1.45	1
47	1	15	0	-1	0	-1	0	-15	1	++-	P I	2.52	0.28	-0.21	0
48	5	10	0	-5	0	-7	0	-10	7	++-	0	2.51	0.41	0.94	1
49	2	1	0	-1	11	0	-1	-12	0	+-+	P I	2.50	0.81	0.63	1
50	2	1	0	-8	19	0	6	-20	0	+--	P I	2.43	0.63	0.83	1
51	0	16	2	0	-16	2	0	0	-4	+--	0	2.43	0.87	0.83	1
52	8	9	0	-7	12	0	-1	-21	0	+-+	0	2.29†	1.03	1.50	1
53	5	6	0	1	11	0	-6	-17	0	+++	0	2.29†	0.97	0.89	1
54	6	1	0	2	8	0	-8	-9	0	+++	0	2.28	0.30	0.28	1
55	0	18	2	0	-26	3	0	8	-5	+--	0	2.26†	1.50	1.08	1
56	2	1	0	-4	17	0	2	-18	0	+--	P I	2.23	0.65	0.65	-1
57	0	4	3	0	-6	3	0	2	-6	+--	P I	2.22†	1.09	1.66	1
58	9	10	0	-8	11	0	-1	-21	0	+-+	0	2.21	1.49	1.94	1
59	0	23	3	0	-26	3	0	3	-6	+--	P I	2.18	0.37	0.00	1
60	0	22	1	0	-16	2	0	-6	-3	+-+	0	2.17†	1.51	1.83	1

Table 2 (Continued)

SERIAL	$\vec{h}$			$\vec{k}$			$-\vec{h}-\vec{k}$			SIGNS*	TPS*	A**	$\cos(o_{\vec{h}}+\phi_{\vec{k}}+\phi_{-\vec{h}-\vec{k}})$ TPROD	MDKS	TRUE
61	5	6	0	-7	12	0	2	-18	0	+--	PI	2.16	0.55	0.58	1
62	0	16	2	0	7	3	0	-23	-5	+++	PI	2.15†	1.01	1.25	1
63	0	25	2	0	-23	4	0	-2	-6	+-+	PI	2.13	0.19	-0.09	1
64	0	11	1	0	-18	2	0	7	-3	+--	PI	2.08	1.27	1.18	1
65	5	0	3	-10	0	4	5	0	-7	+--	0	2.07	0.81	0.92	1
66	2	1	0	5	21	0	-7	-22	0	+++	PI	2.04	0.70	0.54	1
67	2	1	0	-8	9	0	6	-10	0	+--	PI	2.01†	0.70	0.75	1
68	0	19	2	0	-15	4	0	-4	-6	+-+	PI	2.00†	1.33	1.17	1
69	0	12	1	0	-18	2	0	6	-3	+--	PI	1.99†	1.24	1.86	1
70	0	22	1	0	-18	2	0	-4	-3	+-+	0	1.99†	1.44	2.48	1
71	5	6	0	-1	11	0	-4	-17	0	+-+	0	1.97†	1.19	0.81	1
72	0	18	2	0	-19	2	0	1	-4	+--	PI	1.97	0.78	0.71	1
73	2	10	0	-1	11	0	-1	-21	0	+-+	PI	1.97†	1.51	1.68	1
74	0	6	3	0	-7	3	0	1	-6	+--	0	1.95†	1.14	0.96	1
75	3	6	0	4	6	0	-7	-12	0	+++	PI	1.93	0.68	0.30	1
76	3	1	0	5	10	0	-8	-11	0	+++	0	1.91	0.71	0.12	1
77	2	1	0	8	9	0	-10	-10	0	+++	0	1.91†	0.71	1.07	1
78	0	22	1	0	-23	3	0	1	-4	+--	0	1.90	0.41	-0.07	-1
79	0	16	2	0	-17	2	0	1	-4	+--	PI	1.90	0.52	0.01	1
80	0	11	1	0	-13	5	0	2	-6	+--	0	1.90	1.19	1.20	1
81	0	12	1	0	-2	6	0	-10	-7	+-+	0	1.89†	1.52	1.72	1
82	0	12	1	0	-19	2	0	7	-3	+--	0	1.88†	0.92	0.97	1
83	4	17	0	-4	0	-2	0	-17	2	++-	PI	1.87	0.78	2.94	0
84	0	12	1	0	-15	4	0	3	-6	+--	PI	1.86	0.76	1.06	-1
85	2	6	0	5	6	0	-7	-12	0	+++	PI	1.85	0.78	0.66	1
86	0	18	2	0	-7	3	0	-11	-5	+-+	PI	1.84	0.76	-0.10	1
87	0	9	1	0	6	3	0	-15	-4	+++	PI	1.84	0.75	0.53	1
88	0	9	1	0	-16	2	0	7	-3	+--	PI	1.84	0.95	0.95	1
89	0	11	1	0	-27	1	0	16	-2	+--	0	1.83	0.62	0.14	1
90	0	12	1	0	-26	3	0	14	-4	+--	PI	1.83†	0.75	0.88	1
91	5	7	0	-7	12	0	2	-19	0	+--	0	1.81	0.41	0.44	1
92	10	9	0	-8	11	0	-2	-20	0	+-+	PI	1.81	0.89	0.02	1
93	0	16	2	0	-8	3	0	-8	-5	+-+	PI	1.77†	1.29	1.33	1
94	2	1	0	-6	19	0	4	-20	0	+--	PI	1.76	0.80	0.19	1
95	0	4	3	0	-7	3	0	3	-6	+--	0	1.75†	1.35	1.79	1
96	6	1	0	-11	9	0	5	-10	0	+--	0	1.74	1.13	0.78	-1
97	0	12	1	0	-13	5	0	1	-6	+--	0	1.74	1.02	1.53	1
98	0	18	2	0	-18	2	0	0	-4	+--	0	1.73	1.06	1.25	1
99	2	1	0	-1	12	0	-1	-13	0	+-+	0	1.73†	0.81	0.86	1
100	0	22	1	0	-14	4	0	-8	-5	+-+	0	1.72†	1.31	1.58	1
101	0	18	2	0	-8	5	0	-10	-7	+-+	PI	1.71†	1.66	1.61	1
102	7	4	0	1	15	0	-8	-19	0	+++	0	1.70	0.27	0.29	-1
103	0	12	1	0	-11	5	0	-1	-6	+-+	PI	1.68	1.10	0.87	1
104	1	15	0	-1	0	-6	0	-15	6	++-	PI	1.68	0.26	-0.12	1
105	10	10	0	-7	12	0	-3	-22	0	+-+	0	1.68	1.08	1.02	1
106	1	0	1	4	0	6	-5	0	-7	+++	PI	1.66	0.87	1.37	1
107	2	1	0	-1	15	0	-1	-16	0	+-+	PI	1.66	0.48	-0.03	1
108	0	16	2	0	-24	3	0	8	-5	+--	0	1.66†	1.18	0.72	1
109	8	11	0	-6	12	0	-2	-23	0	+-+	0	1.65	0.87	0.36	1
110	0	12	2	0	-14	4	0	2	-6	+--	0	1.65	0.64	0.95	1
111	0	23	3	0	-24	3	0	1	-6	+--	PI	1.64	0.65	0.34	1
112	0	7	3	0	1	4	0	-8	-7	+++	0	1.63	0.77	0.97	1
113	2	1	0	1	21	0	-3	-22	0	+++	PI	1.63	1.01	0.91	1
114	6	1	0	-2	20	0	-4	-21	0	+-+	0	1.61	0.23	-0.14	-1
115	0	11	1	0	-8	5	0	-3	-6	+-+	0	1.61	1.17	0.90	-1
116	0	17	2	0	-18	2	0	1	-4	+--	0	1.60	0.12	0.05	-1
117	0	6	3	0	-14	4	0	8	-7	+--	PI	1.59†	1.46	1.87	1
118	0	9	1	0	-23	3	0	14	-4	+--	0	1.57	0.51	0.09	-1
119	10	9	0	-2	10	0	-8	-19	0	+-+	0	1.57	0.71	0.69	1
120	8	11	0	-10	12	0	2	-23	0	+--	PI	1.56	0.78	-0.16	1

Table 2 (Continued)

SERIAL	$\vec{h}$			$\vec{k}$			$-\vec{h}-\vec{k}$			SIGNS*	TPS*	A**	$\cos(\phi_{\vec{h}}+\phi_{\vec{k}}+\phi_{-\vec{h}-\vec{k}})$		
													TPROD	MDKS	TRUE
121	6	1	0	-2	19	0	-4	-20	0	+-+	PI	1.56	0.71	0.88	1
122	7	4	0	-6	17	0	-1	-21	0	+-+	O	1.56	1.20	0.85	1
123	0	24	3	0	-26	3	0	2	-6	+--	PI	1.55†	0.92	1.35	1
124	8	0	0	-1	-12	0	-7	12	0	++-	O	1.54†	1.55	1.23	1
125	8	4	0	-7	12	0	-1	-16	0	+-+	O	1.53	0.13	-0.10	-1
126	0	22	1	0	-23	5	0	1	-6	+--	O	1.52†	1.65	1.29	1
127	8	9	0	-1	13	0	-7	-22	0	+-+	PI	1.52†	0.68	1.76	1
128	5	0	1	5	0	3	-10	0	-4	+++	O	1.52	1.10	1.03	1
129	2	1	0	3	6	0	-5	-7	0	+++	PI	1.51	0.33	-0.12	1
130	0	16	2	0	-1	4	0	-15	-6	+-+	O	1.51	-0.01	-0.14	-1
131	0	11	1	0	-3	6	0	-8	-7	+-+	PI	1.51	1.75	1.29	1
132	6	1	0	1	16	0	-7	-17	0	+++	PI	1.51	0.15	0.06	1
133	0	25	2	0	-25	2	0	0	-4	+--	PI	1.51	0.21	-0.12	-1
134	6	10	0	1	12	0	-7	-22	0	+++	PI	1.51†	1.14	1.58	1
135	0	12	1	0	4	3	0	-16	-4	+++	O	1.50†	1.28	2.13	1
136	2	1	0	-6	20	0	4	-21	0	+--	PI	1.50	0.25	-0.70	1
137	0	6	3	0	-8	3	0	2	-6	+--	PI	1.50†	1.43	1.55	1
138	3	6	0	-3	0	-3	0	-6	3	++-	O	1.49	0.19	0.28	1
139	0	18	2	0	-14	4	0	-4	-6	+-+	O	1.48†	1.46	2.14	1
140	0	9	1	0	-8	5	0	-1	-6	+-+	O	1.46	0.86	0.27	-1
141	6	1	0	-1	23	0	-5	-24	0	+-+	PI	1.45	1.33	0.10	1
142	8	9	0	-1	11	0	-7	-20	0	+-+	PI	1.45†	1.04	2.42	1
143	4	6	0	-5	7	0	1	-13	0	+--	PI	1.44	0.51	0.52	1
144	0	11	1	0	-1	6	0	-10	-7	+-+	PI	1.44	1.07	0.72	1
145	0	19	2	0	-16	4	0	-3	-6	+-+	PI	1.44†	1.65	1.37	1
146	0	7	3	0	-8	3	0	1	-6	+--	PI	1.43†	0.43	0.06	-1
147	4	0	0	-2	-10	0	-2	10	0	++-	O	1.43†	-0.39	-0.30	-1
148	5	10	0	1	11	0	-6	-21	0	++-	O	1.43	0.43	0.44	-1
149	5	24	0	-5	0	-3	0	-24	3	++-	O	1.43	0.21	1.09	1
150	1	0	1	-6	0	6	5	0	-7	+--	O	1.43	1.44	2.23	1
151	4	0	0	1	0	1	-5	0	-1	+++	PI	1.43†	1.63	0.78	1
152	2	1	0	4	20	0	-6	-21	0	+++	O	1.42	1.12	0.45	1
153	0	17	2	0	-16	4	0	-1	-6	+-+	PI	1.42†	1.39	1.73	1
154	0	19	2	0	4	3	0	-23	-5	+++	PI	1.42†	1.09	1.94	1
155	7	4	0	-1	13	0	-6	-17	0	+-+	O	1.41	0.68	-0.18	1
156	0	19	2	0	-11	5	0	-8	-7	+-+	O	1.41	0.58	1.13	1
157	7	4	0	-1	15	0	-6	-19	0	+-+	O	1.41	0.52	0.04	1
158	3	6	0	1	11	0	-4	-17	0	+++	O	1.40	0.44	1.01	1
159	0	17	2	0	-14	4	0	-3	-6	+-+	PI	1.40†	1.53	2.16	1
160	1	11	0	-1	15	0	0	-26	0	+--	O	1.39	-0.12	-1.01	1
161	2	8	0	-8	9	0	6	-17	0	+--	O	1.39	0.12	0.00	-1
162	1	16	0	-1	0	-4	0	-16	4	++-	O	1.39	0.01	0.50	0
163	4	0	0	2	10	0	-6	-10	0	+++	O	1.39†	0.96	0.60	1
164	0	11	1	0	12	2	0	-23	-3	+++	'PI	1.39	0.53	0.11	1
165	0	9	1	0	-11	5	0	2	-6	+--	O	1.37	0.99	0.35	-1
166	1	23	0	-1	0	-4	0	-23	4	++-	PI	1.37	0.41	1.52	1
167	8	9	0	-2	10	0	-6	-19	0	+-+	O	1.37	0.48	0.40	1
168	5	0	3	0	0	4	-5	0	-7	+++	PI	1.37	0.97	0.34	1
169	0	19	2	0	-6	3	0	-13	-5	+-+	O	1.37	1.10	0.89	1
170	0	12	1	0	1	4	0	-13	-5	+++	PI	1.36	1.11	1.29	1
171	0	9	1	0	-1	6	0	-8	-7	+-+	PI	1.36	1.01	0.47	1
172	3	1	0	-2	10	0	-1	-11	0	+-+	PI	1.36	0.23	0.26	-1
173	0	4	3	0	-14	4	0	10	-7	+-+	PI	1.36†	1.41	2.13	1
174	11	9	0	-7	12	0	-4	-21	0	+-+	PI	1.35	0.91	-0.03	1
175	4	6	0	-6	17	0	2	-23	0	+--	O	1.35	1.20	0.85	1
176	0	15	1	0	-13	5	0	-2	-6	+-+	O	1.34	0.73	0.00	1
177	2	1	0	-7	20	0	5	-21	0	+--	O	1.34	0.40	0.74	1
178	0	12	1	0	-1	4	0	-11	-5	+-+	O	1.32	0.64	0.89	1
179	8	0	0	-4	-6	0	-4	6	0	++-	O	1.32	0.45	1.42	1
180	5	6	0	1	13	0	-6	-19	0	+++	O	1.32	0.46	0.98	-1

Table 2. (Continued)

Footnotes

* Let $\phi_{\vec{h}'}$, $\phi_{\vec{k}'}$, and $\phi_{-(\vec{h}+\vec{k})'}$ denote the phases for the parent forms of reflections $\vec{h}$, $\vec{k}$, and $-\vec{h}-\vec{k}$ respectively, and $\phi_{\vec{h}}$, $\phi_{\vec{k}}$, and $\phi_{-\vec{h}-\vec{k}}$ denote the phases of the three reflections in the form in which they occur in the Σ_2 triple. The parent form is taken to be that symmetry variant with all positive indices. Then

$$\phi_{\vec{h}}+\phi_{\vec{k}}+\phi_{-\vec{h}-\vec{k}}=s_{\vec{h}}\phi_{\vec{h}'}+s_{\vec{k}}\phi_{\vec{k}'}+s_{-\vec{h}-\vec{k}}\phi_{-(\vec{h}+\vec{k})'}+ \text{TPS}$$

where $s_{\vec{h}} = \pm 1$ according as $\phi_{\vec{h}}=\pm\phi_{\vec{h}'}+\varepsilon$, ($\varepsilon = 0$ or π), respectively. The column headed "SIGNS" gives the signs of $s_{\vec{h}}$, $s_{\vec{k}}$, and $s_{-\vec{h}-\vec{k}}$, and the entries in the column TPS are the "total phase shifts".

** A values associated with "mixed" triples have been doubled in accordance with the discussion in §4 of Chapter V.

† A triple used in the phasing procedure described in the text.

Table 3. 9α-Fluorocortisol initial phases in the order
in which they were determined.

Serial	Reflection	$\lvert E \rvert$	Phase (radians)	Reason
1	6 1 0	4.05	0	Origin
2	7 12 0	3.48	$\pi/2$	Origin
3	1 11 0	2.60	$\pi/2$	Triple 1
4	1 13 0	1.79	$-\pi/2$	Triple 7
5	5 10 0	1.70	$\pi/2$	Triple 24
6	1 21 0	1.98	$-\pi/2$	Pairs with (1,11,0) and (1,13,0)
7	1 23 0	1.43	$\pi/2$	
8	5 6 0	1.85	$\pi/2$	Pair with (5,10,0)
9	8 11 0	4.25	π	Triples 3 and 14
10	7 22 0	2.54	$\pi/2$	Triples 16 and 37
11	7 20 0	1.66	$-\pi/2$	Triple 44; pair with (7,22,0)
12	6 17 0	2.47	π	Triple 53
13	4 17 0	2.12	0	Triple 71; pair to (6,17,0)
14	8 9 0	1.73	0	Triples 52, 127, and 142
15	2 10 0	1.98	π	Triples 2 and 73
16	4 0 0	1.88	π	$\Sigma_1;\Sigma_1$ cosines; 1-dimensional squared-tangent; 400; 800 pair; relationship to paired (g,10,0) and
17	8 0 0	1.44	0	(g,12,0) reflections
18	2 1 0	3.12	π	Triples 9 and 22
19	1 12 0	1.60	$\pi/2$	Triples 49, 99, and 124
20	6 10 0	1.92	0	Triples 5, 67, and 134
21	10 10 0	1.83	π	Triples 8 and 77 Pairs to
22	6 12 0	1.55	π	Triple 15 each other
23	10 12 0	1.47	0	Triple 18
24	0 12 1	2.11	0	Origin
25	7 0 1	1.61	$-\pi/2$	Triple 10
26	1 0 1	2.24	$-\pi/2$	Triple 32
27	5 0 1	1.75	$\pi/2$	Triple 151
28	0 22 1	1.94	0	Triple 28
29	0 10 7	1.60	0	
30	0 8 3	1.40	π	Pairs with (0,22,1)
31	0 8 5	2.17	0	
32	0 8 7	2.03	π	
33	0 24 3	1.31	0	
34	0 16 2	3.02	0	Triples 43, 93, 108
35	0 18 2	2.55	π	Triple 101; pair with (0,16,2)
36	0 16 4	1.77	π	Pairs to 0,16,2 and (0,18,2)
37	0 4 3	2.08	π	Triples 45, 70, 135
38	0 6 3	1.91	0	Triples 60, 69; pair with (0,4,3)
39	0 26 3	2.12	π	Triple 55
40	0 2 6	2.90	0	Triples 46, 57, 81, 123, and 137
41	0 14 4	2.11	0	Triples 20, 90, 100, 117, and 173
42	0 15 4	3.47	$\pi/2$	Enantiomorph
43	0 1 6	2.42	$\pi/2$	Triple 6
44	0 3 6	2.00	$-\pi/2$	Triple 25; pair with (0,1,6)
45	0 17 2	1.71	$-\pi/2$	Triples 27, 159, and 153
46	0 7 3	2.19	$-\pi/2$	Triples 30, 36, 74, 95, and 146
47	0 23 5	1.68	$\pi/2$	Triples 62 and 126
48	0 19 2	2.10	$\pi/2$	Triples 82, 145, and 154; pair with 0,17,2
49	0 4 6	1.42	π	Triples 68 and 139

$$E_{0\,k_1\,\ell_1}\,E_{0\,k_2\,\ell_2} = \left|E_{0\,k_1\,\ell_1}\,E_{0\,k_2\,\ell_2}\right|\cos(\phi_{0\,k_1\,\ell_1} + \phi_{0\,k_2\,\ell_2}) \approx$$

$$\frac{N}{2}\left\langle (-1)^{h+\frac{1}{2}(k_1+k_2)}\,(\left|E_{h,\frac{1}{2}(k_1+k_2),\frac{1}{2}(\ell_1+\ell_2)}\right|^2-1)\,\times \right.$$

$$\left. (\left|E_{h,\frac{1}{2}(k_1-k_2),\frac{1}{2}(\ell_1-\ell_2)}\right|^2-1) \right\rangle_h \approx$$

$$\frac{N}{2}\left\langle (-1)^{h+\frac{1}{2}(k_1-k_2)+\ell_1}\cdot(\left|E_{h,\frac{1}{2}(k_1+k_2),\frac{1}{2}(\ell_1-\ell_2)}\right|^2-1)\,\times \right.$$

$$\left. (\left|E_{h,\frac{1}{2}(k_1-k_2),\frac{1}{2}(\ell_1+\ell_2)}\right|^2-1) \right\rangle_h \qquad\qquad (1)$$

where $k_1 \pm k_2$ and $\ell_1 \pm \ell_2$ are even (i.e. k_1 and k_2 have the same parity, and ℓ_1 and ℓ_2 have the same parity). Clearly, the contributions from the two parts of this formula may be combined. Analogous formulas for the $h0\ell$ and $hk0$ reflections may be found through a cyclic permutation of the indices. Because of the cumbersome notation occurring in formulas of this sort, it is convenient to introduce the notation $\vec{h}_1=(0k_1\ell_1)$, $\vec{h}_2=(0k_2\ell_2)$, $\vec{h}_3=(h,\frac{1}{2}(k_1+k_2),\frac{1}{2}(\ell_1+\ell_2))$, $\vec{h}_4=(h,\frac{1}{2}(k_1-k_2),\frac{1}{2}(\ell_1-\ell_2))$, and $E_{\vec{h}_1}=E_1$, etc. In the Type I formulas (e.q. (1)), reflections $\vec{h}_3$ and $\vec{h}_4$ may be, and generally are, three-dimensional, but in the Type II formulas, of which

$$E_{0\,k_1\,\ell_1}\,E_{0\,k_1\,\ell_2} = \left|E_{0\,k_1\,\ell_1}\,E_{0\,k_1\,\ell_2}\right|\cos(\phi_{0\,k_1\,\ell_1} + \phi_{0\,k_1\,\ell_1}) \approx$$

$$\frac{N}{2} \left\langle (-1)^{k+\frac{1}{2}(\ell_1-\ell_2)} (|E_{0,k,\frac{1}{2}(\ell_1+\ell_2)}|^2-1)(|E_{0,k+k_1,\frac{1}{2}(\ell_1-\ell_2)}|^2-1) \right\rangle_k,$$

$$(2)$$

is an example, reflections $\vec{h}_3$ and $\vec{h}_4$ always have at least one zero index. It is apparent that formulas of the second type are applicable only to pairs, $\vec{h}_1$ and $\vec{h}_2$, having a common nonzero component.

Some data for the pairs between the strongest 9α–fluoro-cortisol two-dimensional reflections are presented in Table 4. For example, the scaled average of the products of $|E|^2-1$ for the pairs of (h,17,2) and (h,1,0) reflections and the pairs of (h,17,0) and (h,1,2) reflections yielded -143.7 as the calculated value of $E_{0,16,2} E_{0,18,2}$, indicating that (0,16,2) and (0,18,2) have different phases. The quantity $|E_1E_2|_{calc}*\cos(\phi_1-\phi_2)$ is tabulated rather than $E_1E_2=|E_1E_2|*\cos(\phi_1+\phi_2)$ because the relation-ship between the two phases is immediately apparent from the sign of the former. If this sign is positive, the phases have the same value, but they differ by π radians if the sign is negative. As can be seen from Table 4, the magnitude of the calculated product of the normalized structure factor amplitudes is quite different from the observed value. However, this is not a serious problem because it is known that the phase of each of the paired reflections is restricted to one of two possible values, and the pair formula is only used to determine if these values are the same or differ by π radians. The probability that the correct relationship between the phases is the indicated one increases as the absolute value of the calculated product increases, and one measure of the reliability of the relative phase indication is given by the significance level,

Table 4. Pairing of the two-dimensional reflections having

$$|E|_{obs} > 1.4$$

Reflection 1	Reflection 2	Obs $|E_1E_2|$ *cos($\phi_1-\phi_2$)	Calc $|E_1E_2|$ *cos($\phi_1-\phi_2$)	No. Ctr.	s
0 16 2	0 18 2	− 7.7	−143	18	11.8
0 16 2	0 16 4	− 5.3	−74	17	6.1
0 16 2	0 16 4	− 5.3	−30	36	4.0*
0 18 2	0 16 4	4.5	68	19	5.9
0 22 1	0 8 7	− 3.9	−147	19	12.1
0 4 3	0 6 3	− 3.9	−107	21	9.4
0 4 3	0 8 3	2.9	89	12	5.8*
0 4 3	0 26 3	4.4	−92†	11	5.9*
0 6 3	0 8 3	− 2.6	−71	21	6.3
0 6 3	0 8 3	− 2.6	74†	12	4.9*
0 8 3	0 8 5	− 3.0	−36	43	4.8*
0 8 3	0 8 7	2.8	37	41	4.7*
0 8 5	0 8 7	− 4.4	−52	17	4.3
0 8 5	0 10 7	3.4	56	17	4.5
0 8 7	0 10 7	− 3.2	−76	13	5.3
0 17 2	0 19 2	− 3.6	−183	18	14.6
0 1 4	0 23 4	− 3.0	−81	19	6.8
0 15 4	0 23 4	5.5	−55†	15	4.3
0 1 6	0 3 6	− 4.8	−133	17	10.5
0 9 1	0 23 5	− 2.4	−102	19	8.4
0 7 3	0 23 3	5.8	−148†	11	9.4*
0 11 5	0 13 5	3.0	145	10	8.6*
4 0 0	8 0 0	− 2.7	−268	25	25.0
5 0 1	7 0 1	− 2.8	−66	48	8.9
7 0 1	9 0 1	− 2.5	−33	41	4.2
3 0 3	5 0 3	− 3.3	−41	51	5.8
4 0 0	4 20 0	2.6	131	9	7.4*
2 6 0	2 8 0	2.5	107	19	8.8*
2 6 0	2 18 0	− 2.6	−65	18	5.3*
2 10 0	10 10 0	3.6	32	39	4.1*
6 10 0	10 10 0	− 3.5	−235	36	26.5*
6 10 0	6 12 0	− 2.9	−600	14	41.7*
6 10 0	10 12 0	2.8	633	13	42.4
10 10 0	6 12 0	2.8	633	13	42.4
10 10 0	10 12 0	− 2.6	−785	10	46.0*
6 12 0	10 12 0	− 2.2	−246	34	26.8*
6 12 0	6 14 0	2.4	80	13	5.5
4 20 0	6 20 0	2.2	86	11	5.4
2 1 0	6 1 0	−12.6	−76	18	6.9

Table 4. (continued)

Reflection 1	Reflection 2	Obs $\|E_1E_2\|$ $*\cos(\phi_1 - \phi_2)$	Calc $\|E_1E_2\|$ $*\cos(\phi_1 - \phi_2)$	No. Ctr.	s
6 1 0	10 9 0	6.6	86	14	6.4
6 1 0	4 17 0	8.6	58	16	4.9
6 1 0	6 21 0	6.8	149	14	10.8*
8 9 0	8 11 0	− 7.3	−360	13	24.5*
8 11 0	8 19 0	10.6	−121†	11	8.1
4 17 0	6 17 0	− 5.2	−165	13	11.3
4 17 0	8 19 0	− 5.3	−98	13	6.9
6 17 0	6 19 0	5.1	81	12	5.5
6 17 0	6 19 0	5.1	75	13	5.3*
7 4 0	5 6 0	3.0	−59†	16	4.6
5 6 0	5 10 0	3.1	94	17	7.4
5 10 0	7 12 0	5.9	334	15	24.3
7 20 0	7 22 0	− 4.2	−273	10	16.2*
1 11 0	1 13 0	− 4.6	−479	19	39.0*
1 11 0	1 21 0	5.1	85	15	6.4
1 11 0	1 23 0	− 3.7	−60	15	4.5
1 11 0	1 23 0	− 3.7	−73	17	5.8*
1 13 0	1 21 0	− 3.5	−104	15	7.7
1 13 0	1 23 0	2.5	74	13	5.1
1 21 0	1 23 0	− 2.8	−265	13	17.8*

† Disagreement between observed and calculated $\|E_1E_2\|*\cos(\phi_1 - \phi_2)$.

* Indicates that pair relation of Type II was used.

$$s = 2n^{1/2}(\text{obs}|E_1E_2| + \text{calc}|E_1E_2|)/N, \qquad (3)$$

where n is the number of contributors to the average in equation (1), or (2), and N is the number of atoms in the unit cell. From a study of the results of the application of formulas of this type to the data for several structures having 20-30 nonhydrogen atoms in the asymmetric unit, it appears that the indicated phase relationship is normally correct if s is four or greater. Only those pairs having this minimum significance level are included in Table 4. Inspection of the table shows that only six of the fifty-nine pair relations were incorrectly indicated. An asterisk beside the value of s indicates that the figures were obtained from a Type II formula, and Type I formulas were used in those cases where there is no asterisk.

The pairings among the uu0 reflections proved to be particularly useful in the case of 9α-fluorocortisol. Figure 1 shows that, not only were the signs of all possible pairings among four of the reflections of this type strongly determined, but they were also all internally consistent. Since the phases of two of these reflections, (1,11,0) and (1,13,0), were already known, the symbol α was assigned the value π/2, and (1,21,0) and (1,23,0) could be added to the set of reflections with known phase (Table 3). Reflection (5,6,0) was also strongly paired to the known (5,10,0), and knowledge of the value of this phase (5,6,0) and of the two uu0 phases made seven additional phases accessible from the Σ_2 triples, resulting in a total number of 15 known phases (Table 3).

It is, perhaps, easier to gain a feeling for the true reliability of the phase assignments for (1,21,0) and (1,23,0) by considering the nature of the strongest individual contributors to the pair summations. In order to examine the phase relationship among paired reflections in detail, the case of (1,11,0) and

Figure 1. Pairing among the uu0 reflections. The phase of
           (1,11,0) is assigned a symbolic value α, and s values
           are indicated on the arrows.

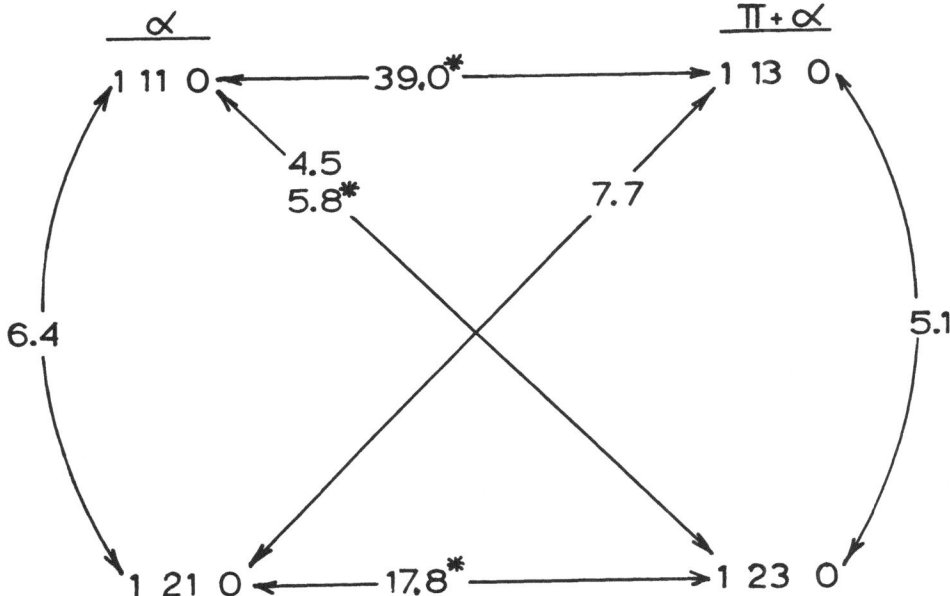

* Indicates that a pair relationship of Type II was used.

(1,21,0) will be considered as an illustration. (Refer to §3.2 of Chapter V for a brief discussion in $P2_1$). The Type I summation for these two reflections is over the products of $(|E|^2-1)$'s for the pairs of (1,5,ℓ) and (0,16,ℓ) reflections and the pairs of (0,5,ℓ) and (1,16,ℓ) reflections. The paired reflections (1,11,0) and (1,21,0) each form a triple of the Σ_2 type with some symmetry variant of each (1,5,ℓ) and (0,16,ℓ) and with each (0,5,ℓ) and (1,16,ℓ). The largest contributor to the summation occurs when $\vec{h}_3 = (1,5,2)$ and $\vec{h}_4 = (0,16,2)$, and the Σ_2 triples corresponding to this contributor are given in Table 5. It is convenient to introduce the notation $\cos(134) = \cos(\phi_{\vec{h}_1} + \phi_{\vec{h}_3}* + \phi_{\vec{h}_4}*)$ and $\cos(234) = \cos(\phi_{\vec{h}_2} + \phi_{\vec{h}_3}* + \phi_{\vec{h}_4}*)$ where the asterisk indicates that symmetry variant of the associated reflection needed to yield a structure invariant. Owing to the space group symmetries, $\cos(134) = \pm \cos(234)$. If $\phi_{\bar{1}\ \bar{5}\ \bar{2}} = \alpha$, $\phi_{0\ \overline{16}\ 2} = \beta$, $\phi_{1\ 11\ 0} = \phi_1$, and $\phi_{1\ 21\ 0} = \phi_2$, then

$$\bar{\Phi}_1 = \phi_{1\ 11\ 0} + \phi_{\bar{1}\ \bar{5}\ \bar{2}} + \phi_{0\ \overline{16}\ 2} = \phi_1 + \alpha + \beta \qquad (4)$$

and

$$\bar{\Phi}_2 = \phi_{1\ 21\ 0} + \phi_{\bar{1}\ 5\ 2} + \phi_{0\ \overline{16}\ 2} = \phi_2 + \alpha + \beta \ . \qquad (5)$$

It follows that

$$\bar{\Phi}_1 = \bar{\Phi}_2 \text{ if } \phi_1 = \phi_2 \qquad (6)$$

and

$$\bar{\Phi}_1 = \pi + \bar{\Phi}_2 \text{ if } \phi_1 = \pi + \phi_2. \qquad (7)$$

Table 5. Σ_2 triples corresponding to a large contributor to the summation for the pair (1,11,0; 1,21,0).

$\vec{h}(\vec{h}_1$ or $\vec{h}_2)$	$\vec{k}(\vec{h}_3$ variant)	$-\vec{h}-\vec{k}(\vec{h}_4$ variant)	Signs	TPS	A
1 21 0	$\bar{1}$ 5 $\bar{2}$	0 $\bar{1}6$ 2	$-+$	0	2.17
1 11 0	$\bar{1}$ 5 2	0 $\bar{1}6$ $\bar{2}$	$+-+$	0	2.86

Hence $\phi_1 = \phi_2$ or $\phi_1 = \pi + \phi_2$ according as $\cos \bar{\Phi}_1 = \cos \bar{\Phi}_2$ or $\cos \bar{\Phi}_1 = -\cos \bar{\Phi}_2$. (If $\cos \bar{\Phi}_1 = \pm \cos \bar{\Phi}_2 \approx 0$, no information is available). If the A values for the two triples are both large, as is true in this example, then it is probable that both cosine seminvariants will have positive values and that the phase relationship as indicated by equation (6), which agrees with the overall indication from equation (1), will be correct. Consistent with the pair equations (1) and (2) is the assumption that the two cosine seminvariants related to a single dominant contributor have the same sign.* It is apparent then that the pair relationship will fail in those cases where there is a single dominant contributor, and that contributor has one positive and one negative associated cosine seminvariant. Thus, erroneous conclusions based on the pair formulas can be avoided in large part by examining the large contributors and checking the values of their cosines which are predicted by the triple product and MDKS formulas. The contributors to the set of strongly paired uu0 reflections of 9α-fluorocortisol for which both of the related Σ_2 triples have A values greater than 1 are presented in Table 6,

---

* The same relationship between the paired phases is not obtained from the sign of the individual contribution to the pair summations (equations 1 and 2) as is obtained from the assumption of equal cosine seminvariants for the two related Σ_2 triples in the case that either, but not both, of $|E_{\vec{h}_3}|$, $|E_{\vec{h}_4}|$ is less than unity. As an example, consider the contribution of (6,17,0) and (6,1,2) to the (0,16,2; 0,18,2) pair as given in Table 6.

Table 6. Σ_2 triples for large contributors (both A>1) to some selected pairs.

$\vec{h}_1$	$\vec{h}_2$	$\vec{h}_3$	$\vec{h}_4$	A(134)	cos(134) TPROD	cos(134) TRUE	A(234)	cos(234) TPROD	cos(234) TRUE	calc $|E_1E_2|$ $*\cos(\phi_1-\phi_2)$
		1 1 1	0 12 1	2.14	0.69	0.99	1.48	0.67	0.99	-583
1 11 0	1 13 0	7 12 0	6 1 0	7.06	0.84	1.00	4.87	0.84	1.00	-9262
		1 12 0	2 1 0	2.50	0.81	1.00	1.73	0.81	1.00	-744
1 11 0	1 21 0	1 5 2	0 16 2	2.86	0.71	0.92	2.17	0.63	0.92	1133
		1 17 3	0 6 3	2.95	0.76	0.93	1.62	1.05	0.93	-1223
1 11 0	1 23 0	6 17 0	5 6 0	2.29	0.97	1.00	1.26	1.34	1.00	-674
		4 17 0	5 6 0	1.97	1.19	1.00	1.08	0.87	1.00	-464
		1 17 3	0 4 3	2.21	0.58	0.93	2.43	0.54	0.93	-1525
1 13 0	1 21 0									
		6 17 0	7 4 0	1.41	0.68	1.00	1.56	1.19	1.00	-484
1 13 0	1 23 0	1 5 2	0 18 2	1.66	0.76	0.92	1.32	0.93	0.92	767
		1 1 1	0 22 1	1.49	0.78	0.99	1.08	1.31	0.99	-466
1 21 0	1 23 0	7 22 0	6 1 0	3.92	1.03	1.00	2.83	1.29	1.00	-4558
		3 22 0	2 1 0	1.63	1.01	1.00	1.18	0.96	1.00	-422
		5 16 3	5 10 0	1.22	0.79	0.89	1.35	0.70	0.89	-285
0 6 3	0 26 3									
		0 16 4	0 10 7	1.05	1.77	1.00	1.16	1.38	1.00	-185
		0 15 4	0 11 1	2.68	1.09	1.00	2.73	0.70	-1.00	-1631
0 4 3	0 26 3	8 15 3	8 11 0	2.04	0.55	-0.96	2.08	0.81	0.96	-409
		1 15 3	1 11 0	1.71	0.85	0.99	1.75	0.87	0.99	534
0 6 3	0 24 3	0 15 4	0 9 1	1.84	0.75	1.00	1.26	0.56	-1.00	-645
		6 7 3	6 1 0	2.43	0.67	0.95	1.78	0.68	0.95	-1391
0 6 3	0 8 3									
		0 7 3	0 1 6	1.95	1.14	1.00	1.43	0.43	-1.00	1002
		2 18 2	2 1 0	2.07	0.70	0.82	2.54	0.47	0.82	-1437
0 17 2	0 19 2	6 18 2	6 1 0	2.42	0.75	0.99	2.96	0.60	0.99	-1893
		0 18 2	0 1 4	1.60	0.12	-1.00	1.97	0.78	1.00	785
		2 17 2	2 1 0	2.28	0.66	0.71	1.92	0.78	0.71	-273
		6 17 2	6 1 0	4.78	0.67	0.99	4.03	0.75	0.99	-2589
0 16 2	0 18 2									
		6 17 0	6 1 2*	1.28	0.81	0.92	1.08	0.87	0.92	57
		0 17 2	0 1 4	1.90	0.53	1.00	1.60	0.12	-1.00	277
		1 8 6	1 0 1	1.18	0.66	0.97	1.11	0.55	0.97	-131
		5 8 6	5 0 1	1.69	0.89	0.76	1.58	0.60	0.76	-485
0 8 5	0 8 7	0 1 6	0 9 1	1.46	0.86	-1.00	1.36	1.01	1.00	283
		0 3 6	0 11 1	1.61	1.17	-1.00	1.51	1.75	1.00	442
		0 4 6	0 12 1	1.26	1.43	1.00	1.18	1.31	1.00	-194
		1 9 6	1 1 1	1.89	1.46	0.96	1.40	1.37	0.96	674
0 8 5	0 10 7									
		0 9 1	0 1 6	1.46	0.86	-1.00	1.07	0.56	-1.00	283
		6 9 7	6 1 0	2.59	0.80	0.99	2.04	0.95	0.99	-1399
0 8 7	0 10 7									
		0 9 1	0 1 6	1.36	1.01	1.00	1.07	0.56	-1.00	283

* $|E_{612}|$ = 0.89<1.00. See footnote on page 344

and it can be verified that not only are all of these individual contributions internally consistent, but the predicted cosine seminvariants for the Σ_2 triples are all strongly indicated to be positive. In order that the phase assignments which were made for (1,21,0) and (1,23,0) be incorrect, the cosine seminvariants for several of the triples in this group would have to be negative, contrary to the entries in Table 6.

An hk0 reflection for which the phase was still unknown, and which has a high normalized structure factor amplitude, was the (2,1,0) reflection, and it was considered to be worth some effort to try to obtain this phase. If the phases of the (4,0,0) and the (8,0,0) reflections, which are paired to each other and which are themselves seminvariants and might be obtainable from the Σ_1 formulas, could be found, then the (2,1,0) phase would be accessible using triples 9 and 22 (Table 2). Triples 9 and 22 also form the largest contributor to the pair summation for (4,0,0; 8,0,0), and they, as well as the overall sum, indicate that these phases should differ by π radians. Unfortunately, the Σ_1 indications for (4,0,0) and (8,0,0), as shown in Table 7, were not definitive, and there was only a weak indication that the phase of (8,0,0) was 0. However, examination of the triple product and MDKS calculated values of the cosine seminvariants for the Σ_2 triples related to the individual Σ_1 contributors to (4,0,0) proved to be more fruitful. As shown in Table 8, there were two relatively large contributors, and since space group symmetry requires that the cosine seminvariants for their related triples have opposite sign, it was encouraging to find that the value of the cosine for one of these triples was calculated to be large (>1) but for the other triple the calculated cosine was found to be very low (<0) by both formulas. Consequently, both large Σ_1 contributors were consistent with a (4,0,0) phase of π. Further evidence for the assignment of this value to the (4,0,0) phase came from the squared-tangent calculations as displayed in

Table 7. Σ_1 indications for reflections
having $|E|_{obs} > 1.4$.

$\vec{h}$	$|E|_{obs}$	E_{calc}	$P_+(E_{\vec{h}})$	True Phase
4 0 0	1.88	0.82	0.59	π rad
		−2.90	0.38	
8 0 0	1.44	0.50	0.54	0
		5.76	0.66	
8 4 0	1.76	7.84	0.75	0
2 6 0	1.49	7.51	0.71	0
4 6 0	2.18	−3.45	0.34	π
2 8 0	1.68	0.09	0.50	0
2 10 0	1.98	−0.57	0.47	π
6 10 0	1.92	2.68	0.60	0
10 10 0	1.83	1.45	0.55	π
6 12 0	1.55	−0.68	0.47	π
10 12 0	1.47	−3.74	0.39	0
6 14 0	1.60	1.01	0.53	π
2 18 0	1.74	−10.85	0.17	π
4 20 0	1.41	1.82	0.55	π
6 20 0	1.62	1.03	0.53	π
0 16 2	3.02	0.58	0.54	0
0 18 2	2.55	0.51	0.53	π
0 24 2	1.41	2.17	0.57	0
10 0 4	2.08	2.47	0.76	0
0 14 4	2.11	0.72	0.54	0
0 16 4	1.77	−0.01	0.49	π
4 0 6	1.61	0.99	0.59	0
0 2 6	2.90	5.41	0.85	0
0 4 6	1.42	0.82	0.53	π

Table 8. Cosine seminvariants for Σ_1 triples having A>1.4. Reflection $\vec{h}$ is a structure seminvariant for which $|E|_{obs}$ >1.4.

$\vec{h}$	$\vec{k}$	$-\vec{h}-\vec{k}$	A	$\cos(\phi_{\vec{h}} + \phi_{\vec{k}} + \phi_{-\vec{h}-\vec{k}})^*$ TPROD	MDKS	If $\phi_{\vec{h}}$ =0
4 0 0	−2 −1 0	−2 1 0	3.54	1.01	1.32	−1
4 0 0	−2 −10 0	−2 10 0	1.43	−0.39	−0.06	+1
4 6 0	−2 −3 −3	−2 −3 3	1.53	0.37	0.11	−1
2 18 0	−1 −9 −6	−1 −9 6	1.69	1.01	0.50	−1
2 18 0	−1 −9 −8	−1 −9 8	1.70	−0.07	−0.72	−1
0 2 6	−11 −1 −3	11 −1 −3	2.80	0.51	0.46	+1

* The values of the cosine seminvariants, $\cos(\phi_{\vec{h}} + \phi_{\vec{k}} + \phi_{-\vec{h}-\vec{k}})$, as computed by the triple product and MDKS formulas are compared to the values they would have if $\phi_{\vec{h}}$ = 0.

Table 9. Squared-tangent indications ((4.4)-(4.6), Chapter VI) for one-dimensional reflections with $|E|_{obs} > 1.0$. K is a positive scaling parameter (10).

1D Reflection	$|E|_{obs}$	2D Contributors No.	2D Contributors $\frac{1}{k} \cos \phi_{1D}$	3D Contributors No.	3D Contributors $\frac{1}{k} \cos \phi_{1D}$	$\frac{1}{k} \cos \phi_{1D}$ (total)	True Phase
4 0 0	1.88	13	-0.29	19	-21.04	-21.33	π rad
8 0 0	1.44	10	6.07	12	7.20	13.28	0
0 8 0	1.00	5	-3.41	6	2.07	-1.34	0
0 20 0	1.03	8	3.72	4	9.92	13.65	0
0 22 0	1.22	6	-20.12	4	-7.79	-27.92	π
0 26 0	1.30	4	0.33	5	9.21	9.54	π
0 0 4	1.37	14	3.79	20	-2.08	1.71	0

Table 9. Cos $\phi_{2h,0,0}$, as given in Table 9, is obtained from equation VI.4.4,

$$\cos \phi_{2h,0,0} = K \left\langle (-1)^{h+k} |E_{hk\ell}|^2 \cos 2\phi_{hk\ell} \right\rangle_{k,\ell} \qquad (10)$$

or a cyclic permutation thereof in the case of 0g0 (equation VI.4.5) or 00g (equation VI.4.6) reflections. In the actual applications of these formulas which have been made to date, K was not evaluated, and only those reflections for which $\frac{1}{K}|\cos \phi_{1D}|$ was large (e.g. 20 or greater) have been considered to be determined. If the sign of the calculated $\cos \phi_{1D}$ is positive, the phase is 0, and if the sign is negative, then the phase has a value of π. Note that each contributor to the summation forms a Σ_2 triple of the type

$$2h\ 0\ 0 \qquad \overline{hk\ell} \qquad \overline{h}k\ell$$

with the one-dimensional reflections. The two-dimensional contribution is equivalent to the Σ_1 indication, but the three-dimensional contribution is based on the squared-tangent values of the three-dimensional phases (see section 4.2 of Chapter VI for further details). The $|E|_{obs}$ minimum for the two-dimensional contributors was 1.0, and the $|E|_{obs}$ minimum for the three-dimensional contributors was 1.5. The squared-tangent calculation for $\phi_{8,0,0}$, although not conclusive, was consistent with the indicated pairing of (4,0,0) and (8,0,0). The (0,22,0) phase, which was strongly, and correctly, indicated to be π was not used because of its relatively low $|E|$ value.

A final piece of evidence supporting the relative phase assignments for (4,0,0) and (8,0,0) came from the pair indications for the (g,10,0) and (g,12,0) reflections. If the pairings among these reflections which are given in Table 4 are correct, and if

Table 10. Σ_2 triples relating (4,0,0) and (8,0,0) to the paired (g,10,0) and (g,12,0) reflections.

$\vec{h}$	$\vec{k}$	$-\vec{h}-\vec{k}$	Signs	TPS	A	$\cos(\phi_{\vec{h}}+\phi_{\vec{k}}+\phi_{-\vec{h}-\vec{k}})$ TPROD	MDKS
400	2 10 0	$\bar{6}$ $\overline{10}$ 0	+++	0	1.39	0.96	0.60
400	6 10 0	$\overline{10}$ $\overline{10}$ 0	+++	0	1.28	1.17	0.58
800	$\bar{2}$ 10 0	$\bar{6}$ $\overline{10}$ 0	+++	0	1.06	0.31	1.12
800	2 10 0	$\overline{10}$ $\overline{10}$ 0	+++	0	1.01	2.17	1.96
400	6 12 0	$\overline{10}$ $\overline{12}$ 0	+++	0	0.82	1.21	1.27

$\phi_{4,0,0} = \pi$ and $\phi_{8,0,0} = 0$, then the only Σ_2 triple among the set of five given in Table 10 which does not have a cosine seminvariant value of +1 is the triple having an A value of 1.06. Consequently, the phases π for (4,0,0) and 0 for (8,0,0) were accepted on the basis of several pieces of evidence, none of which was conclusive by itself, but which when taken together formed a firm basis for these phase assignments. The phases of (2,1,0), (1,12,0), and four gg0 reflections were then found from the Σ_2 triples as recorded in Table 3.

After the values of the phases of 23 hk0 reflections were determined, the third origin specifying phase was assigned. Any reflection with ℓ odd (Chapter I, Table 1) would have served this purpose, but the (0,12,1) reflection was chosen because it was noticed that it would interact well with the hk0 phases which was already known. Four more phases could then be found as shown in Table 3, and these phases included the (0,22,1) phase which formed part of a set of paired 0gu reflections as shown in Table 4 and Figure 2. Five of these 0gu reflections (0,10,7; 0,8,3; 0,8,5; 0,8,7; 0,24,3) were strongly interrelated and were immediately added to the set of reflections with known phase, but conflicts were encountered when three additional reflections (0,26,3; 0,6,3; 0,4,3) were considered. For this reason, phases were first found for three related 0gg reflections because one of these phases, (0,16,2), could be consistently obtained from three Σ_2 triples each of which was strongly indicated to have a cosine seminvariant equal to unity. After these 0gg phases had been found the remainder of the 0gu reflections under consideration were accessible from triples, and their phases, relative to the remainder of the 0gu reflections were found to be as shown in Figure 2. This resulted in contradiction of three of the 0gu pair averages of which the most serious violation involved the (0,4,3; 0,26,3) pair. However, the two largest contributors to the mildly discrepant (0,6,3; 0,26,3) pair disagreed with the overall sum, and

Figure 2.   Pairings among the 0gu reflections. Values of s are
            indicated on the arrows, and in those cases where the
            pair formula indicates a relationship different from
            that shown, the s values are circled.

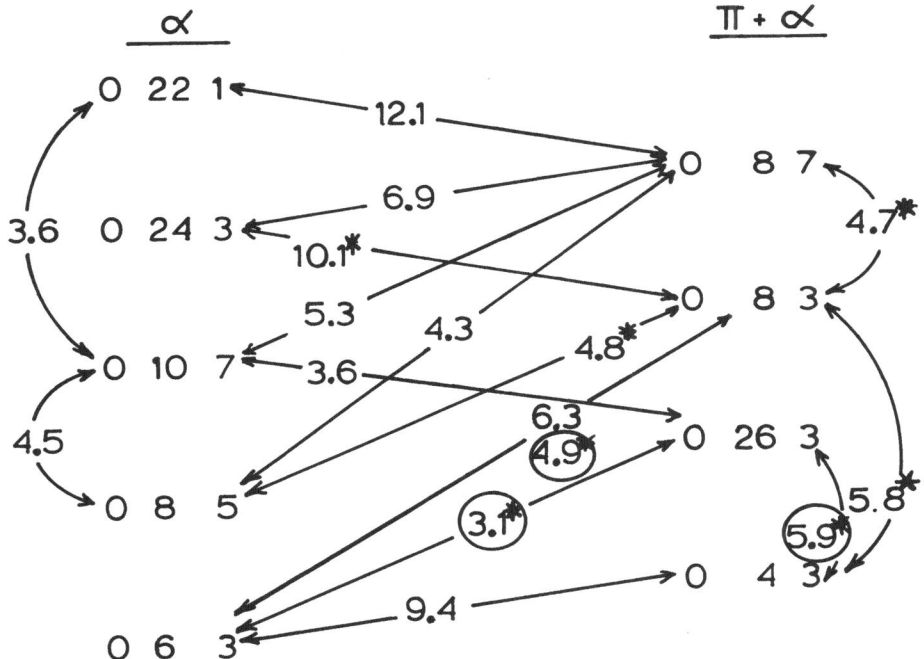

* Indicates that a pair relationship of Type II was used.

these   contributors are listed in Table 6. In addition to the
three pair averages which are incorrect, two large contributors
to the (0,4,3; 0,26,3) pair and a large contributor to each of
the pairs (0,6,3; 0,24,3) and (0,6,3; 0,8,3) are wrong, and these
contributors are also shown in Table 6. The evidence for the
relative 0gu phases being as shown in Figure 2 was quite strong,
however, and any changes in these phases would have resulted in
a greater number of violations.

The phases of (0,2,6) and (0,14,4) were each determinable
from several triples after the phases in the set of 0gu reflections were found. It should be noted that even if the absolute
phases in the 0gu set were unknown, the relative phases of the
(0,4,3; 0,6,3) and (0,6,3; 0,8,3) pairs as given by the pair
averages would in conjunction with triples 57 and 137, determine
the (0,2,6) phase in agreement with the Σ_1 indication. Such
confirmation is also given by the (0,18,2; 0,16,4) and the
(0,15,4; 0,17,2) pairs.

It may be verified from a consideration of the rules for
space group $P2_12_12_1$ given in Chapter I that, given the choice
of origin defining reflections used in this structure determination (gu0, ug0 and 0gu reflections were used), the enantiomorph
can be chosen by selecting one of the two possible values for a
phase of one of the following types: 0ug, u0g, g0u, and 0uu
(e.g. $\cos(\phi_{gu0} + \phi_{0ug}) = 0$). Inspection of Table 1 reveals that
(0,15,4) has the highest normalized structure factor amplitude
of all the reflections in these four classes and, in addition,
Table 2 shows that it interacts, with reflections whose phases
are already known, in triples having well-determined cosines.
Consequently, the enantiomorph was fixed by assigning a value
of $\pi/2$ to $\phi_{0,15,4}$, and the phases for an interrelated set of
seven additional reflections were found by using 17 of the
triples in Table 2. In this way the phases of 49 two-dimensional
reflections, used as input to the tangent procedures, were

determined. The 17 triples are noted in Table 3, and they were, with a single exception (triple 146), strongly indicated to have positive cosine values. Analysis of the (0,6,3; 0,8,3) pair (see Table 6) shows that either triple 74 or 146 must have a negative cosine, and the values calculated by the triple product and MDKS formulas indicated that the negative cosine seminvariant belonged to triple 146.

The above discussion of how the initial phase set for 9α-fluorocortisol was obtained may leave some questions unanswered. One such question, which is of importance whenever an approach similar to that used here is applied to an unknown structure, is how many two-dimensional phases should be determined? A more specific question is why were some triples (e.g. triple 41), for which the cosine values calculated by both formulas passed the acceptance limits, left unused? The answers to these questions are related to some extent. The analyses of several moderately complex noncentrosymmetric structures (25-40 nonhydrogen atoms in the asymmetric unit) have been expedited by determining about 50 two-dimensional phases prior to application of the tangent formula. On the other hand, the two-dimensional techniques should be abandoned if it becomes apparent that they are leading to ambiguities or if new phases cannot be determined with high probability. It is, however, a good idea to expend some extra effort to obtain phases for additional reflections whose indices are of parity types which are poorly represented. Failure to do this, or to make certain that all the independent reflections for which phases are specified arbitrarily interact strongly, often results in unresolved enantiomorphs or fragments of the molecule based on different origins.

In the 9α-fluorocortisol example, an attempt was made to determine phases for blocks of related reflections so that there would be internal checks on virtually all phase assignments. Since it was possible to obtain an adequate starting set without great

difficulty, it was not necessary to include less reliable phases
or phases which did not help to make other phases accessible.
Some examples of reflections whose phases might have been
determined are (0,11,1); (0,1,4) and (0,9,1) and the two-dimensional
triples involving these reflections which have A>1 and $|E_{\vec{h}}|$, $|E_{\vec{k}}|$,
$|E_{-\vec{h}-\vec{k}}|$>1.3 are listed in Table 11. Although $\phi_{0,11,1}$ could have
been found from triple 41, the opposite indication was given for
this phase by triple 38 if it were assumed that the cosine
seminvariant for this triple has a value of unity. The values
of 0.70 and 0.59, calculated for this cosine by the triple
product and MDKS formulas respectively, are only slightly less
than the acceptance limit of 0.75 used for triples with A values
in the range 2-3. Consequently, there was uncertainty as to the
true phase for (0,11,1). Inspection of Table 11 reveals that
two of the four triples involving (0,11,1) which have A values
greater than 1.5, and which have known phases for the other two
reflections, have negative cosines. Three triples with A values
between 1.0 and 1.5 agree with triples 41 and 131 assuming that
these triples with lower A values have cosine seminvariant values
of unity. However, confusion arises because the negative cosines
belong to two of the triples with higher A values, and neither
formula gives a definitive negative indication for either of these
cosines. Under such circumstances, consideration of the internal
relationships among the triples may be helpful. Table 6 shows
that triples 38 and 41 form a large contributor to the (0,4,3;
0,26,3) pair, and triples 115 and 131 form a contributor to the
(0,8,5; 0,8,7) pair. In both cases, the accepted relationship
between the paired reflections requires that the cosines for these
particular contributors have opposite signs, and in both sets of
contributing triples, the triple with a negative cosine had a
lower cosine value as calculated by both formulas. However,
despite the fact that it now seems apparent that the bulk of the
evidence indicated that $\phi_{0,11,1} = -\pi/2$, hindsight is always

Table 11. Two-dimensional triples involving 0,11,1, 0,1,4 or 0,9,1 which have A>1.

Serial	$\vec{h}$			$\vec{k}$			$-\vec{h}-\vec{k}$			SIGNS	TPS	A	$\cos(\phi_{\vec{h}}+\phi_{\vec{k}}+\phi_{-\vec{h}-\vec{k}})$		
													TPROD	MDKS	TRUE
38	0	11	1	0	$\bar{26}$	3	0	15	$\bar{4}$	+--	PI	2.73*	0.70	0.59	-1
41	0	11	1	0	4	3	0	$\bar{15}$	$\bar{4}$	+++	PI	2.68*	1.09	1.77	1
89	0	11	1	0	$\bar{27}$	1	0	16	$\bar{2}$	+--	O	1.83	0.62	0.14	1
115	0	11	1	0	$\bar{8}$	5	0	$\bar{3}$	$\bar{6}$	+-+	O	1.61*	1.17	0.90	-1
131	0	11	1	0	$\bar{3}$	6	0	$\bar{8}$	$\bar{7}$	+-+	PI	1.51*	1.75	1.29	1
144	0	11	1	0	$\bar{1}$	6	0	$\bar{10}$	$\bar{7}$	+-+	PI	1.44*	1.07	0.72	1
198	0	11	1	0	$\bar{17}$	2	0	6	$\bar{3}$	+--	O	1.22*	1.57	1.84	1
228	0	11	1	0	$\bar{19}$	2	0	8	$\bar{3}$	+--	O	1.09*	0.82	1.27	1
72	0	1	4	0	18	$\bar{2}$	0	$\bar{19}$	$\bar{2}$	+-+	PI	1.97*	0.78	0.71	1
78	0	1	4	0	22	$\bar{1}$	0	$\bar{23}$	3	+-+	O	1.90	0.41	-0.07	-1
79	0	1	4	0	16	$\bar{2}$	0	$\bar{17}$	$\bar{2}$	+-+	PI	1.90*	0.52	0.01	1
112	0	1	4	0	7	3	0	$\bar{8}$	$\bar{7}$	+++	O	1.63*	0.77	0.97	1
116	0	1	4	0	17	$\bar{2}$	0	$\bar{18}$	$\bar{2}$	++-	O	1.60*	0.12	0.05	-1
130	0	1	4	0	$\bar{16}$	2	0	15	$\bar{6}$	+--	O	1.51	-0.01	-0.14	-1
170	0	1	4	0	12	1	0	$\bar{13}$	$\bar{5}$	+++	PI	1.36	1.11	1.29	1
178	0	1	4	0	$\bar{12}$	1	0	11	$\bar{5}$	+--	O	1.32	0.64	0.89	1
187	0	1	4	0	$\bar{27}$	$\bar{1}$	0	26	$\bar{3}$	++-	O	1.26	0.23	-0.69	1
196	0	1	4	0	24	$\bar{2}$	0	$\bar{25}$	$\bar{2}$	+-+	PI	1.23	0.58	0.35	1
206	0	1	4	0	22	1	0	$\bar{23}$	$\bar{5}$	+++	PI	1.20*	0.81	1.25	1
219	0	1	4	0	$\bar{9}$	1	0	8	$\bar{5}$	+--	PI	1.14	0.87	1.28	1
87	0	9	1	0	6	3	0	$\bar{15}$	$\bar{4}$	+++	PI	1.84*	0.75	0.53	1
88	0	9	1	0	$\bar{16}$	2	0	7	$\bar{3}$	+--	PI	1.84*	0.95	0.95	1
118	0	9	1	0	$\bar{23}$	3	0	14	$\bar{4}$	+--	O	1.57	0.51	0.09	-1
140	0	9	1	0	$\bar{8}$	5	0	$\bar{1}$	$\bar{6}$	+-+	O	1.46*	0.86	0.27	-1
165	0	9	1	0	$\bar{11}$	5	0	2	$\bar{6}$	+--	O	1.37	0.99	0.35	-1
171	0	9	1	0	$\bar{1}$	6	0	$\bar{8}$	$\bar{7}$	+-+	PI	1.36*	1.01	0.47	1
188	0	9	1	0	$\bar{24}$	3	0	15	$\bar{4}$	+--	PI	1.26*	0.56	0.32	-1
216	0	9	1	0	$\bar{27}$	1	0	18	$\bar{2}$	+--	O	1.15	0.81	1.89	1
219	0	9	1	0	$\bar{1}$	4	0	$\bar{8}$	$\bar{5}$	+-+	PI	1.14	0.87	1.28	1
233	0	9	1	0	7	3	0	$\bar{16}$	$\bar{4}$	+++	O	1.08*	0.57	0.20	-1
236	0	9	1	0	1	6	0	$\bar{10}$	$\bar{7}$	+++	O	1.07*	0.56	-0.45	-1
252	0	9	1	0	17	2	0	$\bar{26}$	$\bar{3}$	+++	O	1.01*	-0.55	0.08	-1

* $\phi_{\vec{k}}$ and $\phi_{-\vec{h}-\vec{k}}$ are both known.

clearer than foresight, and at the time the initial phases were
derived, it did not seem as if it was necessary to introduce this
phase when there was a chance that it might be wrong.

The triples involving (0,1,4) and (0,9,1) also have an
interesting relationship to some of the pairs. The (0,17,2;
0,19,2) pair requires that the cosine seminvariants for triples
72 and 116 have opposite sign, and the 0,16,2; 0,18,2) pair
places the same requirement on triples 79 and 116. Thus, triple
116 probably has a negative cosine since the conflicting contributors to both pairs would then be resolved, and this conclusion is also favored by the triple product and MDKS cosine
seminvariant values. The pair relationships among the (0,9,1)
triples are more complex and are summarized in Table 12. Here,
as in the case of the (0,11,1) triples, the triples with cosine
seminvariant values of unity do have higher calculated values
from both formulas although the distinction is not as clear as
would be desired. Triples 140 and 236 constitute an example of
the relatively infrequent type of pair contributor which gives a
correct indication of the relationship among the paired phases
because both cosine seminvariants are negative. Reflection
(0,9,1) occurs in many more triples having negative cosines than
predicted by probability theory. A few reflections of this type
occasionally occur, but it is wise to verify the intensity of
such a reflection since it might have been mismeasured. If the
normalized structure factor amplitude of (0,9,1) had been
smaller, the A values of these triples would also have been
smaller, and it would then have been more probable that so many
cosines were negative. However, in this case, (0,9,1) was not
mismeasured since $|F_{obs}| = 53.5$ and $|F_{calc}| = 55.1$ (at R = 6.2%
for all data).

Table 12. Relationships among the cosine seminvariants
           for triples involving (0,9,1) as indicated
           by pairs.

Triple 1	Triple 2	Cos_1*Cos_2	Paired reflections
140	171	−1	0,8,5; 0,8,7
140	219	−1	0,1,4; 0,1,6
140	236	+1	0,8,5; 0,10,7
171	236	−1	0,8,7; 0,10,7

CHAPTER XII

VALINOMYCIN; STRONG ENANTIOMORPH DISCRIMINATION

1. Introduction

The most recently devised technique for using calculated
values of the cosine invariants in order to determine the values
of a basic set of phases, so-called strong enantiomorph discrimination, has been described in §3 of Chapter VI. It will be recalled that the central idea (assuming the space group is $P2_1$)
is to determine an integer k and two orthogonal classes, I and
II, of phases $\phi_{hk\ell}$ such that:

1. $|E_{0\ 2k\ 0}|$ is moderately large;
2. every $|E_{hk\ell}|$ corresponding to any phase $\phi_{hk\ell}$ in Class I
   or II is large;
3. any two phases in Class I differ from each other by 0 or
   π, approximately,
4. any two phases in Class II differ from each other by 0
   or π, approximately,
5. (orthogonality property) any phase in Class I differs
   from any phase in class II by $\pi/2$, approximately.

For further details of the method developed to identify two such
classes of phases and their use in decisive enantiomorph selection,
the reader is referred to Chapter VI. The method has been used to
solve three structures, in the space group $P2_1$, thus far. The
present chapter is devoted to a description of the application to
the **actual** phase determination for valinomycin, $C_{54}N_6O_{18}H_{90}$, space

group $P2_1$.

2.   Calculated Values of the Required

Cosine Invariants, $\text{Cos}(\phi_1 + \phi_2 + \phi_3)$

The values of some 3000 cosine invariants

$$\cos(\phi_1 + \phi_2 + \phi_3), \tag{2.1}$$

where

$$\phi_j = \phi_{\vec{h}_j}, \quad j = 1,2,3, \tag{2.2}$$

$$\vec{h}_1 + \vec{h}_2 + \vec{h}_3 = 0, \tag{2.3}$$

and $\phi_{\vec{h}}$ is the phase of the normalized structure factor $E_{\vec{h}}$, were calculated by the modified triple product procedure (§2.1, Chapter V).   Seventy-three of these cosines which constitute the hard core of those needed in the phase determination process are listed in Table 1.   In addition, some eighty cosines played a secondary role in that they provided supporting evidence for the identification of the orthogonal classes and aided in the interpretation of the results of the tangent procedures.   The parameter A is defined by

$$A = \frac{2\sigma_3}{\sigma_2^{3/2}} \, |E_1 E_2 E_3| \; = \; 0.1586 \, |E_1 E_2 E_3|, \tag{2.4}$$

where

$$E_j = E_{\vec{h}_j}, \quad j = 1,2,3, \qquad (2.5)$$

$$\sigma_n = \sum_{j=1}^{N} Z_j^n, \qquad (2.6)$$

Z_j is the atomic number of the j th atom, and N is the number of
atoms in the unit cell. (If there is no heavy atom present, then

$$\frac{\sigma_3}{\sigma_2^{3/2}} \approx \frac{1}{N^{1/2}} \qquad (2.7)$$

where N is the number of nonhydrogen atoms in the unit cell).
Only the cosines with A values greater than 1.5 were calculated.
With rare but important exceptions, only those cosines were used
whose calculated values were close to or greater than one, the
presumption being that the true values of these cosines were then
very likely equal to unity, at least approximately, and the
identification of these cosines was an important feature of the
method. However, the four cosine invariants with serial numbers
636, 1669, 1760, 2315 (Table 1) were calculated to be small or
negative and these played a crucial role in the process of phase
determination. For the purpose of comparison, the last column
of Table 1 contains the true values of the cosine invariants as
obtained from the refined structure. Using identities among the
cosine invariants (Chapter V), improved values for a number of
these cosines were obtained which, however, are not shown in
Table 1. Although the better values were not in fact essential
for the present structure determination, these improvements would
clearly facilitate the solution of more complex structures.

Table 1. Calculated values of cosine invariants, $\cos(\phi_1+\phi_2+\phi_3)$

Serial Number	$\vec{h}_1$			$\vec{h}_2$			$\vec{h}_3$			A	Calculated Cosine	True Cosine
423	$\overline{11}$	$\overline{4}$	1	10	4	$\overline{10}$	$\overline{1}$	0	9	1.544	+2.02	+0.947
557	0	8	0	16	$\overline{4}$	$\overline{9}$	$\overline{16}$	$\overline{4}$	9	1.562	+2.14	+0.992
598	1	4	$\overline{2}$	16	$\overline{4}$	$\overline{9}$	$\overline{17}$	0	11	1.568	+1.34	+0.963
636	4	4	$\overline{3}$	9	$\overline{4}$	$\overline{10}$	$\overline{13}$	0	7	1.572	+0.30	+0.602
977	1	4	$\overline{2}$	$\overline{11}$	$\overline{4}$	1	10	0	1	1.618	+1.03	+0.992
1115	0	8	0	10	$\overline{4}$	$\overline{10}$	$\overline{10}$	$\overline{4}$	10	1.640	+1.61	+0.675
1415	19	4	$\overline{7}$	1	$\overline{4}$	$\overline{2}$	$\overline{20}$	0	9	1.694	+1.87	+0.962
1669	0	8	0	5	$\overline{4}$	$\overline{3}$	$\overline{5}$	$\overline{4}$	3	1.743	−0.23	−0.888
1760	0	8	0	9	$\overline{4}$	$\overline{10}$	$\overline{9}$	$\overline{4}$	10	1.763	+0.16	−0.689
1876	0	8	0	17	$\overline{4}$	$\overline{7}$	$\overline{17}$	$\overline{4}$	7	1.796	+1.06	+0.993
2016	21	3	$\overline{7}$	$\overline{17}$	$\overline{3}$	0	$\overline{4}$	0	7	1.832	+1.44	+0.999
2050	11	3	$\overline{4}$	$\overline{18}$	$\overline{3}$	9	7	0	$\overline{5}$	1.842	+1.16	+0.969
2212	12	1	4	$\overline{5}$	0	$\overline{2}$	$\overline{7}$	$\overline{1}$	$\overline{2}$	1.892	+1.75	+0.999
2223	7	3	2	0	$\overline{2}$	0	$\overline{7}$	$\overline{1}$	$\overline{2}$	1.898	+1.30	+0.963
2230	10	1	4	$\overline{8}$	$\overline{1}$	$\overline{11}$	$\overline{2}$	0	7	1.901	+1.22	+0.944
2267	0	8	0	$\overline{11}$	$\overline{4}$	1	11	$\overline{4}$	$\overline{1}$	1.914	+1.44	+0.983
2279	8	1	11	$\overline{4}$	0	$\overline{7}$	$\overline{12}$	$\overline{1}$	$\overline{4}$	1.917	+1.85	+0.997
2315	0	8	0	$\overline{21}$	$\overline{4}$	$\overline{5}$	21	$\overline{4}$	5	1.931	−0.85	−0.860
2329	14	3	2	$\overline{21}$	0	$\overline{4}$	7	$\overline{3}$	2	1.938	+1.47	+0.915
2534	6	1	$\overline{7}$	$\overline{7}$	0	5	1	$\overline{1}$	2	2.018	+1.34	+0.909
2564	9	1	$\overline{1}$	$\overline{4}$	$\overline{1}$	0	$\overline{5}$	0	1	2.031	+1.27	+0.955
2599	11	4	$\overline{10}$	$\overline{14}$	$\overline{4}$	8	3	0	2	2.059	+1.69	+0.986
2675	6	2	$\overline{11}$	$\overline{13}$	$\overline{1}$	$\overline{13}$	7	$\overline{1}$	2	2.101	+1.52	+0.960
2682	1	4	$\overline{14}$	$\overline{4}$	$\overline{4}$	12	3	0	2	2.106	+1.44	+0.999
2739	18	3	$\overline{9}$	$\overline{25}$	0	7	7	$\overline{3}$	2	2.148	+1.01	+0.999
2753	4	3	0	3	0	2	$\overline{7}$	$\overline{3}$	$\overline{2}$	2.155	+1.09	+0.954
2771	14	4	$\overline{9}$	4	$\overline{1}$	0	$\overline{18}$	$\overline{3}$	9	2.164	+1.12	+0.997
2796	7	3	2	6	$\overline{2}$	11	$\overline{13}$	$\overline{1}$	$\overline{13}$	2.183	+0.94	+0.857
2838	10	7	10	4	$\overline{4}$	$\overline{12}$	$\overline{14}$	$\overline{3}$	2	2.230	+1.73	+0.972
2893	12	7	3	9	$\overline{4}$	$\overline{10}$	$\overline{21}$	$\overline{3}$	7	2.288	+2.64	+0.979
2909	10	1	11	$\overline{13}$	$\overline{1}$	$\overline{13}$	3	0	2	2.297	+1.80	+0.992
2918	5	1	$\overline{2}$	$\overline{10}$	$\overline{1}$	$\overline{4}$	5	0	2	2.309	+1.19	+0.946
2925	9	0	$\overline{15}$	$\overline{4}$	$\overline{4}$	12	$\overline{5}$	4	3	2.323	+0.83	+0.898
2941	11	4	$\overline{11}$	$\overline{16}$	$\overline{4}$	9	5	0	2	2.335	+1.28	+0.999
2960	14	3	$\overline{2}$	$\overline{18}$	$\overline{3}$	9	$\overline{4}$	0	7	2.366	+1.64	+0.990
2963	19	4	$\overline{7}$	$\overline{5}$	0	2	$\overline{14}$	$\overline{4}$	9	2.368	+1.46	+0.995
2981	0	2	0	$\overline{4}$	$\overline{1}$	0	$\overline{4}$	$\overline{1}$	0	2.388	+0.90	+0.837
2995	13	1	$\overline{7}$	$\overline{21}$	0	$\overline{4}$	$\overline{8}$	$\overline{1}$	11	2.407	+1.37	+0.982
2999	5	4	13	16	$\overline{4}$	$\overline{9}$	$\overline{21}$	0	$\overline{4}$	2.413	+1.33	+0.990
3001	9	7	1	$\overline{5}$	$\overline{4}$	3	$\overline{14}$	$\overline{3}$	2	2.414	+1.27	+0.996
3020	6	1	$\overline{7}$	$\overline{4}$	$\overline{1}$	0	$\overline{2}$	0	7	2.445	+1.13	+0.717

Table 1. (Continued)

Serial Number	$\vec{h}_1$	$\vec{h}_2$	$\vec{h}_3$	A	Calculated Cosine	True Cosine
3028	1 0 6	4 0 $\bar{7}$	$\bar{5}$ 0 1	2.463	+0.84	+1.000
3091	5 4 $\bar{3}$	$\bar{9}$ $\bar{4}$ 10	4 0 $\bar{7}$	2.589	+1.22	+0.986
3121	23 3 $\bar{7}$	$\bar{5}$ 0 $\bar{2}$	$\overline{18}$ $\bar{3}$ 9	2.661	+1.67	+0.949
3124	16 4 $\bar{9}$	$\overline{19}$ $\bar{4}$ 7	3 0 $\bar{2}$	2.672	+1.43	+0.999
3130	1 0 9	4 0 $\bar{7}$	$\bar{5}$ 0 $\bar{2}$	2.691	+1.91	+1.000
3131	10 0 $\bar{3}$	$\bar{3}$ 0 $\bar{2}$	$\bar{7}$ 0 5	2.694	+0.74	+1.000
3138	19 0 7	2 0 $\bar{3}$	$\overline{21}$ 0 $\bar{4}$	2.702	+0.59	-1.000
3139	10 7 10	$\bar{9}$ $\bar{7}$ $\bar{1}$	$\bar{1}$ 0 $\bar{9}$	2.703	+1.76	+0.990
3142	17 3 0	$\bar{3}$ 0 $\bar{2}$	$\overline{14}$ $\bar{3}$ 2	2.709	+1.23	+0.982
3153	11 1 $\bar{8}$	$\bar{6}$ $\bar{1}$ 7	$\bar{5}$ 0 1	2.728	+1.12	+0.908
3159	12 7 3	5 4 $\bar{3}$	$\overline{17}$ $\bar{3}$ 0	2.743	+0.97	+0.999
3162	1 1 2	4 $\bar{1}$ 0	$\bar{5}$ 0 $\bar{2}$	2.747	+1.22	+0.942
3170	21 3 $\bar{7}$	$\bar{3}$ 0 $\bar{2}$	$\overline{18}$ $\bar{3}$ 9	2.773	+1.18	+0.999
3171	13 3 $\overline{11}$	$\overline{18}$ $\bar{3}$ 9	5 0 2	2.775	+1.33	+0.969
3182	4 4 $\overline{12}$	$\bar{9}$ $\bar{4}$ 10	5 0 2	2.815	+1.73	+0.812
3184	3 1 9	$\bar{8}$ $\bar{1}$ $\overline{11}$	5 0 2	2.822	+1.53	+0.999
3204	14 4 $\bar{8}$	$\bar{5}$ 0 $\bar{2}$	$\bar{9}$ $\bar{4}$ 10	2.922	+1.67	+0.728
3214	0 2 0	$\bar{8}$ $\bar{1}$ $\overline{11}$	$\bar{8}$ $\bar{1}$ 11	2.974	+0.87	+0.998
3217	9 7 $\bar{1}$	$\bar{9}$ $\bar{4}$ $\overline{10}$	$\overline{18}$ $\bar{3}$ 9	2.999	+2.09	+0.994
3218	17 4 $\bar{7}$	$\bar{3}$ 0 $\bar{2}$	$\overline{14}$ $\bar{4}$ 9	3.013	+0.86	+0.995
3258	16 0 5	$\bar{5}$ 0 $\bar{1}$	$\overline{21}$ 0 $\bar{4}$	3.375	+0.44	+1.000
3259	7 1 2	$\bar{4}$ $\bar{1}$ 0	$\bar{3}$ 0 $\bar{2}$	3.375	+1.20	+0.974
3263	14 7 3	$\overline{10}$ $\bar{7}$ $\overline{10}$	$\bar{4}$ 0 7	3.426	+1.23	+0.996
3267	10 0 1	$\bar{5}$ 0 $\bar{2}$	$\bar{5}$ 0 1	3.478	+0.75	+1.000
3269	8 7 10	4 0 $\bar{7}$	$\overline{12}$ $\bar{7}$ $\bar{3}$	3.493	+1.24	+0.939
3282	9 0 $\bar{5}$	$\bar{5}$ 0 $\bar{2}$	4 0 $\bar{7}$	3.861	+1.23	+1.000
3283	5 0 2	$\bar{2}$ 0 0	$\bar{3}$ 0 $\bar{2}$	3.888	+1.67	+1.000
3289	14 7 3	$\bar{9}$ $\bar{7}$ $\bar{1}$	$\bar{5}$ 0 $\bar{2}$	4.034	+1.80	+0.975
3291	2 0 $\bar{7}$	$\bar{7}$ 0 5	$\bar{5}$ 0 $\bar{2}$	4.190	+1.18	+1.000
3294	13 1 13	$\bar{5}$ 0 $\bar{2}$	$\bar{8}$ $\bar{1}$ $\overline{11}$	4.250	+1.71	+0.953
3299	4 0 $\bar{7}$	$\bar{7}$ 0 5	$\bar{3}$ 0 $\bar{2}$	4.812	+1.16	+1.000
3305	2 0 $\bar{3}$	$\bar{5}$ 0 1	$\bar{3}$ 0 $\bar{2}$	6.470	+0.55	+1.000

3. Determination of Seventeen
Two-Dimensional Phases $\phi_{h\,0\,\ell}$

The values of seventeen two-dimensional phases $\phi_{h\,0\,\ell}$ are
listed in Table 2 in the order in which they were determined.
The method of phase determination is given in column 3 of this
table. Thus $\phi_{3\,0\,2}$ and $\phi_{5\,0\,\bar{1}}$ were arbitrarily set equal to zero
as is permitted by the recipe for origin specification in space
group $P2_1$. Again $\phi_{2\,0\,0} = 180°$ was determined by means of the
Σ_1 formula ((2.10) of Chapter II). The phase $\phi_{21\,0\,4} = 180°$ was
found from $\phi_{3\,0\,2} = 0$ by means of the formula for $\cos(\phi_1 + \phi_2)$
((3.57) of Chapter II) the major contributor to which is related
to the pair of invariants with serial numbers 3303, 3304 (§1.2
and §3.2 of Chapter V). The phases $\phi_{7\,0\,\bar{5}}$ and $\phi_{25\,0\,\bar{7}}$ could not
be directly determined but they were needed in the evaluation of
additional phases. Hence each was permitted to take on its two
possible values and only the correct values (180° for each) are
listed in Table 2. For the remaining phases, the serial number
of the required cosine invariants, as given in Table 1, is listed
in the third column of Table 2. Only the phase $\phi_{19\,0\,7}$ was
incorrectly calculated, a consequence of the negative cosine
invariant 3138 which was calculated to be +0.59 and mistakenly
assumed to be +1 (Table 1).

4. Determination of Thirty-Two Three-Dimensional
Phases Which Are Enantiomorph Independent

In Table 3 are listed the values of thirty-two three-dimen-
sional (i.e. $k \neq 0$) phases, $\phi_{hk\ell}$, in the order in which they were
determined. The phase $\phi_{4\,1\,0}$ was arbitrarily set equal to 180°
in accordance with the procedure for origin-fixing in space group
$P2_1$. For the remaining phases, the serial numbers of the required

Table 2. Values, in degrees, of seventeen two-dimensional phases

h k ℓ	$|E|$	Phase Determined by Means of	Calculated Phase	True Phase
3 0 2	3.732	origin	0	0
5 0 $\bar{1}$	3.448	origin	0	0
2 0 $\bar{3}$	3.170	3305	0	0
2 0 0	1.838	Σ_1	180	180
5 0 2	3.574	3283	180	180
10 0 1	1.779	3267	180	180
21 0 4	3.453	ϕ_{302}, $\cos(\phi_1+\phi_2)$, 3303, 3304	180	180
16 0 5	1.787	3258	180	180
19 0 7	1.556	3138	180	0
7 0 $\bar{5}$	2.866	ambiguous	180	180
4 0 $\bar{7}$	2.836	3299	180	180
2 0 $\bar{7}$	2.579	3291	0	0
9 0 $\bar{5}$	2.401	3282	0	0
10 0 $\bar{3}$	1.588	3131	180	180
1 0 9	1.674	3130	0	0
1 0 6	2.522	3028	180	180
25 0 $\bar{7}$	2.291	ambiguous	180	180

Table 3. Values, in degrees, of thirty-two, enantiomorph
independent, three-dimensional phases

h k ℓ	$|E|$	Phase Determined by Means of	Calculated Phase	True Phase	Error
4 1 0	2.554	origin	180	165	+15
7 1 2	2.233	3259	180	−177	− 3
1 1 2	1.897	3162	180	−175	− 5
6 1 $\bar{7}$	2.340	3020, 2534	180	−149	−31
11 1 $\bar{8}$	2.131	3153	180	−171	− 9
12 1 4	1.495	2212	0	3	− 3
8 1 11	2.850	2279	180	−174	− 6
13 1 13	2.630	3294	0	− 12	+12
3 1 9	1.747	3184	0	− 2	+ 2
13 1 $\bar{7}$	1.542	2995	180	−159	−21
10 1 11	1.476	2909	0	− 14	+14
9 1 $\bar{1}$	1.454	2564	180	−177	− 3
10 1 4	1.630	2230	180	166	+14
5 1 2	2.499	2918	0	4	− 4
6 2 11	2.255	2675	0	− 2	+ 2
7 3 2	2.320	2796, 2223	0	41	−41
4 3 0	1.569	2753	0	5	− 5
14 3 2	1.526	2329	0	45	−45
18 3 $\bar{9}$	2.548	2739	0	28	−28
13 3 $\overline{11}$	1.921	3171	180	−171	− 9
21 3 $\bar{7}$	1.838	2016, 3170	0	16	−16
23 3 $\bar{7}$	1.842	3121	180	−134	−46
14 3 $\bar{2}$	2.064	2960	180	−170	−10
17 3 0	2.217	3142	180	−161	−19
11 3 $\bar{4}$	1.590	2050	180	−179	− 1
0 2 0	2.309	2981, 3214	180	−176	−4

Table 3. (Continued)

h k ℓ	$|E|$	Phase Determined by Means of	Calculated Phase	True Phase	Error
14 4 $\bar{9}$	2.096	2771	0	11	−11
19 4 $\bar{7}$	1.993	2963	180	−168	−12
16 4 $\bar{9}$	2.265	3124	180	−168	−12
17 4 $\bar{7}$	2.428	3218	0	7	− 7
11 4 $\overline{11}$	1.818	2941	0	27	−27
5 4 13	1.945	2999	0	20	−20

cosine invariants, as given in Table 1, are listed in the third
column of Table 3. Since the values of all the cosine invariants
used in this calculation are, in accordance with Table 1, presumed
to have the value +1, every structure invariant $\phi_1 + \phi_2 + \phi_3$ has the
same value, 0, for both enantiomorphs permitted by the observed
structure factor magnitudes $|E|$. Hence the calculated values
(0° or 180°) of all phases listed in Table 3 are enantiomorph
independent. The true values of these thirty-two phases are
listed in Column 5 of the table and the last column gives the
error $(\phi_{Calc} - \phi_{True})$ of the calculated phase.

5. Enantiomorph Selection

In accordance with the procedure for strong enantiomorph
discrimination described in §3 of Chapter VI which employs calculated cosine invariants and a class of structure invariants with
values $\pm\pi/2$, approximately, two orthogonal classes, I and II, of
phases $\phi_{h\ 4\ \ell}$ were tentatively identified. First, $|E_{0\ 8\ 0}| = 1.921$
so that Property 1 is satisfied with k = 4. Second, using the
modified triple product procedure, the values of all cosine
invariants

$$\cos(\phi_{h\ \bar{4}\ \ell} + \phi_{\bar{h}\ \bar{4}\ \bar{\ell}} + \phi_{0\ 8\ 0}), \qquad (5.1)$$

with

$$A = 0.1586 \, |E^2_{h\ 4\ \ell} E_{0\ 8\ 0}| > 1.5, \qquad (5.2)$$

were calculated. Those cosines which were calculated to be greater
than unity led to four phases $\phi_{h\ 4\ \ell}$ which were placed unambiguously
in Class I. Those cosines calculated to be most negative led to

two phases $\phi_{h\ 4\ \ell}$ which were placed unambiguously in Class II.
The results are summarized in Table 4 which lists also, for
comparison, the true values of the cosine invariants as obtained
from the refined structure.

Next, the membership of Classes Ī and II was confirmed and the
classes themselves extended by calculating appropriate cosine
invariants,

$$\cos(\phi_{h\ 4\ \ell} + \phi_{h'\ \bar{4}\ \ell'} + \phi_{-h-h',0,-\ell-\ell'}), \qquad (5.3)$$

in accordance with the procedure described in §3, Chapter VI.
Thus

$$\cos_{calc}(\phi_{\overline{11}\ \bar{4}\ 1} + \phi_{10\ 4\ \overline{10}} + \phi_{1\ 0\ 9}) = +2.02, \text{ A} = 1.544, \qquad (5.4)$$

confirms the presence of $\phi_{11\ 4\ \bar{1}}$ and $\phi_{10\ 4\ \overline{10}}$ in the same class
(I). Again, since $\phi_{16\ 4\ \bar{9}}$ and $\phi_{11\ 4\ \bar{1}}$ are in Class I, the values
of the calculated cosines,

$$\cos_{calc}(\phi_{1\ 4\ \bar{2}} + \phi_{16\ \bar{4}\ \bar{9}} + \phi_{\overline{17}\ 0\ 11}) = +1.34, \text{ A} = 1.568, \qquad (5.5)$$

$$\cos_{calc}(\phi_{1\ 4\ \bar{2}} + \phi_{\overline{11}\ \bar{4}\ 1} + \phi_{10\ 0\ 1}) = +1.03, \text{ A} = 1.618, \qquad (5.6)$$

require that $\phi_{1\ 4\ \bar{2}}$ be placed in Class I, too. Next, since $\phi_{16\ 4\ \bar{9}}$
and $\phi_{1\ 4\ \bar{2}}$ are in Class I,

$$\cos_{calc}(\phi_{19\ 4\ \bar{7}} + \phi_{\overline{16}\ \bar{4}\ 9} + \phi_{\bar{3}\ 0\ \bar{2}}) = +1.43, \text{ A} = 2.672, \qquad (5.7)$$

$$\cos_{calc}(\phi_{19\ 4\ \bar{7}} + \phi_{1\ \bar{4}\ \bar{2}} + \phi_{\overline{20}\ 0\ 9}) = +1.87, \text{ A} = 1.694, \qquad (5.8)$$

Table 4. Tentative definition of the orthogonal classes
I and II of phases, $\phi_{h4\ell}$, via calculated cosine
invariants, $\cos(\phi_{h\bar{4}\ell} + \phi_{\overline{h4}\ell} + \phi_{080})$. $|E_{080}| = 1.921$

Class	h 4 ℓ	$|E|$	A	$\cos(\phi_{h\bar{4}\ell} + \phi_{\overline{h4}\ell} + \phi_{080})$ Calculated	True
	11 4 $\bar{1}$	2.506	1.914	+1.44	+0.961
I	17 4 $\bar{7}$	2.428	1.796	+1.06	+0.975
	10 4 $\overline{10}$	2.320	1.640	+1.61	+0.559
	16 4 $\bar{9}$	2.265	1.562	+2.14	+0.999
II	21 4 5	2.518	1.931	−0.85	−0.899
	5 4 $\bar{3}$	2.392	1.743	−0.23	−0.809

imply that $\phi_{19\ 4\ \bar{7}}$ likewise belongs to Class I. Continuing in this way the membership of Class I was extended and firmly established. The final membership of Class II was firmly established in a similar way. The ten phases, $\phi_{h\ 4\ \ell}$, finally placed in Class I and the six phases, $\phi_{h\ 4\ \ell}$, placed in Class II are shown in Table 5.

Only one cosine invariant directly connecting the Classes I and II had a sufficiently large A value to warrant calculating the cosine. This was

$$\cos_{calc}(\phi_{4\ 4\ 3} + \phi_{9\ \bar{4}\ \overline{10}} + \phi_{\overline{13}\ 0\ 7}) = +0.30, \ A = 1.572, \qquad (5.9)$$

sufficiently close to 0 to serve as additional confirmation of orthogonality.

In defining the orthogonal classes I and II, great stress has been placed on the calculated values of the special cosine invariants (5.1) and (5.3). However, the cosines (5.3) can be calculated accurately only if the corresponding A values are at least moderately large. Again, if $\phi_{h_1\ 4\ \ell_1}$ belongs to Class I and $\phi_{h_2\ 4\ \ell_2}$ belongs to Class II, then, because of the orthogonality property (5),

$$\cos(\phi_{h_1\ 4\ \ell_1} + \phi_{h_2\ \bar{4}\ \ell_2} + \phi_{-h_1-h_2,0,-\ell_1-\ell_2}) \approx 0 \qquad (5.10)$$

and, in view of the known probability distribution of the cosine invariants, there is no requirement that the value of A associated with (5.10) be large (although an occasional large value for A is not ruled out). These observations suggest that a convenient way to facilitate the tentative extension of the Classes I and II, after the initial definition of Table 4, would be to compare intraclass interactions (as measured by A values rather than

Table 5. Final definition of the orthogonal
          classes I and II of phases, $\phi_{h4\ell}$, using
          Table 4 and calculated cosine invariants,
          $\cos(\phi_{h4\ell} + \phi_{h'\bar{4}\ell'} + \phi_{-h-h',0,-\ell-\ell'})$.

Class	h 4 ℓ	$|E|$
	11 4 $\bar{1}$	2.506
	4 4 3	2.497
	17 4 $\bar{7}$	2.428
	10 4 $\overline{10}$	2.320
I	1 4 $\bar{2}$	2.289
	16 4 $\bar{9}$	2.265
	14 4 $\bar{9}$	2.096
	19 4 $\bar{7}$	1.993
	5 4 13	1.945
	11 4 $\overline{11}$	1.818
	24 4 4	3.627
	21 4 5	2.518
II	9 4 $\overline{10}$	2.406
	5 4 $\bar{3}$	2.392
	14 4 $\bar{8}$	2.143
	4 4 $\overline{12}$	2.064

calculated cosines as done previously) with interclass interactions. From the very definition of the classes it is known
that the former interactions must be strong (i.e. many large A
values) but the interclass interaction is expected to be weak
(i.e. mostly small A values). The suggested study was carried
out and the results summarized in Tables 6, 7, 8, and 9. The ten
indices $\vec{h}$ corresponding to the phases $\phi_{\vec{h}} = \phi_{h\ 4\ \ell}$ of Class I are
shown (together with $|E|$ values, all of which are large) in the
first row of Table 6. The first column gives the same information
with the label $\vec{h}' = (h'\ 4\ \ell')$ instead of $\vec{h}$. In each box above
the main diagonal of Table 6 are displayed the vector $(h+h',0,\ell+\ell')$
where $(h\ 4\ \ell)$ and $(h'\ 4\ \ell')$ head the column and row, respectively,
of the box; $|E_{h+h',0,\ell+\ell'}|$; and the value of Λ associated with
the triple

$$(\bar{h}\ 4\ \bar{\ell}) + (\bar{h}'\ \bar{4}\ \bar{\ell}') + (h+h',0,\ell+\ell') = 0. \qquad (5.11)$$

In each box below the main diagonal are displayed the vector
$(-h+h',0,-\ell+\ell')$, where $(h\ 4\ \ell)$ and $(h'\ 4\ \ell')$ head the column and
row, respectively, of the box; $|E_{-h+\ell',0,-\ell+\ell'}|$; and the value of
Λ associated with the triple

$$(h\ 4\ \ell) + (\bar{h}'\ \bar{4}\ \bar{\ell}') + (-h+h',0,-\ell+\ell') = 0. \qquad (5.12)$$

If $|E_{\pm h+h',0,\pm\ell+\ell'}|$ was not in the observable range, the entry xxx
was made.

Table 7 was constructed in the same way as Table 6, but the
members of Class II were used instead of Class I.

Table 8 was constructed in order to exhibit the relatively
weak interaction between the Classes I and II. In the first
column are listed the vectors $\vec{h}_1 = (h_1\ 4\ \ell_1)$ of the ten phases
$\phi_{\vec{h}_1}$ in Class I. In the first row are listed the vectors

Table 6. Class I of phases, $\phi_{h4\ell}$, showing strong interactions.

In each cell the three stacked quantities are, respectively, $\to \hbar_3$, $\to |E^3|$, $\to A^3$. Column headers are $\vec{h} = (h4\ell)$ with $|E|$; row headers are $\vec{h}' = (h'4\ell')$ with $|E'|$.

$\vec{h}'=(h'4\ell')$, $|E'|$ \ $\vec{h}=(h4\ell)$, $|E|$	14 4 9̄ 2.09	19 4 7̄ 1.99	16 4 9̄ 2.27	17 4 7̄ 2.43	11 4 1̄1̄ 1.82	5 4 13 1.95	10 4 1̄0̄ 2.32	11 4 1̄ 2.51	1 4 2̄ 2.29	4 4 3 2.50
14 4 9̄ 2.10		33 0 1̄6̄ x x x	30 0 1̄8̄ x x x	31 0 1̄6̄ x x x	25 0 2̄0̄ x x x	19 0 4 0.61 0.40	24 0 1̄9̄ x x x	25 0 1̄0̄ 1.46 1.21	15 0 1̄1̄ 0.68 0.52	18 0 6̄ 0.19 0.16
19 4 7̄ 1.99	5 0 2̄ 3.57 2.37		35 0 1̄6̄ x x x	36 0 1̄4̄ x x x	30 0 1̄8̄ x x x	24 0 6 1.05 0.65	29 0 1̄7̄ x x x	30 0 8 x x x	20 0 9̄ 2.34 1.69	23 0 4̄ 0.90 0.71
16 4 9̄ 2.27	2 0 0 1.84 1.38	3̄ 0 2̄ 3.73 2.67		33 0 1̄6̄ x x x	27 0 2̄0̄ x x x	21 0 4 3.45 2.41	26 0 1̄9̄ x x x	27 0 1̄0̄ x x x	17 0 1̄1̄ 1.91 1.57	20 0 6̄ 0.77 0.69
17 4 7̄ 2.43	3 0 2̄ 3.73 3.01	2̄ 0 0 1.84 1.41	1 0 2 1.31 1.14		28 0 1̄8̄ x x x	22 0 6 0.92 0.69	27 0 1̄7̄ x x x	28 0 8 x x x	18 0 9̄ 0.58 0.51	21 0 4̄ 1.36 1.30
11 4 1̄1̄ 1.82	3̄ 0 2̄ 3.73 2.26	8̄ 0 4̄ 0.42 0.24	5̄ 0 2̄ 3.57 2.33	6̄ 0 4̄ 1.69 1.18		16 0 2 1.30 0.73	21 0 2̄1̄ x x x	22 0 1̄2̄ 1.30 0.94	12 0 1̄3̄ 0.33 0.21	15 0 8̄ 1.00 0.72
5 4 13 1.95	9̄ 0 22 x x x	1̄4̄ 0 20 x x x	1̄1̄ 0 22 x x x	1̄2̄ 0 20 x x x	6̄ 0 24 x x x		15 0 3 0.11 0.08	16 0 12 1.03 0.80	6 0 11 1.33 0.94	9 0 16 x x x
10 4 1̄0̄ 2.32	4̄ 0 1̄ 2.45 1.89	9̄ 0 3̄ 1.70 1.24	6̄ 0 1̄ 0.27 0.23	7̄ 0 3̄ 1.11 0.99	1̄ 0 1 0.20 0.13	5 0 2̄3̄ x x x		21 0 1̄1̄ 0.50 0.46	11 0 1̄2̄ 1.64 1.39	14 0 7̄ 0.23 0.21
11 4 1̄ 2.51	3̄ 0 8 1.60 1.33	8̄ 0 5 0.36 0.28	5̄ 0 8 0.46 0.41	6̄ 0 6 1.13 1.09	0 0 10 0.30 0.22	6 0 1̄4̄ 0.56 0.43	1 0 9 1.67 1.54		12 0 3̄ 0.38 0.35	15 0 2 1.44 1.43
1 4 2̄ 2.29	1̄3̄ 0 7 1.65 1.26	1̄8̄ 0 5 0.21 0.15	1̄5̄ 0 7 0.46 0.38	1̄6̄ 0 5 0.24 0.56	1̄0̄ 0 9 0.56 0.37	4̄ 0 1̄5̄ 1.08 0.76	9̄ 0 8 1.09 0.91	1̄0̄ 0 4 1.78 1.62		5 0 1 1.31 1.18
4 4 3 2.50	1̄0̄ 0 12 1.02 0.85	1̄5̄ 0 10 1.58 1.24	1̄2̄ 0 12 2.10 1.89	1̄3̄ 0 10 1.67 1.61	7̄ 0 14 1.11 0.80	1̄ 0 1̄0̄ 0.13 0.10	6̄ 0 13 0.36 0.33	7̄ 0 4 1.02 1.02	3 0 5 0.32 0.29	

Table 7. Class II of phases, $\phi_{h4\ell}$, showing strong interactions.

$\vec{h} = \frac{(h4\ell)}{|E|} \rightarrow$ $\vec{h}' = \frac{(h'4\ell')}{|E'|} \downarrow$	9 4 $\overline{10}$ 2.41	4 4 $\overline{12}$ 2.06	5 4 $\overline{3}$ 2.39	14 4 $\overline{8}$ 2.14	21 4 5 2.52	24 4 4 3.63
↑ h / ↓ |E³| / ↓ A						
9 4 $\overline{10}$ 2.41		13 0 $\overline{22}$ / X X X	14 0 $\overline{13}$ / 0.36 / 0.33	23 0 $\overline{18}$ / X X X	30 0 $\overline{5}$ / X X X	33 0 $\overline{6}$ / X X X
4 4 $\overline{12}$ 2.06	$\overline{5}$ 0 2 / 3.57 / 2.82		9 0 $\overline{15}$ / 2.97 / 2.32	18 0 $\overline{20}$ / X X X	25 0 $\overline{7}$ / 2.29 / 1.89	28 0 $\overline{8}$ / X X X
5 4 $\overline{3}$ 2.39	$\overline{4}$ 0 7 / 2.84 / 2.59	1 0 9 / 1.67 / 1.31		19 0 $\overline{11}$ / 1.25 / 1.01	26 0 2 / 0.64 / 0.61	29 0 $\overline{1}$ / X X X
14 4 $\overline{8}$ 2.14	5 0 2 / 3.57 / 2.92	10 0 4 / 0.88 / 0.62	9 0 $\overline{5}$ / 2.40 / 1.95		35 0 $\overline{3}$ / X X X	38 0 $\overline{4}$ / X X X
21 4 5 2.52	12 0 15 / X X X	17 0 17 / X X X	16 0 8 / 0.29 / 0.27	7 0 13 / 0.84 / 0.72		45 0 9 / X X X
24 4 4 3.63	15 0 14 / X X X	20 0 16 / X X X	19 0 7 / 1.56 / 2.14	10 0 12 / 0.24 / 0.29	3 0 $\overline{1}$ / 2.04 / 2.96	

Table 8. Weak interaction between the classes I and II of phases, $\phi_{h4\ell}$

Header for each Class II column: $\hbar_2 = (h_2 4\ell_2)$, $|E_2|$
Header for each Class I row: $\hbar_1 = (h_1 4\ell_1)$, $|E_1|$
Value rows within each cell correspond to $\hbar_3$ / $|E_3|$ / A^3 (with $+$ $+$ $+$).

Class I \ Class II	9 4 10 2.41	4 4 12 2.06	5 4 3 2.39	14 4 8 2.14	21 4 5 2.52	24 4 4 3.63
14 4 9 2.10	23 0 19 x x x 5 0 1 1.31 1.04	18 0 21 x x x 10 0 3 0.66 0.45	19 0 12 0.53 0.42 9 0 6 0.30 0.24	28 0 17 x x x 0 0 1 1.18 0.84	35 0 4 x x x 7 0 14 0.17 0.15	38 0 5 x x x 10 0 13 0.35 0.42
19 4 7 1.99	28 0 17 x x x 10 0 3 0.66 0.50	23 0 19 x x x 15 0 5 0.62 0.41	24 0 10 0.64 0.48 14 0 4 0.08 0.06	33 0 17 x x x 5 0 1 1.30 0.88	40 0 2 x x x 2 0 12 0.52 0.41	43 0 3 x x x 5 0 11 0.51 0.59
16 4 9 2.27	25 0 19 x x x 7 0 1 0.35 0.30	20 0 21 x x x 12 0 3 0.49 0.36	21 0 12 0.99 0.85 11 0 6 0.27 0.24	30 0 17 x x x 2 0 1 0.19 0.15	37 0 4 x x x 5 0 14 0.42 0.38	40 0 5 x x x 8 0 13 0.70 0.91
17 4 7 2.43	26 0 17 x x x 8 0 3 0.12 0.11	21 0 19 x x x 13 0 5 0.23 0.19	22 0 10 0.57 0.53 12 0 4 0.41 0.38	31 0 15 x x x 3 0 1 0.68 0.56	38 0 2 x x x 4 0 12 0.32 0.31	41 0 3 x x x 7 0 11 0.28 0.39
11 4 11 1.82	20 0 21 x x x 2 0 1 0.19 0.13	15 0 23 x x x 7 0 1 0.35 0.21	16 0 14 0.51 0.35 6 0 8 0.26 0.18	25 0 19 x x x 3 0 3 1.34 0.83	32 0 6 x x x 10 0 16 x x x	35 0 7 x x x 13 0 15 x x x
5 4 13 1.95	14 0 3 0.37 0.27 4 0 23 x x x	9 0 1 0.40 0.26 1 0 25 x x x	10 0 10 0.61 0.45 0 0 16 0.31 0.23	19 0 5 0.32 0.21 9 0 21 x x x	26 0 18 x x x 16 0 8 0.45 0.35	29 0 17 x x x 19 0 9 0.92 1.02
10 4 10 2.32	19 0 20 x x x 1 0 0 0.50 0.44	14 0 22 x x x 6 0 2 0.92 0.70	15 0 13 0.95 0.84 5 0 7 1.05 0.93	24 0 18 x x x 4 0 2 0.53 0.42	31 0 5 x x x 11 0 15 0.34 0.31	34 0 6 x x x 14 0 14 0.51 0.34
11 4 1 2.51	20 0 11 1.05 1.01 2 0 9 0.60 0.57	15 0 13 0.95 0.78 7 0 11 0.28 0.23	16 0 4 0.11 0.10 6 0 2 0.92 0.88	25 0 9 1.03 0.87 3 0 7 1.52 1.29	32 0 4 x x x 10 0 6 0.20 0.20	35 0 3 x x x 13 0 5 0.23 0.34
1 4 2 2.29	10 0 12 1.02 0.89 8 0 8 0.09 0.08	5 0 14 0.42 0.32 3 0 10 0.29 0.22	6 0 5 0.19 0.17 4 0 1 0.80 0.70	15 0 10 1.58 1.23 5 0 10 0.31 0.24	22 0 3 0.43 0.39 20 0 7 0.60 0.54	25 0 2 0.50 0.66 23 0 6 0.68 0.90
4 4 3 2.50	13 0 7 1.65 1.57 5 0 13 0.32 0.30	8 0 9 0.40 0.33 0 0 15 0.49 0.40	9 0 0 0.72 0.69 1 0 6 0.74 0.70	18 0 5 0.21 0.17 10 0 11 0.07 0.06	25 0 8 x x x 17 0 2 0.18 0.18	28 0 7 x x x 20 0 1 0.51 0.73

$\vec{h}_2 = (h_2 \ 4 \ \ell_2)$ of the six phases $\phi_{\vec{h}_2}$ in Class II. The body of the table contains the vectors $\vec{h}_3 = (h_3 \ 0 \ \ell_3)$ satisfying

$$(\bar{h}_1 \ 4 \ \bar{\ell}_1) + (\bar{h}_2 \ \bar{4} \ \bar{\ell}_2) + (h_3 \ 0 \ \ell_3) = 0 \qquad (5.13)$$

or

$$(\bar{h}_1 \ 4 \ \bar{\ell}_1) + (h_2 \ \bar{4} \ \ell_2) + (h_3 \ 0 \ \ell_3) = 0 \qquad (5.14)$$

where $\vec{h}_1$ and $\vec{h}_2$ head the row and column, respectively, of the box; $|E_3|$; and A. In marked contrast to Tables 6 and 7, the $|E|$ and A entries in the body of Table 8 show a preponderance of small values, thus demonstrating the weak interaction between the Classes I and II, consistent with orthogonality.

The two columns labeled "Intra-class interaction (strong)" and "Inter-class interaction (weak)" of Table 9 summarize the information contained in Tables 6, 7, and 8. Each entry of the first of these columns for Class I (or Class II, respectively) identifies the vector (h 4 ℓ) at the head of the row and contains the average value of A appearing in the row or column of Table 6 (or Table 7, respectively) headed by the vector (h 4 ℓ). Each entry in the second of these columns for Class I (or Class II, respectively) identifies the vector (h 4 ℓ) at the head of the row and contains the average value of A appearing in the row (or column, respectively) of Table 8 headed by the vector (h 4 ℓ). Comparison of the entries in these two columns shows again that the intra-class interaction is strong but that the inter-class inter-action is weak.

In light of the theoretical basis for the definition of the orthogonal Classes I and II, it is instructive to study further the values of those cosine invariants (5.1) which were actually calculated. Only those cosines corresponding to A values greater than 1.5 were calculated and these are shown in the penultimate

Table 9. Comparison of the strong intra-class and the weak inter-class interactions.

Class	h 4 ℓ	|E|	Intra-class interaction (strong)	Inter-class interaction (weak)	φ assigned (degrees)	φ true (degrees)	Error |Δφ| (degrees)	A	$\cos(\phi_{h4\ell}+\phi_{h4\bar{\ell}}+\phi_{080})$ Cos calc.	Cos true
I	14 4 5̄	2.096	1.386	0.509	0	11	11	1.335	xxx	+0.998
	19 4 7̄	1.993	1.151	0.476	180	-168	12	1.207	xxx	+0.999
	16 4 5̄	2.265	1.373	0.456	180	-168	12	1.562	+2.14	+0.999
	17 4 7̄	2.428	1.227	0.352	0	7	7	1.796	+1.06	+0.978
	11 4 1̄1̄	1.818	0.845	0.340	0	27	27	1.005	xxx	+0.883
	5 4 13	1.945	0.726	0.399	0	20	20	1.150	xxx	+0.970
	10 4 1̄0̄	2.320	0.785	0.606	xxx	-15	xxx	1.640	+1.61	+0.559
	11 4 1̄	2.506	0.875	0.627	xxx	5	xxx	1.914	+1.44	+0.961
	1 4 2̄	2.289	0.814	0.527	xxx	-178	xxx	1.597	+0.26	+0.927
	4 4 3	2.497	0.854	0.513	xxx	-152	xxx	1.900	+0.33	+0.866
Average			0.982	0.492					+1.14	+0.914
II	9 4 1̄0̄	2.406	2.164	0.556	90	71	19	1.763	+0.16	-0.438
	4 4 1̄2̄	2.064	1.790	0.373	-90	-62	28	1.295	xxx	-0.866
	5 4 3	2.392	1.394	0.470	-90	-95	5	1.743	-0.23	-0.809
	14 4 8̄	2.143	1.252	0.597	-90	-141	51	1.396	xxx	+0.616
	21 4 5	2.518	1.289	0.322	xxx	90	xxx	1.931	-0.85	-0.899
	24 4 4	3.627	1.798	0.662	xxx	89	xxx	4.008	+0.39	-0.883
Average			1.797	0.492			19		-0.13	-0.546

column of Table 9. In accordance with the theory, the calculated
values of those cosine invariants (5.1) associated with Class I
(average calculated cosine = +1.14) were observed to be significantly larger than those associated with Class II (average calculated
cosine = -0.13). (For comparison, the true values of the cosine
invariants (5.1) are shown in the last column of Table 9). It
should, however, be emphasized that the initial, tentative definition of the classes (Table 4) was based on the six calculated
values of the cosines (5.1) which were very large and positive or
very large and negative so that their true values were not much
in doubt. The final definition (Table 5) depended on those
cosines (5.3) which were definitely calculated to be +1 and, owing,
in addition, to the great redundancy, the resulting classes
possessed the desired properties, in particular that of orthogonality. Tables 6-9 serve to facilitate the search and confirm
the results implied by the calculated values of the special cosine
invariants (5.1) and (5.3).

The orthogonality property (5) of Classes I and II implies
that

$$\phi_{h_1 \, 4 \, \ell_1} + \phi_{h_2 \, \bar{4} \, \ell_2} + \phi_{-h_1-h_2,0,-\ell_1+\ell_2} \approx \pm 90° \qquad (5.15)$$

and

$$\phi_{h_1 \, 4 \, \ell_1} + \phi_{\bar{h}_2 \, \bar{4} \, \bar{\ell}_2} + \phi_{-h_1+h_2,0,-\ell_1+\ell_2} \approx \pm 90° \qquad (5.16)$$

where $\phi_{h_1 \, 4 \, \ell_1}$ is an arbitrary phase in Class I and $\phi_{h_2 \, 4 \, \ell_2}$ is
an arbitrary phase in Class II. In short the invariants (5.15)
and (5.16) constitute a class of structure invariants each member
of which is approximately equal to ±90° and is therefore suitable
for decisive enantiomorph selection.

Table 10.   Values, in degrees, of twelve phases,
mostly enantiomorph dependent.

h	k	ℓ	$|E|$	Phase Determined by Means of	Calculated Phase	True Phase	Error
9	4	$\overline{10}$	2.406	Enantiomorph	+90	71	+19
14	4	$\overline{8}$	2.143	3204 (Enan.)	-90	-140	+50
4	4	$\overline{12}$	2.064	3182 (Enan.)	-90	- 62	-28
5	4	$\overline{3}$	2.392	3091 (Enan.)	-90	- 95	+ 5
9	0	$\overline{15}$	2.925	2925	0	0	0
1	4	$\overline{14}$	1.724	2682	-90	- 57	-33
11	4	$\overline{10}$	1.623	2599	-90	-149	+59
9	7	1	3.084	3001, 3217	+90	+101	-11
12	7	3	3.262	2893, 3159	+90	+109	-19
10	7	10	3.301	3139, 2838	+90	+113	-23
14	7	3	2.370	3263, 3289	-90	- 56	-34
8	7	10	2.380	3269	-90	- 62	-28

Inspection of the last six rows of Table 3 shows that the ten phases $\phi_{h\,4\,\ell}$ of Class I must have values equal to 0 or 180 degrees, approximately. Hence, because of the orthogonality property, the values of the six phases $\phi_{h\,4\,\ell}$ of Class II must be equal to ±90 degrees, approximately. As it turned out, values could be unambiguously assigned to only six phases of Class I (Table 3) and to four phases of Class II (§6), as shown in Column 6 of Table 9. However, Column 7 gives the true values (obtained from the refined structure) of all sixteen phases in the Classes I and II and these show good agreement with the predicted values (0 or 180 degrees for Class I, ±90 degrees for Class II). In fact, as is readily verified, the average deviation of the sixteen phases in Classes I and II from the appropriate one of 0°, 180°, 90°, -90° is only 15°.

The stage was now set for the determination of the remaining phases, mostly enantiomorph dependent, to be used as input to the tangent procedures.

6. Determination of Twelve Phases Which Are Mostly Enantiomorph Dependent

The values of the remaining twelve phases (all but one of which are enantiomorph dependent), needed for use in the tangent procedures, are listed in Table 10. In view of §5 and the description of Table 3, Table 10 is self-explanatory.

7. Determination of Remaining Phases by the Tangent Procedures

The 61 phases thus far determined were used as input to the tangent formula and modified tangent procedure (Chapter VII) to

determine the values of 452 phases. Modified and simple tangent
figures of merit on the four runs required, due to the h0ℓ reflections of ambiguous phase, showed no preferred solution. Inspection
of phasing of some Σ_1 type reflections and the 020 and certain
h1ℓ phase values, strongly indicated by the phasing process,
suggested the most promising solution. The 51 strongest peaks in
that E-map (using modified tangent phases) formed an oval. Structure factor calculations based upon those peaks gave an R value
(defined as $\Sigma||F_o|-|F_c||/\Sigma|F_o|$) of 36% for 5982 reflections. Two
subsequent difference maps led to the unambiguous identification
of the 78 nonhydrogen atoms of the structure. Three cycles of
isotropic refinement on all nonhydrogen atoms reduced the R factor
to 15%.

When comparing the true phases, as determined from the
structure, with the phases that led to the solution, it was
found that the average magnitude of the error in the 43 three-
dimensional phases of the basis set was 18°. This good agreement
suggested that an initial E-map with more than 452 terms might
give the full structure. To test this hypothesis, the 61 basis
phases were again used as input to the modified tangent procedure
in order to calculate 552 phases. In the initial E-map using
these 552 phases, 57 of the 60 strongest peaks were found to be
at an average distance of 0.15Å from the refined atomic positions.
Seventeen of the remaining 21 atoms were also present in this
initial 552 phase E-map at an average distance of 0.27Å from
refined positions. The 4 atoms not in the map were terminal
carbons on isopropyl groups exhibiting greatest thermal motion.

CHAPTER XIII*

9-*t*-BUTYL-9,10-DIHYDROANTHRACENE;

MODIFIED TANGENT PROCEDURE

The modified tangent procedure (VII, (2.4)) has been shown
to result in more accurate phasing than the simple tangent formula
(VII, (1.15)) when applied to the data for three solved steroid
crystal structures (VII, §3 and §4). However, the simple tangent
formula had been adequate to solve these structures and, based on
these results alone, it was questionable whether the modified
tangent procedure would allow for a structure solution in a case
where the tangent formula itself was unsuccessful. The analysis
of 9-*t*-butyl-9,10-dihydroanthracene ($C_{18}H_{20}$), a hydrocarbon
crystallizing in space group $P2_12_12_1$, provided an example where
this situation occurred.

A base set of 29 two-dimensional phases was obtained for
9-*t*-butyl-9,10-dihydroanthracene through the use of a variety of
techniques, and these starting phases are recorded in Table 1.
Phases of $\pi/2$ were assigned to reflections (0,9,1), (1,9,0),
(0,3,14), and (1,0,5) in order to specify the origin and enantiomorph, and the Σ_1 and squared-tangent formulas indicated the
phases of (0,10,0), (0,0,6), and (2,0,0) to be π. It should be
noted that, while specifying arbitrarily the values of $\phi_{0,9,1}$,

_____

* This chapter was written by Charles Weeks.

$\phi_{1,9,0}$, $\phi_{0,3,14}$, $\phi_{1,0,5}$ does serve to fix the origin and
enantiomorph, so that all phases are uniquely determined, the
greatest common divisor of the fifty-six 3 x 3 subdeterminants of
the 8 x 3 matrix

$$\begin{pmatrix} 0 & 9 & 1 \\ 0 & 9 & \bar{1} \\ 1 & 9 & 0 \\ 1 & \bar{9} & 0 \\ . & . & . \end{pmatrix}$$

is 3, not unity. Hence this matrix is not primitive and not all
phases are accessible from the initial quadruple of phases.
However, if the single phase $\phi_{0,10,0}$, determined by means of
Σ_1, is adjoined to the original quadruple, the g.c.d. of all
3 x 3 subdeterminants of the associated matrix is unity, the
matrix is primitive, and all phases are then accessible.

The Σ_2 triples with A values greater than 1.3 which involved
only two-dimensional reflections with $|E|$ greater than 1.2 were
generated, and the cosine seminvariants, $\cos(\phi_{\vec{h}}+\phi_{\vec{k}}+\phi_{-\vec{h}-\vec{k}})$, for
these triples were computed by both the modified triple product
and MDKS formulas. The triples were then divided into four
groups each having approximately constant A values, and within
each group, they were first ranked according to the modified
triple product computed cosines and then according to the MDKS
cosines. Acceptance criteria, as shown in Table 2, were then
assigned to each of the A ranges, and a triple was assumed to have
a cosine of unity and was used to determine a new phase only if
its cosine, as computed by both formulas, ranked among the top
T% for triples in that range. Similarly, triples were assumed to
have cosine values of -1 and were used to determine new phases if
their computed cosines ranked among the bottom B% for the A range
in question.

Table 1. Initial two-dimensional phases for
9-*t*-butyl-9,10-dihydroanthracene.

Serial	Reflection	$|E|$	Phase (radians)	Reason
1	0 9 1	4.02	$\pi/2$	
2	1 9 0	3.25	$\pi/2$	Define origin
3	0 3 14	2.90	$\pi/2$	and enantiomorph
4	1 0 5	3.09	$\pi/2$	
5	0 10 0	4.71	π	$\Sigma_1\ P_+=0.00$; squared-tangent $\cos\phi_{1D}/K=-16$
6	0 0 6	4.08	π	$\Sigma_1\ P_+=0.01$; squared-tangent $\cos\phi_{1D}/K=-35$
7	2 0 0	1.68	π	Squared-tangent $\cos\phi_{1D}/K=-43$
8	0 9 5	2.80	$-\pi/2$	
9	0 7 14	2.66	$-\pi/2$	
10	1 0 1	1.75	$\pi/2$	Cycle 1
11	0 10 6	1.63	0	
12	0 9 7	1.42	$-\pi/2$	
13	1 0 11	1.66	$-\pi/2$	
14	0 2 13	1.85	0	Cycle 2
15	0 2 15	1.80	0	
16	1 0 3	2.05	$\pi/2$	(1,0,5; 1,0,3) pair
17	2 0 2	2.11	0	Cycle 3
18	2 0 8	2.89	π	
19	3 0 3	1.33	$\pi/2$	
20	0 3 8	1.41	$-\pi/2$	Cycle 4
21	0 1 1	1.21	$\pi/2$	
22	0 8 0	1.30	0	Σ_2 triple with (0,9,1) and (0,1,1) confirms squared-tangent $\cos\phi_{1D}/K=12$
23	0 5 5	2.77	$-\pi/2$	Cycle 5, $\cos(\phi_{0\,2\,\bar{1}3}+\phi_{0\,3\,8}+\phi_{0\,5\,5})=-1$
24	0 2 3	2.80	π	
25	0 5 11	2.90	$\pi/2$	Cycle 6
26	0 8 3	1.41	π	
27	0 11 4	1.91	$-\pi/2$	
28	0 4 10	1.80	0	Cycle 7
29	0 5 17	2.35	$-\pi/2$	

Table 2.   Acceptance criteria for the two-dimensional triples.

A Range	Average A	No. triples in range	T %	B %
1.3- 1.5	1.40	69	50%	5%
1.5- 2.0	1.73	69	60	2
2.0- 3.0	2.38	60	80	2
3.0-10.5	5.00	46	90	0

At A values of 2.0, 12% of the cosine seminvariants for triples of two-dimensional reflections have values of -1. Using only those triples with A values of two or greater whose cosines were among the top 80% when computed by both formulas, six additional reflections (reflections with serial numbers 8 to 13 in Table 1) could be related to the origin and Σ_1 reflections, and a second cycle of this procedure yielded two more phases. Reflection (1,0,3) was then introduced into the known set because it was strongly paired to (1,0,5), and the inclusion of this phase allowed four additional phases to be determined from only those triples having A>2. Reflection (0,1,1) was included despite its relatively low normalized structure factor amplitude $(|E_{0\ 1\ 1}|=1.21)$ because it could be obtained from a triple with an A value of 5.2 for which the cosine values calculated by the triple product and MDKS formulas were 1.35 and 1.43 respectively.

It was considered to be important to obtain $\phi_{0,5,5}$ because knowledge of this phase would make several additional phases accessible, but it was necessary to use triples with lower A values in order to reach this phase. The triple product value for $\cos(\phi_{0,10,6}+\phi_{0,\overline{3},8}+\phi_{0,\overline{7},\overline{14}})$ was 1.99, and this triple was used to find $\phi_{0,3,8}$ even though the A value was only 1.4. Reflection (0,3,8) provided a bridge which led to (0,5,5) by means of $\cos(\phi_{0,\overline{2},\overline{13}}+\phi_{0,\overline{3},8}+\phi_{0,5,5})$ for which the computed values were

-0.27 (triple product) and -0.64 (MDKS). This was the only
triple from the bottom B% in any A range which was actually used
to find a new phase. Two more cycles were then based on these
two-dimensional triples, and six more phases were obtained.

Although all of the phase assignments based on the two-dimensional triples proved to be correct, it is apparent, in retro-
spect, that the phase assignments would have had a more substantial basis if related cosine seminvariants had been considered
in the groups suggested by the pair relationship. This was not
done because the necessary computer programs were not available
at the time that this structure was analyzed. The phase of reflection (1,0,3) had been found using the pair relationship with
(1,0,5) only because this pair was dominated by a single very
large contributor, and the triples corresponding to this contributor were easily found in the triple list.

Table 3 shows the three major contributors to the pair (0,2,3;
0,2,13). Two of these contributors, as well as the overall sum,
were negative indicating that (0,2,3) and (0,2,13) have different
phases. The third large contributor, which involved the reflections (0,3,8) and (0,5,5) gave a positive contribution. However,
one of the Σ_2 triples related to this contributor [(0,2,13),
(0,3,$\bar{8}$), (0,$\bar{5}$,$\bar{5}$)] was the triple with the negative cosine which
had been used to find $\phi_{0,5,5}$. Since the other cosine in this pair
contributor is positive, this pair makes an incorrect contribution
to the pair summation for (0,2,3) and (0,2,13). As is shown in
one of the footnotes to Table 3, knowledge only of the relationship between the cosine seminvariants is sufficient to indicate
the relationship between the phases of (0,2,3) and (0,2,13), and
if the cosines have opposite signs, the phases have different
values. This example shows how consideration of a group of cosine
seminvariants, in this case the set of cosines forming contributions to the pair (0,2,3; 0,2,13), may indicate when it is

Table 3. Σ_2 triples for large contributors ($|E_{\vec{h}_3}|$, $|E_{\vec{h}_4}|$, and both A>1) to the (0,2,3; 0,2,13) pair*.

$\vec{h}(\vec{h}_1$ or $\vec{h}_2)$	$\vec{k}(\vec{h}_3)$	$-\vec{h}-\vec{k}(\vec{h}_4)$	A	$\cos(\phi_{\vec{h}}+\phi_{\vec{k}}+\phi_{-\vec{h}-\vec{k}})$			Calc $\|E_1E_2\|$ $*\cos(\phi_1-\phi_2)$
				TPROD	MDKS	TRUE	
0 2 3	$\bar{1}$ $\bar{2}$ $\bar{8}$	1 0 5	3.42	0.92	1.27	0.92	-677
0 2 13	$\bar{1}$ $\bar{2}$ $\bar{8}$	1 0 $\bar{5}$	2.26	0.71	0.60	0.92	
0 2 3	0 $\overline{12}$ $\bar{8}$	0 10 5	1.84	0.51	0.43	1.00	-140
0 2 13	0 $\overline{12}$ $\bar{8}$	0 10 $\bar{5}$	1.22	0.54	1.46	1.00	
0 2 3	0 3 $\bar{8}$	0 $\bar{5}$ 5	2.49	0.66	0.37	1.00	263†
0 2 13	0 3 $\bar{8}$	0 $\bar{5}$ $\bar{5}$	1.65	-0.27	-0.64	-1.00	

* See Chapter XI for a more extensive discussion of individual pair contributors.

† Knowledge of the cosine values for the two triples comprising any contributor determines the relationship between the paired phases. If

$$\cos(\phi_{0,2,3}+\phi_{0,3,\bar{8}}+\phi_{0,\bar{5},5}) = -\cos(\phi_{0,2,13}+\phi_{0,3,\bar{8}}+\phi_{0,\bar{5},5})$$

and α and β designate $\phi_{0,3,\bar{8}}$ and $\phi_{0,\bar{5},5}$ respectively, then

$$\phi_{0,2,3}+\alpha+\beta = \phi_{0,2,13}+\alpha+\beta+\pi$$

since space group symmetry requires that $\phi_{0,5,5}$ also equal β. Thus,

$$\phi_{0,2,3} = \phi_{0,2,13}+\pi.$$

necessary that certain cosines be negative. Conflicts should be
resolved by making as few cosines as possible negative, especially
when the A values involved are large. The conflict between the
(0,2,3; 0,2,13) pair contributors can be resolved by assigning a
negative cosine to one of the triples related to the contributor
composed of (0,3,8) and (0,5,5). This course of action was
strongly favored since the triple product and MDKS values for
$\cos(\phi_{0,\bar{2},\overline{13}} + \phi_{0,\bar{3},8} + \phi_{0,5,5})$ were much lower than the values for
any other triple in this set of six.

Figure 1 summarizes the pair indications for the strongest
0uu and u0u reflections. If it is required that the significance
level s, as defined in Chapter XI (equation 3) be at least 4.0,
then a set of six 0uu reflections, including (0,5,5), are strongly interrelated, and weakly coupled to the origin reflection
(0,9,1). However, the reflection (0,9,5) which is in this set of
six reflections was readily obtained from $\cos(\phi_{0,9,1} + \phi_{0,\bar{9},5} + \phi_{0,0,6})$
for which the A value was 10.45 and the calculated cosine values
were 0.89 (triple product) and 1.82 (MDKS). Thus, the (0,5,5)
phase could have been reached through a path which did not depend
on any specific low A triples. Similarly, the (1,0,3) and (1,0,5)
reflections are part of a strongly interrelated set of four u0u
reflections.

The 29 phases found by means of this analysis of the two-dimensional triples were used as input to both the simple tangent
(VII, (1.15)) and modified tangent formulas (VII, (2.4)), and 221
more phases were found during the course of ten tangent cycles.
The total number of phases known at the end of each cycle and the
modified tangent cycle figures of merit, R cycle (equation (2.7),
Chapter VII) are given in Table 4. Although the figures of merit
for the final two or three cycles seemed to be a little high, the
modified tangent phases yielded an excellent E map nevertheless.
The highest 18 peaks on this E map had a rms deviation of 0.10Å
from the refined positions for the 18 carbon atoms, and the

Figure 1. Pairings among u0u and 0uu reflections having $|E|>1.4$.
Significance levels, s (equation 3, Chapter XI),
which are greater than four are indicated on the
arrows.

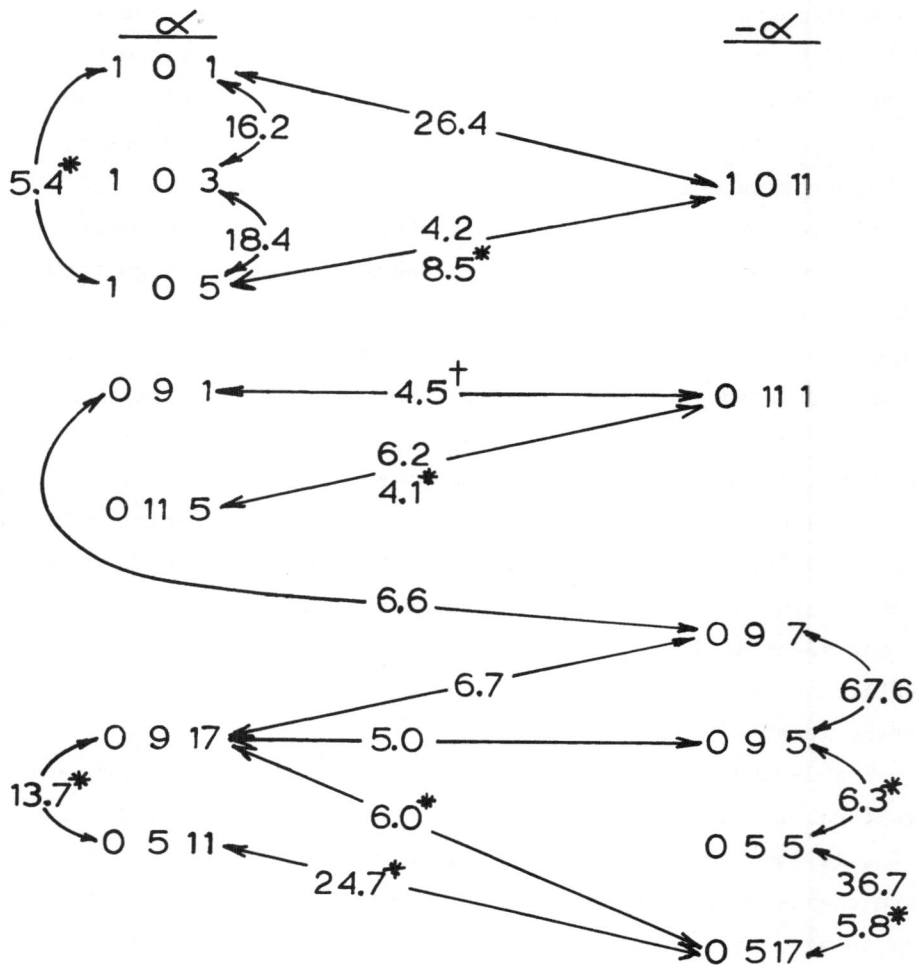

* An indication from a Type II formula.

† The only incorrect indication shown on this figure.

Table 4. Modified tangent cycle figures of merit, R cycle
          (equation (2.7), Chapter VII).

Cycle	No. phases determined	R cycle
8	35	0.79
9	45	0.81
10	60	0.81
11	80	0.82
12	105	0.85
13	135	0.88
14	170	0.90
15	210	0.92
16	250	0.93
17	250	0.93

initial R value was 22%. In contrast, the simple tangent phases
produced an E map in which 13 of the top 50 peaks were within
0.50Å of the carbon atom positions, and formed a recognizable
fragment (after the true solution was known!) of the enantiomorph
of the original molecule referred to a different origin.

Table 5 shows the average absolute deviations of the 250
modified and simple tangent calculated phases from their true
values for all 16 possible positions of the two enantiomorphs of
the molecule. The average absolute deviations of the phases
corresponding to the position of the molecule consistent with the
original origin and enantiomorph selection was 0.27 radians for
the modified tangent formula and 0.98 radians for the simple
tangent formula. The average absolute deviation of the simple
tangent phases from the true values for the position at which the
enantiomorph fragment appeared was 1.04 radians. If the phases
issuing from the tangent program were randomly distributed, these

Table 5.   Average absolute deviations of the 250 modified and
           simple tangent phases calculated in the 17th cycle
           from their true values.

Origin position	1st enantiomorph		2nd enantiomorph	
	Modified tangent	Simple tangent	Modified tangent	Simple tangent
0 0 0	0.27 rad	0.98 rad	1.55 rad	1.56 rad
$\frac{1}{2}$ 0 0	1.54	1.56	1.49	1.56
0 $\frac{1}{2}$ 0	1.57	1.58	1.29	1.52
0 0 $\frac{1}{2}$	1.50	1.56	1.19	1.28
$\frac{1}{2}$ $\frac{1}{2}$ 0	1.51	1.51	1.23	1.04*
$\frac{1}{2}$ 0 $\frac{1}{2}$	1.52	1.46	1.52	1.55
0 $\frac{1}{2}$ $\frac{1}{2}$	1.54	1.58	1.54	1.52
$\frac{1}{2}$ $\frac{1}{2}$ $\frac{1}{2}$	1.44	1.36	1.54	1.52

* A fragment of the molecule was found at this position as
  described in the text.

deviations would be 1.57 radians, and it can be seen that the
deviations are close to this value for most of the positions of
the molecules.   It is clear that such a movement of the molecule
as was seen on the tangent map is not a phenomenon reproducible
in other sets of data, and must be attributed to the fact that
the various positions of the molecules have some phases in common.
In this case, the tangent formula retained a majority of the

phases common to the two positions and altered many of the other phases from the values consistent with the original origin and enantiomorph selection, but it is doubtful that a recognizable molecule would have resulted had the structure been very complex. After the structure was solved, several other phase buildups differing in E minima and rate of new phase pickup were investigated, and in one of these cases both tangent procedures worked, but no case has yet been found where the simple tangent formula yielded the solution and the modified tangent procedure did not.

BIBLIOGRAPHY

CHAPTER I

Woolfson, M. M. (1970). An Introduction to X-Ray Crystallography. Cambridge:University Press. 112.

Harker, D. & Kasper, J. S. (1948). Phases of Fourier Coefficients Directly from Crystal Diffraction Data. Acta Cryst. $\underline{1}$, 70-75.

Karle, J. & Hauptman, H. (1950). The Phases and Magnitudes of the Structure Factors. Acta Cryst. $\underline{3}$, 181-187.

Goedkoop, J. A. (1950). Remarks on the Theory of Phase Limiting Inequalities and Equalities. Acta Cryst. $\underline{3}$, 374-378.

Von Eller, G. (1955). Inégalités de Karle-Hauptman et Géometrie Euclidienne. Acta Cryst. $\underline{8}$, 641-645.

Bouman, J. (1956). A General Theory of Inequalities. Acta Cryst. $\underline{9}$, 777-780.

Toguchi, I. & Naya, S. (1958). Matrix Theoretical Derivation of Inequalities. Acta Cryst. $\underline{11}$, 543-545.

Karle, J. & Hauptman, H. (1953). Application of Statistical Methods to the Naphthalene Structure. Acta Cryst. $\underline{6}$, 473-476.

Christ, C. L. & Clark, J. R. (1956). The Structure of Meyer-hofferite, $2CaO.3B_2O_3.7H_2O$, a $P\bar{1}$ Crystal, Determined by the Direct Method of Hauptman and Karle. Acta Cryst. $\underline{9}$, 830.

Karle, J.; Hauptman, H. & Christ, C. (1958). Phase Determination for Colemanite, $CaB_3O_4(OH)_3.H_2O$. Acta Cryst. $\underline{11}$, 757-761.

Karle, I.; Hauptman, H.; Karle, J. & Wing, A. (1958). Crystal and Molecular Structure of p,p'-Dimethoxybenzophenone by the Direct Probability Method. Acta Cryst. $\underline{11}$, 257-263.

Hauptman, H. & Karle, J. (1954). Solution of the Phase Problem for Space Group $P\bar{1}$. Acta Cryst. $\underline{7}$, 369-374.

Hauptman, H. & Karle, J. (1953). Solution of the Phase Problem I. The Centrosymmetric Crystal. ACA Monograph No. 3. Ann Arbor: Edwards Brothers, Inc. $\underline{34}$, 25.

397

Hauptman, H. & Karle, J. (1956). Structure Invariants and Seminvariants for Non-Centrosymmetric Space Groups. Acta Cryst. 9, 45-55.

Hauptman, H. & Karle, J. (1959). Seminvariants for Centro-symmetric Space Groups with Conventional Centered Cells. Acta Cryst. 12, 93-97.

Karle, J. & Hauptman H. (1961). Seminvariants for Non-Centro-symmetric Space Groups with Conventional Centered Cells. Acta Cryst. 14, 217-223.

Hauptman, H.; Fisher, J.; Hancock, H. & Norton, D. (1969). Phase Determination for the Estriol Structure. Acta Cryst. B25, 811-814.

International Tables for X-Ray Crystallography (1952). Vol. 1. Birmingham:Kynoch Press.

CHAPTER II

Ott, H. (1927). Structure Analysis. Z. f. Krist. 66, 136-153.

Banerjee, K. (1933). Determination of the Signs of the Fourier Terms in Complete Crystal Structure Analysis. Proc. Roy. Soc., A141, 188-193.

Hauptman, H. & Karle, J. (1953). Solution of the Phase Problem I. The Centrosymmetric Crystal. ACA Monograph No. 3. Ann Arbor: Edwards Brothers, Inc. 46, 50, 55, 63 et seq.

Sayre, D. (1952). The Squaring Method: A New Method for Phase Determination. Acta Cryst. 5, 60-65.

Hughes, E. W. (1953). The Signs of Products of Structure Factors. Acta Cryst. 6, 871.

Cochran, W. (1954). The Determination of Signs of Structure Factors from the Intensities. Acta Cryst. 7, 581-583.

Bullough, R. K. & Cruickshank, D. W. J. (1955). Some Relations between Structure Factors. Acta Cryst. 8, 29-31.

Hauptman, H. & Karle, J. (1955). A Relationship among the Structure Factor Magnitudes for P1. Acta Cryst. 8, 355.

Hauptman, H. & Karle, J. (1957). A Unified Algebraic Approach to the Phase Problem. I. Space Group P$\bar{1}$. Acta Cryst. 10, 267-270.

Karle, J. & Hauptman, H. (1957). A Unified Algebraic Approach
to the Phase Problem. II. Space Group P1. Acta Cryst. $\underline{10}$,
515-524.

Vaughan, P. A. (1958). A Phase-Determining Procedure Related to
the Vector-Coincidence Method. Acta Cryst. $\underline{11}$, 111-115.

Hauptman, H. (1964). The Role of Molecular Structure in the Direct
Determination of Phase. Acta Cryst. $\underline{17}$, 1421-1433.

Hauptman, H.; Fisher, J.; Hancock, H. & Norton, D. (1969). Phase
Determination for the Estriol Structure. Acta Cryst. $\underline{B25}$,
811-814.

Hauptman, H. (1970). The Squared-Tangent Formula. Acta Cryst.
$\underline{B26}$, 531-536.

CHAPTERS III AND IV

Watson, G. N. (1958). A Treatise on the Theory of Bessel
Functions. Cambridge at the University Press.

Wilson, A. J. C. (1949). The Probability Distribution of X-Ray
Intensities. Acta Cryst. $\underline{2}$, 318-321.

Hauptman, H. & Karle, J. (1952). Crystal-Structure Determination
by Means of a Statistical Distribution of Interatomic Vectors.
Acta Cryst. $\underline{5}$, 48-59.

Karle, J. & Hauptman, H. (1953). The Probability Distribution of
the Magnitude of a Structure Factor. I. The Centrosymmetric
Crystal. Acta Cryst. $\underline{6}$, 131-135.

Hauptman, H. & Karle, J. (1953). The Probability Distribution of
the Magnitude of a Structure Factor. II. The Non-Centrosymmetric Crystal. Acta Cryst. $\underline{6}$, 136-141.

Hauptman, H. & Karle, J. (1954). Solution of the Phase Problem
for Space Group P$\bar{1}$. Acta Cryst. $\underline{7}$, 369-374.

Hauptman, H. & Karle, J. (1953). Solution of the Phase Problem I.
The Centrosymmetric Crystal. ACA Monograph No. 3. Ann Arbor:
Edwards Brothers, Inc., 30-43.

Woolfson, M. M. (1954). The Statistical Theory of Sign Relationships. Acta Cryst. $\underline{7}$, 61-64.

Cochran, W. (1955). Relations between the Phases of Structure Factors. Acta Cryst. $\underline{8}$, 473-478.

Hauptman, H. & Karle, J. (1958). Phase Determination from New Joint Probability Distributions: Space Group P$\bar{1}$. Acta Cryst. $\underline{11}$, 149-157.

Karle, J. & Hauptman, H. (1958). Phase Determination from New Joint Probability Distributions: Space Group P1. Acta Cryst. $\underline{11}$, 264-269.

Klug, A. (1958). Joint Probability Distributions of Structure Factors and the Phase Problem. Acta Cryst. $\underline{11}$, 515-543.

Cochran, W. (1958). Structure Factor Relations and the Phase Problem. Acta Cryst. $\underline{11}$, 579-585.

Bertaut, E. F. (1960). Ordre Logarithmique des Densités de Repartition I. Acta Cryst. $\underline{13}$, 546-552.

Naya, S.; Nitta, I. & Oda, T. (1965). A Theory of the Joint Probability Distribution of Complex-Valued Structure Factors. Acta Cryst. $\underline{19}$, 734-747.

Hauptman, H. (1971). The Probability Distributions of Several Structure Factors and Their Magnitudes. Z. Krist. $\underline{134}$, 28-43.

Karle, J. & Hauptman, H. (1950). The Phases and Magnitudes of the Structure Factors. Acta Cryst. $\underline{3}$, 181-187.

Tsoucaris, G. (1970). A New Method for Phase Determination. The "Maximum Determinant Rule". Acta Cryst. $\underline{A26}$, 492-499.

CHAPTERS V - VII

Woolfson, M. (1954). The Statistical Theory of Sign Relationships. Acta Cryst. $\underline{7}$, 61-64.

Cochran, W. & Woolfson, M. (1955). The Theory of Sign Relations between Structure Factors. Acta Cryst. $\underline{8}$, 1-12.

Sayre, D. (1952). The Squaring Method: A New Method for Phase Determination. Acta Cryst. $\underline{5}$, 60-65.

Hauptman, H. & Karle, J. (1959). Rational Dependence and the Renormalization of Structure Factors for Phase Determination. Acta Cryst. $\underline{12}$, 846-850.

Hauptman, H.; Karle, I. and Karle, J. (1960). Crystal Structure of Spurrite, $Ca_5(SiO_4)_2CO_3$. I. Determination by the Probability Method. Acta Cryst. $\underline{13}$, 451-453.

Grant, D.; Howells, R. and Rogers, D. (1957). A Method for the Systematic Application of Sign Relations. Acta Cryst. $\underline{10}$, 489-497.

Hauptman, H.; Fisher, J.; Hancock, H. and Norton, D. (1969). Phase Determination for the Estriol Structure. Acta Cryst. $\underline{B25}$, 811-814.

Hauptman, H. (1970). The Calculation of Structure Invariants. Chapter A5, Crystallographic Computing. Proceedings of the 1969 International Summer School on Crystallographic Computing, pages 45-51. Ed. F. Ahmed, S. R. Hall and C. P. Huber. Copenhagen:Munksgaard.

Hauptman, H.; Fisher, I. and Weeks, C. (1971). Phase Determination by Least-Squares Analysis of Structure Invariants: Discussion of This Method as Applied to Two Androstane Derivatives. Acta Cryst. $\underline{B27}$, 1550-1561.

Cochran, W. (1955). Relations between the Phases of Structure Factors. Acta Cryst. $\underline{8}$, 473-478.

Karle, J. (1970). An Alternative Form for $B_{3,0}$, a Phase Determining Formula. Acta Cryst. $\underline{B26}$, 1614-1617.

Karle, J. & Hauptman, H. (1957). A Unified Algebraic Approach to the Phase Problem. II. Space Group P1. Acta Cryst. $\underline{10}$, 515-524.

Karle, J. & Hauptman, H. (1956). A Theory of Phase Determination for the Four Types of Non-Centrosymmetric Space Groups 1P222, 2P22, $3P_12$, $3P_22$. Acta Cryst. $\underline{9}$, 635-651.

Coulter, C. and Dewar, R. (1971). Tangent Formula Applications in Protein Crystallography: An Evaluation. Acta Cryst. $\underline{B27}$, 1730-1740.

Zacharissen, W. (1952). A New Analytical Method for Solving Complex Crystal Structures. Acta Cryst. $\underline{5}$, 68-73.

Karle, J. & Karle, I. (1966). The Symbolic Addition Procedure for Phase Determination for Centrosymmetric and Non-Centrosymmetric Crystals. Acta Cryst. $\underline{21}$, 849-859.

Germain, G. & Woolfson, M. (1968). On the Application of Phase Relationships to Complex Structures. Acta Cryst. $\underline{B24}$, 91-96.

Germain, G.; Main, P. and Woolfson, M. (1970). On the Application of Phase Relationships to Complex Structures. II. Getting a Good Start. Acta Cryst. $\underline{B26}$, 274-285.

CHAPTER VIII

Hauptman, H.; Fisher, J. and Weeks, C. (1971). Phase Determination by Least-Squares Analysis of Structure Invariants: Discussion of This Method as Applied to Two Androstane Derivatives. Acta Cryst. $\underline{B27}$, 1550-1561.

Fisher, J.; Hancock, H. and Hauptman, H. (1970). The Conditional Probability Distribution of Crystal Structure Invariants. Naval Research Laboratory Report 7132.

Fisher, J.; Hancock, H. and Hauptman, H. (1970). Computer Program for the Calculation of the Crystal Structure Invariants Based on Certain Conditional Distributions. Naval Research Laboratory Report 7157.

Hancock, H.; Fisher, J. and Hauptman, H. (1970). Computer Program for the Least-Squares Determination of the Phases of the Crystal Structure Factors. Naval Research Laboratory Report 7167.

INDEX